MICRO IRRIGATION MANAGEMENT

Technological Advances and Their Applications

MICRO IRRIGATION MANAGEMENT

Technological Advances and Their Applications

Edited by

Megh R. Goyal, PhD, PE, Senior Editor-in-Chief

Apple Academic Press Inc.	Apple Academic Press Inc.
3333 Mistwell Crescent	9 Spinnaker Way
Oakville, ON L6L 0A2	Waretown, NJ 08758
Canada	USA

©2017 by Apple Academic Press, Inc.

First issued in paperback 2021

Exclusive worldwide distribution by CRC Press, a member of Taylor & Francis Group
No claim to original U.S. Government works

ISBN 13: 978-1-77463-707-4 (pbk)
ISBN 13: 978-1-77188-390-0 (hbk)

Library and Archives Canada Cataloguing in Publication

Micro irrigation management : technological advances and their applications / edited by Megh R. Goyal, PhD, PE, senior editor-in-chief.

(Innovations and challenges in micro irrigation ; volume 5)
Includes bibliographical references and index.
Issued in print and electronic formats.
ISBN 978-1-77188-390-0 (hardback).--ISBN 978-1-77188-391-7 (pdf)

1. Microirrigation--Technological innovations. 2. Microirrigation--Management--Technological innovations. I. Goyal, Megh Raj, editor II. Series: Innovations and challenges in micro irrigation; v. 5

S619.T74M52 2016	631.5'87	C2016-905462-4	C2016-905463-2

Library of Congress Cataloging-in-Publication Data

Names: Goyal, Megh Raj, editor.
Title: Micro irrigation management : technological advances and their applications / editor: Megh R. Goyal.
Other titles: Innovations and challenges in micro irrigation ; [v. 5]
Description: Waretown, NJ : Apple Academic Press, 2017. | Series: Innovations and challenges in micro irrigation ; [volume 5] | Includes bibliographical references and index.
Identifiers: LCCN 2016035506 (print) | LCCN 2016037178 (ebook) | ISBN 9781771883900 (hardcover : alk. paper) | ISBN 9781771883917 (ebook)
Subjects: LCSH: Microirrigation--Technological innovations. | Microirrigation--Design and construction.
Classification: LCC S619.T74 M525 2017 (print) | LCC S619.T74 M525 2017 (ebook) | DDC 631.587--dc23
LC record available at https://lccn.loc.gov/2016035506

Apple Academic Press also publishes its books in a variety of electronic formats. Some content that appears in print may not be available in electronic format. For information about Apple Academic Press products, visit our website at **www.appleacademicpress.com** and the CRC Press website at **www.crcpress.com**

CONTENTS

LIST OF CONTRIBUTORS

Richard C. Beeson, PhD
Associate Professor (Plant Physiologist) University of Florida, USA, Agricultural, Research and Education Center, Apopka 2725 S. Binion Road, Apopka, FL 32703–8504 Phone: (407) 814–6172, Fax: (407) 814–6186; E-mail: rcbeeson@ufl.edu

R. B. Biniwale, PhD
Associate Professor, Soil Water Conservation Engineering Agricultural College, Dr.PDKV, Nagpur

Mohamed E. El-Hagarey, PhD
Researcher, Desert Research Center (DRC), Soil Conservation and Water Resources Department, Irrigation and Drainage Unite, Cairo, Egypt. E-mail: elhagarey@gmail.com

Rajesh Gajbhiye, M.Tech.
Former Student, Department of Irrigation and Drainage Engineering, Dr. Panjabrao Deshmukh Krishi Vidyapeeth, Akola 444104 (MS), India

Manoj K. Ghosal, PhD
Professor, Department of Farm Machinery and Power, College of Agricultural Engineering and Technology, Orissa University of Agriculture and Technology, Bhubaneswar–751003, Odisha. E-mail: mkghosal1@rdiffmail.com

Victor A. Gillespie, PhD
Retired Professor and Extension Specialist, Department of Molecular Biosciences and Bioengineering, University of Hawaii at Mañoa, Honolulu, Hawaii; Mailing address: 3050 Maileway, Honolulu, Hawaii 96822, USA

Megh R. Goyal, PhD, PE
Retired Professor in Agricultural and Biomedical Engineering, University of Puerto Rico – Mayaguez Campus; and Senior Technical Editor-in-Chief in Agriculture Sciences and Biomedical Engineering, Apple Academic Press Inc., PO Box 86, Rincon – PR – 00677, USA. E-mail: goyalmegh@gmail.com

Eric W. Harmsen, PhD
Professor, Department of Agricultural and Biosystems Engineering, University of Puerto Rico – Mayaguez Campus, Mayaguez, Puerto Rico 00681 USA. Phone: 787 955 5102, E-mail: eric.harmsen@upr.edu, harmsen1000@gmail.com

Ayman A. A. Ibrahim, PhD
Researcher, Agricultural Research Center (ARC), Agricultural Engineering Research Institute (AEnRI), 12311 Dokki, Giza, Cairo, Egypt. E-mail: eng_ayman288@yahoo.com

Deepak Jhajharia, PhD
Associate Professor, Department of Agricultural Engineering, North Eastern Regional Institute of Science & Technology (Deemed University under Ministry of Human Resource Development, Govt. of India), Nirjuli, Itanagar–791109, Arunachal Pradesh, India. Mobile: +91 9774990242, E-mail: jhajharia75@rediffmail.com

R. N. Katkar, PhD
Associate Professor, Soil Water Conservation Engineering Agricultural College, Dr.PDKV, Nagpur

Yogesh Mahalle, M.Tech.
Former student at Soil Water Conservation Engineering Agricultural College, Dr.PDKV, Akola; Postal address: A. T. Mahendra Colony, Near Sharda Agency, P.O. VMV Amravati – 444604 (MS), India. E-mail: yrmahalle1@gmail.com; Mobile: +91 9326279798 or 9420124699

Hani A. Mansour, PhD
Researcher, National Research Centre (NRC), Agricultural Division, Water Relations and Field Irrigation Department, El-Buhouth St., El-Dokki, Giza, Cairo, Egypt. E-mail: mansourhani2011@gmail.com

Ashok Mhaske, PhD
Associate Professor, Soil Water Conservation Engineering Agricultural College, Dr.PDKV, Nagpur. Mobile: +91 9422767788; E-mail: mhaskear@gmail.com

Subu Monia, M. Tech.
Research Scholar, Department of Agricultural Engineering, North Eastern Regional Institute of Science & Technology (Deemed University Under Ministry of Human Resource Development, Govt. of India), Nirjuli, Itanagar–791109, Arunachal Pradesh, India. Mobile: +91 9402495940, E-mail: munisubu23@gmail.com

Hamed Parsiani, PhD
Professor, Director & PI of NOAA-CREST Center at UPRM, University of Puerto Rico at Mayaguez (UPRM), Department of Electrical and Computer Engineering, Call Box 9000, Mayaguez, PR 00681–9000. E-mail: parsiani@ece.uprm.edu

Allan L. Phillips, PhD
Former Professor and Director, Department of Agricultural and Biosystems Engineering, University of Puerto Rico – Mayaguez Campus, Mayaguez, Puerto Rico 00681 USA. Current Address: 7602 Kilgore Rd, Avoca –MI – 48006. E-mail: allan@greatlakes.net

Kevin Rodriguez, BSc
Former Technical Assistant, School of Business and Computer Science, Mailing Address: 49 Alenore Gardens Phase-2, Off De Gannes Street, Arima, Trinidad and Tobago. Mobile: 868–372–3448 / 868–667–5457 E-mail: starkisuke7@gmail.com

Abdelgawad Saad, PhD
Researcher, Agricultural Research Center (ARC), Agricultural Engineering Research Institute (AEnRI), 12311 Dokki, Giza, Cairo, Egypt. E-mail: en_gawad2000@yahoo.com

Ajai Singh, PhD
Associate Professor and Head, Centre for Water Engineering and Management, Central University of Jharkhand, Brambe, Ranchi 834205 India. E-mail: ajai_jpo@yahoo.com; ajai.singh@cuj.ac.in

Dnyaneshwar Tathod, PhD
Former Student, Department of Irrigation and Drainage Engineering, Dr. Panjabrao Deshmukh Krishi Vidyapeeth, Akola 444104 (MS), India; Now Assistant Professor, College of Agricultural Engineering and Technology, Warwat Road, Jalgaon Jamod, Dist. Buldhana, Maharashtra, India. Mobile: +91 9604818220; E-mail: dnyanutathod@gmail.com

Maritza Torres, MSc
Former Student, Electrical and Computer Engineering Department, University of Puerto Rico at Mayagüez, Call Box 9000, Mayaguez, PR 00681–9000, USA

Salvio Torres Justiniano, MSc (Agronomy)
Former Student, Agroambiental Sciences Department, University of Puerto Rico, Call Box 9000, Mayaguez, PR 00681–9000, USA

I. Pai Wu, PhD, PE
Retired Professor, Department of Molecular Biosciences and Bioengineering, University of Hawaii at Mañoa, Honolulu, Hawaii; Mailing address: 3050 Maileway, Honolulu – Hawaii – 96822, USA. Tel. 808–956–8809. E-mail: ipaiwu@hawaii.edu

LIST OF ABBREVIATIONS

ace	acre
Agri.	agriculture
Agril.	agricultural
ANOVA	analysis of variance
ASAE	American Society of Agricultural Engineers
ASTT	Agricultural Society of Trinidad and Tobago
BCR	benefit–cost ratio
BFID	Belle Fourche Irrigation District
BGREI	bringing green revolution in eastern India
BU	Brabender unit
C.A.E.T.	College of Agricultural Engineering and Technology
CARDI	Caribbean Agricultural Research and Development Institute
CAZRI	Central Arid Zone Research Institute
CEC	cation exchange capacity
CH4	methane
cm	centimeter
CPE	cumulative pan evaporator
CPRL	Conservation and Production Research Laboratory
CV	coefficient of variation
CVm	manufacturer's coefficient of variation
CWSI	crop water stress index
Dec	December
Drain.	drainage
DSR	direct seeded rice
DU	distribution uniformity
EC	eddy covariance
EC	electrical conductivity
ET	evapotranspiration
et al.	et alidi (and other)
ETA	actual evapotranspiration

ETc	crop water consumptive use
etc.	etcetera
ETo	reference evapotranspiration
EU	emission uniformity
FAO	Food and Agriculture Organization
FCV	flue cured Virginia
GIS	geographic information system
GPR	ground penetrating radar
GUI	graphical user interface
H	heat flux
hr.	hour
HYVs	high-yielding varieties
i.e.	that is
iCSWI	integrated crop water stress index
Int.	International
IPCC	Intergovernmental Panel on Climate Change
IR	infrared
IRR	internal rate of return
Irrig.	irrigation
IW	irrigation water
IWUE	irrigated water use efficiency
J.	Journal
Ja.	Jamod
JISL	Jain Irrigation System Limited
K	potassium
kPa	kilopascal
LCC	life cycle cost
LE	latent heat flux
lph	liter per hour
LSD	least significant difference
MK	Mann-Kendall test
N	nitrogen
NEH	North-East Hill
NFSM	National Food Security Mission
NP	net profit
NPC	non-pressure compensating
NPV	net present value

Oct	October
OFR	on farm reservoir
OUAT	Orissa University of Agriculture and Technology
P	phosphorus
PBP	payback period
PDPP	Plasticulture Development Project Centers
PE	probability of exceedence
PET	potential evapotranspiration
PETHS	potential evapotranspiration (PET) using the Hargreaves-Samani method
PLC	programmable logic controller
PM	Penman-Monteith
PV	photovoltaic
PVPS	photovoltaic pumping system
Sci.	science
SCS	soil conservation service
SFR	small farm reservoirs
SI	supplemental irrigation
SPV	solar photovoltaic
SPVWPS	solar photovoltaic water pumping system
SR	surface runoff
SRI	system of rice intensification
TC	total cost
TDR	time domain reflectometry
TOC	total organic carbon
TP	Theta Probe™
TPR	transplanted rice
Trans.	transactions
UC	uniformity coefficient
USA	United States of America
USDA	United States Department of Agriculture
VPD	vapor pressure deficits
W&S	Wang and Schmugge
WP	water productivity
WUE	water use efficiency

OU	Europe
OFR	default fallover
OUAT	Orissa University of Agriculture and Technology
P	phosphere
PBP	payback period
PDPT	Plantforative Development Project Centers
PE	probability of exceedance
PET	potential evapotranspiration
PET/HS	potential evapotranspiration (PET) using the Hargreaves-Samani method
PLC	programmable logic controller
PM	Penman & Kostak
PV	photovoltaic
LVPS	photovoltaic pumping system
...	sensor
...	Soil Conservation Service
S-R	small inter reservoirs
SI	supplemental irrigation
SP	Solar photovoltaic
SPVWPS	Solar photovoltaic water pumping system
...	sub first runoff
SRI	system of rice intensification
TC	total cost
TDR	time domain reflectometry
TOC	total organic carbon
...	total flowrate
T8	... m
...	...
DSA	United States of America
USD	United States Department of Agriculture
VPD	vapor pressure deficit
W&S	Water and Resources
WP	water productivity
WUE	water use efficiency

LIST OF SYMBOLS

°C	degree Celsius
A	crop area (m^2)
amsl	above mean sea level
a_o, a_1, a_2	constants
b_o, b_1, b_2	constants
C	a roughness coefficient
C	clay content in percent of dry soil
C	roughness coefficient
c_o, c_1, c_2	constants
CV	discharge coefficient of variation (%)
CV_m	manufacturer's coefficient of variation
D	diameter of emitters, in inches (millimeters)
D	diameter of lateral or sub principal line, in feet (meters)
D	inside diameter, in feet (meters)
D	inside diameter, inches (centimeters)
DV	dependent variable
E	hydraulic energy required (kWh/day)
ea	saturation vapor pressure at temperature T (kPa)
ed	saturation vapor pressure at dew point (kPa)
ET	evapotranspiration
ET_0	grass reference evapotranspiration (mm d^{-1})
ET_A	actual evapotranspiration
ETc	crop water consumptive use
ETo	reference evapotranspiration
E_u	emission uniformity of drip irrigation (Approx. 0.9)
EU	emission uniformity of emitters (%)
f	frequency
g	gravitational acceleration (= 9.81 m/s^2)
G	soil heat flux heat density (MJ m^{-2} d^{-1})
H	operating pressure at the emitter in kPa
H	total hydraulic head

H	total pressure head or operating pressure head, in feet (meters)
$h(\theta_v)$	pore water pressure
H_f	friction drop, in feet (meters)
H_f	frictional head loss, feet (meters)
H_{max}	maximum pressure head, in feet (meters)
H_{min}	minimum pressure head, in feet (meters)
H_p	pressure charge at distance p, from the initial point (inlet), in feet (meters)
H_{var}	pressure head variation, in feet (meters)
IV	independent variable
K	coefficient (a constant value)
$K(\theta_v)$	variably saturated hydraulic conductivity
K_1	constant equal to 3.023 for British units and 0.0837 for SI units
K_1, K_2, K_3	coefficients for system of units
K_2	constant equal to 10.45 for British units and 2.264×10^7 for SI units
K_3	constant equal to 3.6642 for British units and 7.94×10^6 for SI units
K_c	crop coefficient
K_e	emitter discharge coefficient that characterize the emitter dimensions
Kg/cm^2	kilogram per centimeter square
Ks	saturated hydraulic conductivity
L	pipe length, feet (meters)
L	total length of the lateral or sub-main line, in feet (meters)
m	meter
m	number of groups of tied ranks
MJ	Mega-Joule
n	pore size distribution index $m = 1 - n^{-1}$ for $n > 1$
n	number of observations
P	length of the lateral line measured from the header (inlet), in feet (meters)
P_c	Pan coefficient (0.7 to 0.9)
PE	Pan evaporation rate (mm/day) converted to m/day

PET	potential evapotranspiration
q	average emitter flow, in gallons per minute (liter per second)
q	emitter discharge (lph)
Q	expressed as gallons per minute (liters per second)
q_a	average discharge of all emitters in lph
q_{avg}	mean of emitter discharge rate (lph)
Q_e	discharge of one emitter, in gallons per minute (liters per second)
q_{max}	maximum flow of an emitter, in gallons per minute (liters per second)
q_{min}	minimum flow of an emitter, in gallons per minute (liters per second)
q_n	average discharge from emitters in lowest 25% of discharge range (lph)
q_{var}	variation in flow of an emitter, in gallons per minute (liters per second)
RH_{max}	maximum relative humidity
RH_{mean}	mean relative humidity
RH_{min}	minimum relative humidity
R_n	net radiation (MJ m^{-2})
R_{nl}	net long wave radiation (MJ m^{-2} d^{-1})
R_{ns}	net short wave radiation (MJ m^{-2} d^{-1})
Rs	solar radiation
S	sand content in percent of dry soil
S	slope of energy line
sgn(.)	sign function
S_o	Slope of the lateral or sub principal line
S_o	land slope
S_o	line slope
S_p	distance between the emitters, in feet (meters)
Sp	emitter spacing, in feet (meter)
Sq	standard deviation of discharge rates of the emitters in the sample (lph)
T	mean temperature (°C)
T_{dew}	dew point temperature

t_i	tied observations
T_{max}	maximum temperature
T_{mean}	mean temperature
T_{min}	minimum temperature
U_2	average wind speed at 2 m height (m s^{-1})
V	average velocity, in feet per second (meters per second)
V	mean velocity, in feet per second (meters per second)
V	volume of water required (9.5 m^3/day)
Var (S)	variance of statistic S
w_a	wetted area (%, 95% for drip irrigation)
Wind	wind speed
WP	moisture content at the wilting point (pore water pressure, 15 bars)
WP	wilting point
W_r	peak water requirement (m^3/day)
W_t	transition moisture content
x	emitter discharge exponent
X	multiply
x_j	jth observation
α	inverse of the air-entry value (or bubbling pressure)
γ	fitting parameter which is related to WP psychometric constant (kPa °C^{-1})
ΔH	total friction drop at the end of a lateral line or submain, in feet (meters)
ΔH	total frictional head loss at a side of the lateral or sub principal line, in feet (meters)
$\Delta H'$	energy gain or loss due to slope over the total length of the line
$\Delta H'_p$	energy gain or loss due to slopes at a length P measured from the inlet
ΔH_p	total friction drop at a distance P, from the inlet
ε	apparent dielectric constant
ε_1	apparent dielectric constant for moisture content less than or equal to W_t
ε_2	apparent dielectric constant for moisture content greater than Wt

ε_a	dielectric constant of air (1)
ε_i	dielectric constant of ice (3.2)
$\varepsilon_i, \varepsilon_w, \varepsilon_a, \varepsilon_s$	dielectric constants of ice, water, air and solids, respectively
ε_w	dielectric constant of pure water (81)
θr	residual moisture content
θ_v	volumetric moisture content
ρ_b	bulk density
ρ_s	particle density

PREFACE

Water is a scarce, precious natural resource and a most crucial element; it must be planned, developed, conserved, and managed in a sustainable manner. Optimum management of available water resources at the farm level is needed because of increasing demands, limited resources, water table variations in space and time, and soil combination. It is important to increase the crop yield under limited water sources for optimum crop production to meet the future needs of food production. Limited water supply should be used efficiently to irrigate more areas with the same quantity of water. Pressurized irrigation techniques are a very reliable solution for obtaining higher uniformity and application efficiency.

Drip irrigation can potentially provide high application efficiency and application uniformity. Both are important in producing uniformly high crop yields and preserving water quality when both water and chemicals are applied through the irrigation system. In many irrigation methods, drip irrigation is popular for a plant crop because water can be applied directly to the root zone of the crop. Trickle irrigation is gaining importance in the world, especially in areas with limited and expensive water supplies, since it allows efficient use of limited resources. Ideally, all emitters in the system should discharge the same amount of water, but flow differences between two supposedly identical emitters exist due to: manufacturing variations, pressure differences, emitter plugging, aging, friction head losses throughout the pipe network, emitter sensitivity to pressure, and irrigation water temperature changes. Accurate emitter manufacturing is necessary in order to achieve a high degree of system uniformity. However, the complexity of emitter and their individual components make it difficult to maintain precision during production. Along with above factors, variation in pressure causes change in the discharge rate of two identical drippers and non-uniformity of water application, ultimately leading to reduction in crop yield. Variation in discharge causes uneven wetting pattern of the soil due to which plant root development restricted.

In most micro irrigation systems, the plant root development is restricted to the wetted soil volume near each emitter or along each lateral because water is supplied to a concentrated portion of the total soil volume. Excessive restriction of plant root development has the potential to decrease plant growth and yield. Therefore, it is more rational if irrigation water is applied according to the change of required soil wetted patterns during plant growing period.

Modern irrigation technologies have high water savings under well management. Due to high irrigation efficiency, size of the irrigated land with current water supply is higher in comparison with surface irrigated areas and it is possible to obtain high crop yield as well as more income with better management. A poorly managed pressurized irrigation system results in non - uniform water distribution. In such systems, the most valuable outcome of evaluation process is irrigation uniformity. Uniformity of drip irrigation system is usually a combination of measuring the variability of emission from individual emitter and pressure variation within the entire system.

Drip irrigation system in
Rose Gaden, Columbus - OH

Fully drip irrigated garden, Paris

Fully automated drip irrigation
in vine orchards, Paris, France

Irrigation manifold for landscape irrigation
at Airforce Base in Dayton - OH

Fotox Coutesy of Book editor, 2005

Since 1971, I have observed micro irrigation systems in almost all vegetable and fruit crops; in urban and home landscaping; in fields for research and farming; in non-automated and fully automated farms. During January of 2015, I and my wife had a stop-over in Mumbai International Airport of India while heading to New Delhi, capital of India. I was surprised to see *Fully Automated Vertical Micro Irrigation*, for irrigating potted flowers in the form of a heart. What a welcome to the waiting exhausted passengers. I felt relaxed. At our wedding anniversary on February 14, we visited flower festival in the Rose Garden of Chandigarh. I was astonished to see *Vertical Hydroponic Irrigation System* to grow lettuce. Not only I enjoy my tours but I also keenly watch and taste new technological advances in drip irrigation. The enthusiasm in my heart gave birth to this book volume *Micro Irrigation Management: Technological Advances and Their Applications*. WOW..... I am surprised at these advances!

http://www.domyownpestcontrol.com/my-garden-post-vertical-growing-system-with-drip-irrigation-p-8315.html comments, *"My Garden Post Vertical Growing System With Drip Irrigation is a new and innovative vertical gardening system ideal for patios, balconies, and decks. It is the perfect solution for the urban gardeners who want to grow their own herbs, vegetables, fruits or flowers in a small space. Includes drain hole plugs for indoor use. No more weeding, no bending and never a reason to get the knees dirty! The plants can thrive year round! The system provides carefree, consistent moisture for optimal growth and yield and takes the worry out of watering. Just set the fully automated programmable timer and the system does the rest. Easily assembled with the provided step-by-step instructions. This system is specially designed for My Garden Post and is not for indoor use, adds My Garden Post LLC."*

http://www.homeimprovementpages.com.au/article/choosing_irrigation_systems_for_vertical_gardens cites Jaclyn Fitzgerald's comments: *"For a vertical garden to succeed and thrive, it needs to be well watered, especially if it is attached to a wall that is exposed to plenty of sun and wind. The easiest way to ensure that the plants are getting the amount of water that they need is to use an irrigation system. There are three types of available systems: gravity fed drip irrigation system; comprehensive drip irrigation system; and a network of perforated plastic pipes runs throughout the entirety of the vertical garden system."*

The mission of this book volume is to serve as a reference manual for graduate and undergraduate students of agricultural, biological and civil engineering; horticulture, soil science, crop science and agronomy. I hope that it will be a valuable reference for professionals who work with micro irrigation and water management; for professional training institutes, technicals agricultural centers, irrigation centers, agricultural extension service, and other agencies that work with micro irrigation programs. I cannot guarantee the information in this book series will be enough for all situations.

After my first textbook, *Drip/Trickle or Micro Irrigation Management,* was published by Apple Academic Press Inc., and response from international readers, I was motivated to bring out for the world community a ten-volume series on *Research Advances in Sustainable Micro Irrigation.* The website http://www.appleacademicpress.com gives details on these ten book volumes.

This book is published under book series *Innovations and Challenges in Micro Irrigation.* Both book series are a must for those interested in irrigation planning and management, namely, researchers, scientists, educators and students.

The contributions by the cooperating authors to this book series have been most valuable in the compilation of this volume. Their names are mentioned in each chapter and in the list of contributors. This book would not have been written without the valuable cooperation of these investigators, many of whom are renowned scientists who have worked in the field of micro irrigation throughout their professional careers.

I will like to thank editorial staff, Sandy Jones Sickels, Vice President, and Ashish Kumar, Publisher and President at Apple Academic Press, Inc., for making every effort to publish the book when the diminishing water resources are a major issue worldwide. Special thanks are due to the AAP Production Staff for the quality production of this book.

We request that the reader offer us your constructive suggestions that may help to improve the next edition.

I express my deep admiration to my wife, Subhadra Devi Goyal, for her understanding and collaboration during the preparation of this book. As an educator, there is a piece of advice to one and all in the world:

"Permit that our almighty God, our Creator and excellent Teacher, irrigate the life with His Grace of rain trickle by trickle, because our life must continue trickling on..."

—Megh R. Goyal, PhD, PE, Senior Editor-in-Chief

December 25, 2015

ABOUT SENIOR EDITOR-IN-CHIEF

Megh R. Goyal, PhD, PE, is, at present, a Retired Professor in Agricultural and Biomedical Engineering from the General Engineering Department in the College of Engineering at University of Puerto Rico – Mayaguez Campus; and Senior Acquisitions Editor and Senior Technical Editor-in-Chief in Agricultural and Biomedical Engineering for Apple Academic Press Inc.

He received his BSc degree in Engineering in 1971 from Punjab Agricultural University, Ludhiana, India; his MSc degree in 1977 and PhD degree in 1979 from the Ohio State University, Columbus; his Master of Divinity degree in 2001 from Puerto Rico Evangelical Seminary, Hato Rey, Puerto Rico, USA.

Since 1971, he has worked as Soil Conservation Inspector (1971); Research Assistant at Haryana Agricultural University (1972–1975) and the Ohio State University (1975–1979); Research Agricultural Engineer/ Professor at Department of Agricultural Engineering of UPRM (1979– 1997); and Professor in Agricultural and Biomedical Engineering at General Engineering Department of UPRM (1997–2012). He spent one-year sabbatical leave in 2002–2003 at Biomedical Engineering Department, Florida International University, Miami, USA.

He was first agricultural engineer to receive the professional license in Agricultural Engineering in 1986 from College of Engineers and Surveyors of Puerto Rico. On September 16, 2005, he was proclaimed as *"Father of Irrigation Engineering* in Puerto Rico for the twentieth century" by the ASABE, Puerto Rico Section, for his pioneer work on micro irriga-tion, evapotranspiration, agroclimatology, and soil and water engineering. During his professional career of 45 years, he has received awards such as: Scientist of the Year, Blue Ribbon Extension Award, Research Paper Award, Nolan Mitchell Young Extension Worker Award, Agricultural Engineer of the Year, Citations by Mayors of Juana Diaz and Ponce,

Membership Grand Prize for ASAE Campaign, Felix Castro Rodriguez Academic Excellence, RashtryaRatan Award and Bharat Excellence Award and Gold Medal, Domingo Marrero Navarro Prize, Adopted son of Moca, Irrigation Protagonist of UPRM, Man of Drip Irrigation by Mayor of Municipalities of Mayaguez/Caguas/Ponce and Senate/Secretary of Agriculture of ELA, Puerto Rico.

He has authored more than 200 journal articles and textbooks, including: *Elements of Agroclimatology* (Spanish) by UNISARC, Colombia; two *Bibliographies on Drip Irrigation*. Apple Academic Press Inc. (AAP) has published his books, namely: *Management of Drip/Trickle or Micro Irrigation*, and *Evapotranspiration: Principles and Applications for Water Management*. During 2014–2015, AAP has published his ten-volume set on *Research Advances in Sustainable Micro Irrigation*. During 2016–2017, AAP will be publishing book volumes on emerging technologies/issues/challenges under the book series, *Innovations and Challenges in Micro Irrigation*, and *Innovations in Agricultural and Biological Engineering*.

Readers may contact him at: goyalmegh@gmail.com.

WARNING/DISCLAIMER

USER MUST READ IT CAREFULLY

The goal of this compendium, *Micro Irrigation Management: Technological Advances and Their Applications*, is to guide the world engineering community on how to efficiently design for economical crop production. The reader must be aware that dedication, commitment, honesty, and sincerity are important factors in a dynamic manner for success.

The editor, the contributing authors, the publisher and the printer have made every effort to make this book as complete and as accurate as possible. However, there still may be grammatical errors or mistakes in the content or typography. Therefore, the contents in this book should be considered as a general guide and not a complete solution to address any specific situation in irrigation. For example, one size of irrigation pump does not fit all sizes of agricultural land and to all crops.

The editor, the contributing authors, the publisher and the printer shall have neither liability nor responsibility to any person, any organization or entity with respect to any loss or damage caused, or alleged to have caused, directly or indirectly, by information or advice contained in this book. Therefore, the purchaser/reader must assume full responsibility for the use of the book or the information therein.

The mention of commercial brands and trade names are only for technical purposes. This does not mean that a particular product is endorsed over to another product or equipment not mentioned. Author, cooperating authors, educational institutions, and the publisher Apple Academic Press Inc. do not have any preference for a particular product.

All weblinks that are mentioned in this book were active on December 31, 2015. The editors, the contributing authors, the publisher and the printing company shall have neither liability nor responsibility, if any of the weblinks is inactive at the time of reading of this book.

BOOK SERIES ENDORSEMENTS

This book series is user-friendly and is a must for all irrigation planners to minimize the problem of water scarcity worldwide. Each book volume is concise and covers topics in depth related to the main theme. The *Father of Irrigation Engineering in Puerto Rico of 21st Century and World pioneer on micro irrigation*, Dr. Goyal (my longtime colleague) has done an extraordinary job in the presentation of this book series.

—Miguel A Muñoz, PhD
Ex-President of University of Puerto Rico,
University of Puerto Rico

I congratulate Dr. Megh Goyal on his professional contributions and the distinction in micro irrigation. I believe that this innovative book series on micro irrigation will aid the irrigation fraternity throughout the world.

—A. M. Michael, PhD
Former Professor and Project Director, Water Technology Center,
IARI, New Delhi,
Director, IARI, New Delhi; and
Ex-Vice-Chancellor, Kerala Agricultural University, Trichur, Kerala

In providing these book volumes under the book series on micro irrigation, Prof. Megh Raj Goyal as well as the Apple Academic Press have rendered an important service to the irrigators/investigators. Dr. Goyal, *Father of Irrigation Engineering in Puerto Rico*, and his colleagues have done an unselfish job in the presentation of this book volume that focuses on technological adances and their applications in micro irrigation. Micro irrigation is our future to solve the water scarcity problems throughout the world.

—Gajendra Singh, PhD
Ex-President (2010–2012) of ISAE
Former Deputy Director General (Engineering) of ICAR, and
Former Vice-President/Dean/Professor and Chairman, Asian Institute of Technology, Thailand

OTHER BOOKS on MICRO IRRIGATION TECHNOLOGY from APPLE ACADEMIC PRESS, INC.

Management of Drip/Trickle or Micro Irrigation
Megh R. Goyal, PhD, PE, Senior Editor-in-Chief

Evapotranspiration: Principles and Applications for Water Management
Megh R. Goyal, PhD, PE, and Eric W. Harmsen, Editors

Book Series: Research Advances in Sustainable Micro Irrigation
Senior Editor-in-Chief: Megh R. Goyal, PhD, PE
  Volume 1: Sustainable Micro Irrigation: Principles and Practices
  Volume 2: Sustainable Practices in Surface and Subsurface Micro
              Irrigation
  Volume 3: Sustainable Micro Irrigation Management for Trees and Vines
  Volume 4: Management, Performance, and Applications of Micro
              Irrigation Systems
  Volume 5: Applications of Furrow and Micro Irrigation in Arid and
              Semi-Arid Regions
  Volume 6: Best Management Practices for Drip Irrigated Crops
  Volume 7: Closed Circuit Micro Irrigation Design: Theory and
              Applications
  Volume 8: Wastewater Management for Irrigation: Principles and
              Practices
  Volume 9: Water and Fertigation Management in Micro Irrigation
  Volume 10: Innovation in Micro Irrigation Technology

Book Series: Innovations and Challenges in Micro Irrigation
Senior Editor-in-Chief: Megh R. Goyal, PhD, PE

• Principles and Management of Clogging in Micro Irrigation
• Sustainable Micro Irrigation Design Systems for Agricultural Crops:
  Methods and Practices

- Performance Evaluation of Micro Irrigation Management: Principles and Practices
- Potential Use of Solar Energy and Emerging Technologies in Micro Irrigation
- Micro Irrigation Management: Technological Advances and Their Applications
- Micro Irrigation Engineering for Horticultural Crops: Policy Options, Scheduling, and Design
- Micro Irrigation Scheduling and Practices

PART I

ESTIMATIONS OF EVAPOTRANSPIRATION

CHAPTER 1

REFERENCE EVAPOTRANSPIRATION ESTIMATIONS USING THE PENMAN-MONTEITH METHOD: PUERTO RICO

ERIC W. HARMSEN[1] and SALVIO TORRES JUSTINIANO[2]

[1]*Professor, Department of Agricultural and Biosystems Engineering, University of Puerto Rico - Mayaguez Campus, Mayaguez, Puerto Rico 00681 USA. Phone: 787 955 5102 E-mail: eric.harmsen@upr.edu, harmsen1000@gmail.com*

[2]*Former Student, Agroambiental Scienes Department, University of Puerto Rico, Call Box 9000, Mayaguez, PR 00681-9000*

CONTENTS

Edited version of: *"Eric W. Harmsen and Salvio Torres Justiniano, 2001. Estimating island-wide reference evapotranspiration for Puerto Rico using the Penman-Monteith Method. Unpublished Paper Presentation at the 2001 ASAE Annual International Meeting, Sacramento, California, USA July 30–August 1"*; *"E. W. Harmsen, Megh R. Goyal, and S. Torres Justiniano, 2002. Estimating Evapotranspiration in Puerto Rico. J. Agric. Univ. P.R. 86(1–2):35–54."*

1.1 INTRODUCTION

Estimates of long-term average daily evapotranspiration (ET) and reference evapotranspiration (ET_o) have been made for numerous agricultural crops in Puerto Rico [3]. These data are essential for determining monthly irrigation volumes, sizing of pumps and water conveyance devices, and for determining irrigation system fixed and operating costs. Most of the estimates previously made were based on the Soil Conservation Service (USDA-SCS) Blaney-Criddle method [12] and the Hargreaves-Samani method [7]. Harmsen et al. [9] reported large differences between the SCS Blaney-Criddle method [estimates obtained from Goyal, 4] and the Penman-Monteith method in a study that compared seasonal consumptive use for pumpkin and onion at two locations in Puerto Rico. The maximum observed differences were in the order of 100 mm per season. No comparisons have been made between the Hargreaves-Samani [7] and Penman-Monteith methods at locations in Puerto Rico. Inaccurate predictions of ET for an irrigated crop can lead to inefficient use of water and energy, increased potential for surface and groundwater contamination, and reduced profits for the grower.

In 1990 a committee of the United Nations Food and Agriculture Organization (FAO) recommended that the Penman-Monteith method should be the single approach for calculating ET_o. This recommendation was based on comprehensive studies, which compared twenty ET calculation methods with weighing lysimeter data [10]. These studies found that the Penman-Monteith method gives superior results compared to all other methods (including the SCS Blaney-Criddle [12] and Hargreaves-Samani [7]).

Therefore, it is imperative that improved estimates of long-term average daily reference evapotranspiration should be made available for Puerto Rico at this time.

The objectives of this study were 1) to present a simplified procedure for estimating long-term average daily reference evapotranspiration for any location in Puerto Rico; and 2) to compare previous estimates of reference evapotranspiration using the Hargreaves-Samani [7] method with the Penman-Monteith method.

1.2 MATERIALS AND METHODS

Harmsen and Torres Justiniano [8] presented procedures for estimating climate parameters to be used in the Penman-Monteith method for Puerto Rico. Their methodology was based on methods presented in the literature, which were then calibrated for Puerto Rico conditions. The study compared estimates of reference evapotranspiration at four locations (San Juan, Aguadilla, Mayagüez and Ponce) using measured and estimated climate data as input to the Penman-Monteith method. Figure 1.1 shows the results of their comparison of ET_o based on measured and estimated climate input data.

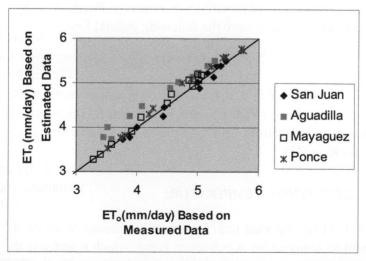

FIGURE 1.1 Comparison of long-term average daily reference evapotranspiration (ET_o) calculated with measured and estimated data by Harmsen and Torres Justiniano [8].

Input to the Penman-Monteith method includes: maximum daily air temperature (T_{max}), minimum daily air temperature (T_{min}), dew point temperature (T_{dew}), wind speed measured at 2 meters above the ground (U_2), and solar radiation (R_s). Although the methodology tended to overestimate slightly (Figure 1.1), yet it appears to provide reasonably good results for estimation purposes. As noted by Harmsen and Torres Justiniano [8], from an irrigation design standpoint, the fact that ETo (based on all parameters being estimated) overestimates slightly is not a serious problem. The procedures presented by Harmsen and Torres Justiniano [8] were used in this study for estimating Island-wide reference evapotranspiration. ET is related to the ETo by the following relation:

$$ET = K_c \, ET_o, \text{ where Kc is a crop coefficient} \qquad (1)$$

The procedure is outlined in the following sub-sections.

1.2.1 MINIMUM AND MAXIMUM AIR TEMPERATURE

Goyal et al. [5] developed regression equations for minimum and maximum long-term average daily air temperatures for Puerto Rico based on surface elevation. Table 1.1 lists the regression coefficients for the daily average minimum and maximum temperatures in Puerto Rico by month. The regression equations have the following general form:

$$T = A + BZ \qquad (2)$$

where, T is temperature (°C), A and B are regression coefficients, and Z is elevation (m) above mean sea level. Regression equations were derived with temperature data from Climatography of the United States No. 86–45 for Puerto Rico [11].

1.2.2 DEW POINT TEMPERATURE

The FAO [1] has reported that T_{dew} can be estimated based on the daily minimum air temperature. A correction factor, which is added to the minimum temperature, is recommended based on local conditions. Therefore, T_{dew} can be estimated in Puerto Rico from the following equation:

TABLE 1.1   Relationship Among Temperature (T) and Elevation (Z) for Puerto Rico [5]

Month	Mean daily maximum temperatures, °C			Mean daily minimum temperatures, °C		
	A	B, -10^{-5}	r^2	A	B, -10^{-5}	r^2
Jan.	29.24	770	0.73	18.58	544	0.44
Feb.	29.37	752	0.72	18.37	558	0.46
Mar.	30.08	711	0.71	18.71	590	0.48
Apr.	30.59	687	0.71	19.9	686	0.63
May	31.16	707	0.76	21.23	608	0.63
Jun.	31.76	686	0.73	21.92	577	0.59
Jul.	32.07	717	0.64	22.14	591	0.58
Aug.	32.12	682	0.75	22.21	585	0.58
Sep.	32.12	696	0.79	21.95	586	0.62
Oct.	31.84	705	0.79	21.48	553	0.59
Nov.	30.89	706	0.75	20.68	562	0.55
Dec.	29.83	744	0.73	19.52	547	0.47

* T = A + BZ, where T = temperature, °C; Z = elevation above mean sea level, m; A and B are regression coefficients and r^2 is the square of the coefficient of correlation.

$$T_{dew} = T_{min} + K_{corr} \qquad (3)$$

where, K_{corr} is a temperature correction factor in degrees °C, and the other variables have been previously defined. The correction factors (K_{corr}) are presented in Table 1.2. Figure 1.2 shows the Climatic Divisions for Puerto Rico [6].

1.2.3   WIND SPEED

For Puerto Rico, daily average wind speeds measured at 2 meters above the ground surface (U_2) were estimated based on averaging station data within the Climatic Divisions established by NOAA. Data are presented in Table 1.3.

1.2.4   RADIATION

The FAO recommends that solar radiation be estimated using the following equation for islands:

TABLE 1.2 Temperature Correction Factor Kcorr used in Eq. (2) for Climatic Divisions Within Puerto Rico [7]

Climatic Division	1	2	3, 4, 5, 6
K_{corr} (°C)	0.5 if Tdew is estimated using estimated Tmin data	−2.9	0
	−1.5 if Tdew is estimated using measured Tmin data		

* See Figure 1.1 for Climate Divisions.

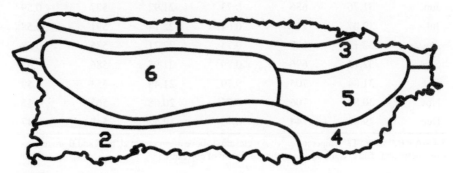

FIGURE 1.2 Climatic Divisions of Puerto Rico: 1 North Coastal, 2 South Coastal, 3 Northern Slopes, 4 Southern Slopes, 5 Eastern Interior, and 6 Western Interior.

$$R_s = (0.7\, R_a - b) \tag{3}$$

where, R_s is solar radiation, b is an empirical constant = 4 MJ.m^{-2}.day^{-1} and R_a is the incoming extraterrestrial radiation. Table 1.4 lists values of R_a by month and for latitudes applicable to Puerto Rico. The equations used to develop Table 1.4 are presented by Allen et al. [1]. Equation (3) is limited to elevations less than 100 m above sea level. Therefore, for higher elevations, in the interior areas of Puerto Rico where the ocean does not moderate air temperatures as much as along the low altitude coastal areas, the Hargreaves' radiation formula can be used:

$$R_s = K_{Rs}\, R_a\, (Tmax - Tmin)^{1/2} \tag{4}$$

where, K_{Rs} is an adjustment factor = 0.19, and the other variables have been previously defined.

TABLE 1.3 Average Daily Wind Speeds by Month and Climatic Division* with Puerto Rico [7]

Climatic Division*	Average daily wind speeds (m/s)**											
	Jan	Feb	Mar	Apr	May	June	July	Aug	Sept	Oct	Nov	Dec
1	2.7	2.8	3.0	2.9	2.6	2.6	2.9	2.7	2.1	1.9	2.2	2.6
2	1.8	2.0	2.2	2.1	2.2	2.4	2.4	2.1	1.7	1.5	1.4	1.5
3	2.2	2.4	2.6	2.4	2.2	2.4	2.7	2.5	2.0	1.8	2.0	2.3
4	1.8	2.0	2.1	2.1	2.0	2.0	2.0	1.8	1.6	1.6	1.6	1.6
5	1.1	1.3	1.4	1.5	1.6	1.7	1.6	1.3	1.1	0.9	0.9	0.9
6	1.3	1.5	1.5	1.5	1.6	1.8	1.8	1.5	1.2	1.1	1.0	1.0

* See Figure 1.1 for Climate Divisions.

** Averages are based on San Juan and Aguadilla for Div. 1; Ponce, Aguirre, Fortuna and Lajas for Div. 2; Isabela and Rio Piedras for Div. 3; Mayagüez, Roosevelt Rd. and Yabucoa for Div. 4; Gurabo for Div. 5; and Corozal and Adjuntas for Div. 6.

TABLE 1.4   Monthly Extraterrestrial Radiation and Latitude Within Puerto Rico

	Extraterrestrial Radiation, R_a, MJ.m^{-2}.day^{-1}						
	Latitude (decimal degrees N)						
Month	17.90	18.00	18.10	18.20	18.30	18.40	18.50
Jan	27.90	27.85	27.80	27.74	27.69	27.64	27.58
Feb	31.36	31.32	31.27	31.23	31.19	31.14	31.10
Mar	35.33	35.30	35.28	35.25	35.23	35.20	35.18
Apr	38.03	38.02	38.02	38.02	38.01	38.01	38.01
May	39.02	39.03	39.04	39.06	39.07	39.09	39.10
Jun	39.07	39.09	39.12	39.14	39.16	39.19	39.21
Jul	38.91	38.93	38.95	38.97	38.99	39.01	39.03
Aug	38.30	38.31	38.31	38.32	38.32	38.33	38.33
Sep	36.38	36.36	36.35	36.33	36.32	36.31	36.29
Oct	32.91	32.88	32.84	32.81	32.77	32.74	32.70
Nov	29.10	29.05	29.01	28.96	28.91	28.86	28.81
Dec	26.89	26.84	26.78	26.73	26.67	26.61	26.56

1.3   EXAMPLE APPLICATION

To illustrate the use of the climate estimation procedures for calculating long-term average daily reference evapotranspiration, an example is presented. The following conditions are used:

Location: Dos Bocas, Arecibo County, PR;

Elevation: 60 m;

Latitude: 18°20';

Climatic Division: 6

The temperature correction factor (K_{corr}) = 0 °C for Climatic Division 6

Estimated dew point temperature = the minimum air temperature

The estimated climate data and reference evapotranspiration for January through December are given in Table 1.5. Minimum and maximum temperatures were calculated with data from Table 1.1. Wind speeds were obtained from Table 1.3 for Climate Division 6. From Table 1.4 and Eq. (3), Rs values were determined. The latitude in decimals (required to use Table 1.4) is determined as follows: 18 degrees + (20 minutes/60 minutes per degree) = 18.33 decimal degrees. Estimated R_s, along with R_a values were obtained from Table 1.4 and are presented in Table 1.5. Reference

TABLE 1.5 Estimated Climate Data and Long-Term Average Daily Reference Evapotranspiration for Dos Bocas, PR

Month	Jan	Feb	Mar	Apr	May	June	July	Aug	Sept	Oct	Nov	Dec
T_{max}, °C	29.2	29.4	30.1	30.6	31.2	31.8	32.1	32.1	32.1	31.8	30.9	29.8
T_{min}, °C	18.6	18.4	18.7	19.9	21.2	21.9	22.1	22.2	21.9	21.5	20.7	19.5
T_{dew}, °C	18.6	18.4	18.7	19.9	21.2	21.9	22.1	22.2	21.9	21.5	20.7	19.5
U_2, m/s	1.1	1.3	1.4	1.5	1.6	1.7	1.6	1.3	1.1	0.9	0.9	0.9
R_a, MJ.m^{-2}.day^{-1}	27.7	31.2	35.2	38.0	39.1	39.2	39.0	38.3	36.3	32.8	28.9	26.7
R_s, MJ.m^{-2}.day^{-1}	15.4	17.8	20.7	22.6	23.4	23.4	23.3	22.8	21.4	18.9	16.2	14.7
ET_o, P-M, mm/day	3.1	3.6	4.3	4.7	5.0	5.1	5.1	4.9	4.6	3.9	3.3	2.9

Definitions: maximum daily air temperature (T_{max}), minimum daily air temperature (T_{min}), dew point temperature (T_{dew}), wind speed measured at 2 meters above the ground (U_2), extraterrestrial radiation (R_a), solar radiation (R_s) and long-term daily average reference evapotranspiration (ET_o).

evapotranspiration was calculated using the Penman-Monteith method as described by Allen et al. [1]. The calculation procedure was implemented via an Excel spreadsheet.

Alternatively, the reference evapotranspiration could have been calculated using the computer program CROPWAT [2]. This program is available free of charge on the Internet. Currently, a computer program has been developed by the Senior Author, which will perform the above procedure directly on the World Wide Web. This will increase accessibility to the public and greatly reduce the number of calculations required by the user [6].

1.4 COMPARISON OF ESTIMATED REFERENCE EVAPOTRANSPIRATION AT THIRTY-FOUR LOCATIONS IN PUERTO RICO

Goyal et al. [5] estimated reference evapotranspiration at thirty-four locations in Puerto Rico using the Hargreaves-Samani method [7]. In this section, estimates will be presented based on the Penman-Monteith method and a comparison of the two approaches will be discussed. The locations where estimates were made are shown in Figure 1.3.

Table 1.6 lists the Penman-Monteith-estimated reference evapotranspiration for the thirty-four locations considered by Goyal et al. [5]. Table 1.6 indicates the Climatic Division for each site, upon which the K_{corr} and U_2 values were selected from Tables 1.2 and 1.3. For locations with elevations

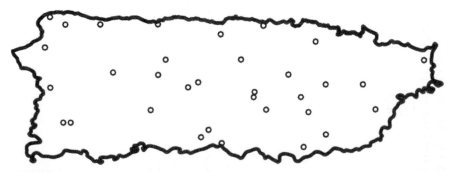

FIGURE 1.3   Locations within Puerto Rico, where estimates of reference evapotranspiration have been made (see Table 1.6 for estimated values).

TABLE 1.6 Reference Evapotranspiration Estimates using the Penman-Monteith (P-M) and Hargreaves-Samani (H-S) Methods for Thirty-Four Locations in Puerto Rico*

Location	Latitude (decimal degrees)	Longitude (decimal degrees)	Elevation (m)	Climatic Division	ET$_o$ Method**	Jan	Feb	Mar	Apr	May	Jun	Jul	Aug	Sep	Oct	Nov	Dec
						Reference Evapotranspiration (mm/day)											
Aguirre	17.97	66.48	15.0	2	P-M	3.8	4.3	4.9	5.3	5.5	5.6	5.7	5.6	5.1	4.5	3.9	3.6
					H-S	3.6	4.0	4.5	4.8	4.9	4.9	5.0	5.0	4.9	4.4	3.9	3.5
Aibonito	18.13	66.27	690.0	5	P-M	2.6	3.1	3.7	4.3	4.4	4.5	4.5	4.4	4.1	3.6	2.9	2.6
					H-S	2.5	2.9	3.6	4.5	4.2	4.2	4.1	4.0	3.9	3.4	2.7	2.5
Arecibo 2 ESE	18.47	66.70	4.5	1	P-M	3.8	4.2	4.9	5.2	5.4	5.6	5.7	5.5	5.1	4.6	4.1	3.8
					H-S	3.6	4.0	4.7	5.1	5.3	5.4	5.4	5.2	5.1	4.5	3.9	3.5
Barranquitas	18.18	66.32	540.0	6	P-M	2.7	3.2	3.8	4.2	4.4	4.5	4.5	4.4	4.1	3.6	3.0	2.6
					H-S	2.9	3.2	3.9	4.3	4.4	4.4	4.3	4.3	4.1	3.7	3.2	2.7
Caguas	18.23	66.03	75.0	5	P-M	3.1	3.7	4.3	4.8	5.0	5.1	5.1	5.0	4.6	4.0	3.3	3.0
					H-S	3.9	4.4	5.1	5.5	5.5	5.5	5.6	5.5	5.3	4.9	4.1	3.8
Canovanas 2N	18.40	65.08	9.0	3	P-M	3.3	3.8	4.4	4.8	5.0	5.0	5.0	4.9	4.6	4.0	3.4	3.1
					H-S	3.5	4.0	4.6	5.0	5.2	5.1	5.0	4.8	4.8	4.4	3.7	3.3
Carite Camp Tunnel	18.07	66.10	600.0	6	P-M	2.9	3.4	3.9	4.3	4.4	4.5	4.5	4.5	4.2	3.7	3.1	2.7
					H-S	3.1	3.5	4.1	4.3	4.2	4.1	4.1	4.2	4.2	3.8	3.4	3.0
Cayey 1 NW	18.12	66.15	420.0	4	P-M	3.5	4.2	4.9	5.2	5.1	5.0	5.0	5.0	4.8	4.3	3.8	3.4
					H-S	3.6	4.1	4.8	5.2	5.1	5.1	5.1	5.0	5.0	4.4	3.9	3.5
Cidra 3 E	18.18	66.13	420.0	4	P-M	3.2	3.8	4.4	4.6	4.7	4.7	4.8	4.8	4.3	4.1	3.6	3.1

TABLE 1.6 Continued

Location	Latitude (decimal degrees)	Longitude (decimal degrees)	Elevation (m)	Climatic Division	ET$_0$ Method**	Reference Evapotranspiration (mm/day)											
						Jan	Feb	Mar	Apr	May	Jun	Jul	Aug	Sep	Oct	Nov	Dec
Coloso	18.38	67.15	15.0	3	H-S	3.3	3.8	4.8	4.8	4.8	4.8	4.8	4.7	4.7	4.3	3.7	3.2
					P-M	3.6	4.1	4.7	5.1	5.1	5.2	5.3	5.2	4.7	4.1	3.6	3.4
Comerio Falls	18.27	66.18	150.0	5	H-S	4.0	4.5	5.1	5.5	5.5	5.7	5.7	5.5	5.3	4.8	4.2	3.8
					P-M	3.0	3.5	4.2	4.5	4.6	4.7	5.2	4.5	4.2	3.7	3.2	2.8
Corozal 4 W	18.33	66.37	120.0	6	H-S	3.3	3.7	4.4	4.8	4.8	4.9	5.4	4.7	4.5	4.0	3.5	3.2
					P-M	3.3	3.9	4.6	5.0	5.1	5.2	5.1	4.9	4.7	4.2	3.5	3.1
Dorado 4 W	18.47	66.28	7.5	1	H-S	3.6	4.1	4.8	5.2	5.4	5.4	5.2	5.0	5.0	4.5	3.8	3.5
					P-M	3.4	3.9	4.6	4.9	5.0	5.2	5.1	5.1	4.9	4.3	3.7	3.4
Dos Bocas	18.33	66.67	60.0	3	H-S	3.2	3.6	4.3	4.6	4.7	4.8	4.6	4.6	4.6	4.1	3.5	3.1
					P-M	3.5	4.0	4.7	5.1	5.1	5.3	5.8	5.2	4.7	4.1	3.6	3.4
Fajardo	18.33	65.65	12.0	4	H-S	3.8	4.4	5.1	5.6	5.6	5.9	6.6	5.5	5.5	4.8	4.1	3.8
					P-M	3.2	3.6	4.2	4.6	4.8	4.9	4.9	4.9	4.6	4.1	3.5	3.0
Garzas Dam	18.13	66.73	745.5	2	H-S	3.1	3.5	4.0	4.3	4.4	4.4	4.4	4.4	4.4	4.0	3.4	3.0
					P-M	3.1	3.5	4.1	4.6	4.5	4.7	4.8	4.6	4.3	3.8	3.3	2.9
Guayama	17.98	66.12	58.5	4	H-S	3.1	3.6	4.1	4.6	4.5	4.7	4.8	4.6	4.4	3.9	3.4	3.0
					P-M	3.4	3.9	4.5	4.9	5.0	5.1	5.1	5.0	4.7	4.2	3.6	3.2
Guineo Reservoir	17.98	66.12	900.0	4	H-S	3.4	3.8	4.4	4.7	4.8	4.8	4.8	4.8	4.7	4.1	3.6	3.2
					P-M	2.7	3.1	3.7	4.0	4.0	4.0	4.1	4.1	3.9	3.4	2.9	2.6

TABLE 1.6 Continued

Location	Latitude (decimal degrees)	Longitude (decimal degrees)	Elevation (m)	Climatic Division	ET₀ Method**	Reference Evapotranspiration (mm/day)											
						Jan	Feb	Mar	Apr	May	Jun	Jul	Aug	Sep	Oct	Nov	Dec
Humacao 1 SW	18.13	65.83	30.0	4	H-S	2.7	3.1	3.7	4.0	4.0	4.1	4.2	4.1	3.9	3.5	3.0	2.7
					P-M	3.2	3.8	4.4	4.8	4.9	4.9	5.0	4.9	4.6	4.1	3.5	3.1
Isabela 4 SW	18.47	67.07	126.0	3	H-S	3.5	3.9	4.6	4.8	4.7	4.7	4.8	4.8	4.6	4.2	3.7	3.3
					P-M	3.4	3.9	4.5	4.9	4.9	4.9	5.0	4.9	4.7	4.2	3.5	3.3
Jayuya	18.22	66.58	420.0	6	H-S	3.5	3.4	4.6	5.0	5.0	5.2	5.1	5.0	4.9	4.4	3.8	3.4
					P-M	2.9	3.6	4.5	4.8	4.8	5.0	5.0	5.0	4.6	4.1	3.3	3.0
Juana Diaz Camp	18.05	66.50	60.0	2	H-S	3.2	3.7	4.6	5.0	5.0	5.2	5.2	5.1	4.9	4.4	3.6	3.3
					P-M	3.7	4.2	4.8	5.1	5.2	5.4	5.5	5.3	4.8	4.2	3.7	3.4
Juncos 1 E	18.23	65.88	81.0	5	H-S	4.0	4.5	5.1	5.5	5.4	5.5	5.6	5.4	5.2	4.6	4.1	3.8
					P-M	3.2	3.8	4.4	4.9	5.1	5.2	5.2	5.0	4.6	3.9	3.3	3.0
Lajas	18.03	67.08	30.0	2	H-S	4.1	4.6	5.3	5.8	5.7	5.8	5.9	5.7	5.4	4.8	4.3	3.9
					P-M	3.7	4.2	4.9	5.2	5.3	5.5	5.6	5.4	4.8	4.2	3.7	3.4
Lares	18.28	66.88	360.0	6	H-S	4.1	4.6	5.4	5.6	5.6	5.7	6.0	5.7	5.2	4.7	4.3	4.0
					P-M	3.6	4.2	4.9	5.3	5.5	5.7	5.6	5.5	5.1	4.2	3.8	3.3
Maniti	18.43	66.45	75.0	1	H-S	3.9	4.4	5.1	5.5	5.7	5.8	5.8	5.6	5.4	4.5	4.2	3.7
					P-M	3.7	4.3	4.9	5.3	5.5	5.6	5.7	5.5	5.1	4.6	4.0	3.7
					H-S	3.5	4.1	4.8	5.2	5.4	5.5	5.4	5.2	5.1	4.6	3.8	3.4

TABLE 1.6 Continued

Location	Latitude (decimal degrees)	Longitude (decimal degrees)	Elevation (m)	Climatic Division	ET₀ Method**	Reference Evapotranspiration (mm/day)											
						Jan	Feb	Mar	Apr	May	Jun	Jul	Aug	Sep	Oct	Nov	Dec
Mayaguez	18.22	67.13	24.0	4	P-M	3.6	4.1	4.8	5.2	5.2	5.3	5.3	5.2	4.9	4.3	3.7	3.4
					H-S	3.9	4.5	5.2	5.6	5.7	5.8	5.8	5.6	5.4	4.7	4.1	3.8
Patillas Dam	18.03	66.03	72.0	4	P-M	3.2	3.8	4.4	4.7	4.9	4.9	4.9	4.9	4.6	4.0	3.4	3.1
					H-S	3.3	3.8	4.4	4.7	4.8	4.7	4.7	4.8	4.7	4.1	3.8	3.1
Ponce 4 E	18.02	66.53	12.0	2	P-M	3.6	4.0	4.6	4.9	5.1	5.2	5.3	5.2	4.8	4.2	3.6	3.3
					H-S	3.8	4.3	4.8	5.1	5.1	5.1	5.2	5.2	5.0	4.5	4.0	3.7
Quebradillas	18.47	66.93	111.6	1	P-M	3.7	4.2	4.9	5.1	5.1	5.3	5.3	5.3	5.0	4.5	4.0	3.7
					H-S	3.4	3.9	4.5	4.9	5.0	5.1	5.1	4.9	4.8	4.3	3.7	3.3
Ramey Air Force Base	18.50	67.13	71.1	1	P-M	3.2	3.6	4.2	4.5	4.7	4.8	4.9	4.9	4.6	4.1	3.5	3.1
					H-S	2.8	3.2	3.8	4.0	4.1	4.2	4.2	4.2	4.0	3.6	3.1	2.8
Rio Piedras	18.40	66.07	30.0	3	P-M	3.3	3.8	4.4	4.8	4.9	5.0	5.0	5.0	4.6	4.0	3.4	3.2
					H-S	3.5	4.0	4.7	5.1	5.1	5.2	5.1	5.0	4.9	4.4	3.8	3.4
San German	18.08	67.05	114.0	4	P-M	4.1	4.7	5.3	5.6	5.5	5.6	5.8	5.7	5.2	4.7	4.2	4.0
					H-S	4.1	4.6	5.2	5.6	5.6	5.7	5.9	5.7	5.3	4.8	4.2	4.0
Utuado	18.27	66.70	129.0	6	P-M	3.9	4.5	5.4	5.6	5.7	6.0	6.1	5.8	5.3	4.7	4.0	3.6
					H-S	4.2	4.8	5.6	5.9	5.9	6.2	6.2	5.9	5.7	5.1	4.4	4.0

* Hargreaves-Samani values of reference evapotranspiration were obtained from Goyal et al. [5]. ** P-M Penman-Monteith method; H-S Hargreaves-Samani method.

less than or equal to 100 m and greater than 100 m, R_s was calculated using Eqs. (3) and (4), respectively. Figure 1.4 shows the results of the comparison.

Figure 1.4 indicates positive and negative differences. The maximum positive difference (i.e., H-S minus P-M) was 0.92 mm/day during the month of November at the Juncos-1E station. On a monthly basis, this is equal to 27.5 mm or 1.1 inches of water. The minimum difference (i.e., negative difference) was –0.75 mm/day during the month of June at Aguirre. On a monthly basis this is –22.5 mm or –0.88 inches of water. Figure 1.4 indicates that while there is agreement between the two methods during many months at many locations, there were also many estimates, which were not in agreement. One could reasonably ask the question: "Which method is more correct?" FAO recommends to use the Penman-Monteith method over

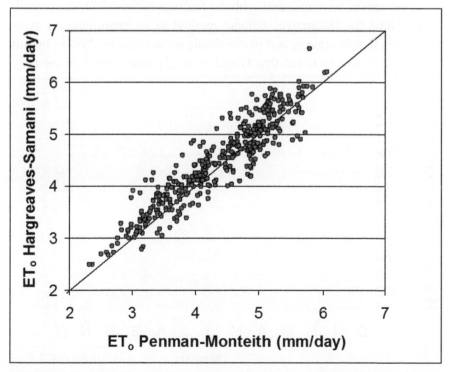

FIGURE 1.4 Comparison of reference evapotranspiration estimated by the Penman-Monteith (P-M) and Hargreaves-Samani (H-S) Methods for thirty-four locations in Puerto Rico.

all other methods, even when local data is missing. Studies have shown that using estimation procedures for missing data with the Penman-Monteith equation will generally provide more accurate estimates of ET_o than will other available methods requiring less input data [1].

Figure 1.5 shows a plot of the differences between ET_o calculated by the two methods (H-S minus P-M) by month, for the Juncos 1E and Aguirre stations. Maximum positive and negative differences were observed at these sites, respectively. If the Penman-Monteith method is taken as the standard ("correct") ET_o, then it can be stated that the Hargreaves-Samani [7] method overestimated ETo at Juncos 1E and underestimated ET_o at Aguirre. Juncos 1E is in Climate Division 5, which is humid, while Aguirre, in Climate Division 2, is semi-arid. The maximum underestimate of –0.75 mm/day at Aguirre (semi-arid) is equal to a 13% error, and the maximum overestimate of 0.92 mm/day at Juncos 1E (humid) is equal to a 28% error.

These results are consistent with the findings of the ASCE study [10], which found the Hargreaves-Samani method to underestimate on average by 9% in arid regions and overestimate on average by 25% in humid regions. It should be noted that Goyal et al. [5] used estimated monthly

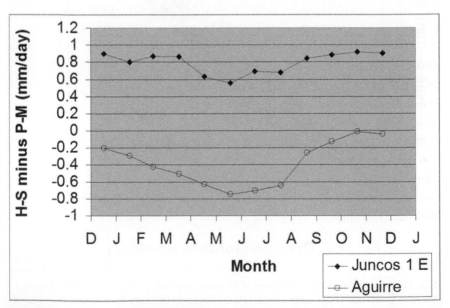

FIGURE 1.5 Estimated difference between ETo calculated by the Hargreaves-Samani (H-S) and Penman-Monteith (P-M) methods at the Juncos 1E and Aguirre stations.

values of R_a based on a single latitude equal to 18 degrees, which may account for some of the differences. In this study, actual site latitudes were used to obtain R_a.

1.4.1 METHOD LIMITATIONS

The approach presented in this paper should be considered only approximate for estimating long-term average daily reference evapotranspiration. Some potential limitations are:

- The data presented in this chapter and Tables 1.1–1.3 are only valid for Puerto Rico.
- The approach has not been validated using measured Tdew data from Climatic Divisions 3, 5 and 6, nor has Eq. (4) been verified to be accurate for areas within Puerto Rico, where elevations exceed 100 m [7].
- The data in Tables 1.1, 1.2 and 1.4 are based on monthly averages of daily data. Therefore, it should be understood that the method presented in this paper, provides a monthly average of the daily value for reference evapotranspiration.

1.5 CONCLUSIONS

This chapter presents a simple method for estimating long-term average daily reference evapotranspiration in Puerto Rico. The only data needed to use the method is the site latitude and surface elevation. With these two parameters, it is possible to estimate all other inputs to the Penman-Monteith method. Comparisons of long-term average daily reference evapotranspiration calculated using the Penman-Monteith method were compared with estimates made using the Hargreaves-Samani method for thirty-four locations in Puerto Rico. Maximum and minimum differences between the two methods (H-S minus P-M) were 0.92 and –0.75 mm/day, respectively.

1.6 SUMMARY

This chapter presents the application of a simple procedure for estimating long-term average daily ETo in Puerto Rico. With only two parameters, site latitude and surface elevation, it is possible to estimate all other inputs to the Penman-Monteith method. Minimum and maximum air temperatures are estimated from surface elevation data. Dew point temperature is estimated from the minimum temperature plus or minus a temperature correction factor. Temperature correction factors and average wind speeds are associated with six climatic divisions for Puerto Rico. Solar radiation is estimated from a simple equation for island settings or by the Hargreaves' radiation equation, based on air temperature differences. Estimated ETo is presented for thirty-four locations within Puerto Rico. Comparisons are made with values of ETo previously made using the Hargreaves-Samani method for the same locations.

ACKNOWLEDGMENTS

This project was funded by the USDA Higher Education Program-Hispanic Serving Institutions (Undergraduate and High School Experiences in Environmental Issues, University of Puerto Rico – Center for Investigation and Development No. 533521).

KEYWORDS

- **Climatology**
- **Crop water use**
- **Daily reference evapotranspiration**
- **Dew point**
- **Evapotranspiration**
- **Extraterrestrial radiation**
- **Hargreaves-Samani method**
- **Latitude**
- **Penman-Monteith**
- **Puerto Rico**
- **Solar radiation**
- **Wind speed**

REFERENCES

1. Allen, R. G., Pereira, Dirk Raes, L. S., & Smith, M. (1998). *Crop Evapotranspiration Guidelines for Computing Crop Water Requirements.* FAO Irrigation and Drainage Paper 56, Food and Agriculture Organization of the United Nations, Rome.
2. Clarke, D. (1998). *CropWat for Windows*: User Guide. Version 4.2.0013.
3. Goyal, M. R. (Ed.) (1989a). *Irrigation Research and Extension Progress in Puerto Rico.* Prepared for the First Congress on Irrigation in Puerto Rico. March 8th, 1989, Ponce, PR. UPR-Ag. Experiment Station.
4. Goyal, M. R. (1989b). *Estimation of Monthly Water Consumption by Selected Vegetable Crops in the Semiarid and Humid Regions of Puerto Rico.* AES Monograph 99–90, June, Agricultural Experiment Station, University of Puerto Rico Rio Piedras, PR.
5. Goyal, M. R., Gonzalez, E. A., & Chao de Baez, C. (1988). Temperature versus elevation relationships for Puerto Rico. *Journal of Agriculture U.P.R.,* 72(3), 440–67.
6. Goyal, M. R., & Harmsen, E. W. (Eds.) (2014). *Evapotranspiration: Principles and Application for Water Management.* Oakville, ON, Canada: Apple Academic Press Inc.,
7. Hargreaves, G. H., & Samani, Z. A. (1985). Reference crop evapotranspiration from temperature. *Appl. Eng. Agric., ASAE.,* 1(2), 96–9.
8. Harmsen, E. W., & Torres Justiniano, S. (2001). Evaluation of prediction methods for estimating climate data to be used with the Penman-Monteith method in Puerto Rico. ASAE Paper No. 01–2048 at *ASAE Annual International Meeting,* Sacramento Convention Center, Sacramento, CA, July 30–August 1.
9. Harmsen, E. W., Caldero, J., & Goyal, M. R. (2001). Consumptive Water Use Estimates for Pumpkin and Onion at Two Locations in Puerto Rico. *Proceedings of the Sixth Caribbean Islands Water Resources Congress.* Editor: Walter F. Silva Araya. University of Puerto Rico, Mayagüez, PR 00680.
10. Jensen, M. E., Burman, R. D., & Allen, R. G. (1990). *Evapotranspiration and irrigation water requirements.* ASCE Manuals and Reports on Engineering Practice No. 70, 332 pp.
11. National Climate Data Center (1992). *International Station Meteorological Climate Summary (ISMCS),* Version 2.
12. SCS (1970). *Irrigation Water Requirements.* Technical Release No. 21. USDA Soil Conservation Service, Engineering Division.

CHAPTER 2

REFERENCE EVAPOTRANSPIRATION: TRENDS AND IDENTIFICATION OF ITS METEOROLOGICAL VARIABLES IN ARID CLIMATE

SUBU MONIA[1] and DEEPAK JHAJHARIA[2]

[1]*Research Scholar, Department of Agricultural Engineering, North Eastern Regional Institute of Science & Technology (Deemed University under Ministry of Human Resource Development, Govt. of India), Nirjuli, Itanagar-791109, Arunachal Pradesh, India. Mobile: +91 9402495940, E-mail: munisubu23@gmail.com; [2]jhajharia75@rediffmail.com*

CONTENTS

2.1   INTRODUCTION

Evapotranspiration (ET) is frequently used in the analysis of regional water studies, field irrigation practices, weather forecasting, etc., owing to the results of impacts of ET changes, which is seriously affecting water use efficiency. The ET, one of the basic components of the hydrologic cycle, is combination of two processes (evaporation and transpiration), which balances the nature. For estimating the ET these two processes need to be checked upon, i.e., amount of water lost to the atmosphere from soil surface by the process of evaporation, as well as the water contained in the stomatal tissues of the sparse vegetation available in arid region that is released to the atmosphere as water vapor by the process of transpiration.

Several terms related to the ET are often used in the literature, and thus these terms may be individually understood first. First term is the potential evapotranspiration (PET), which is defined as the maximum quantity of water capable of being lost as water vapor, under a given climate, by a continuous, extensive stretch of vegetation covering the ground when there is no shortage of water. Another term is reference evapotranspiration (ET_0) which is the measure of water vaporized from a reference crop surface that is actively growing and not deficient of water at all times. The ET_0 is the standard term used for any measurements regarding the ET rates. The ET_0 can be evaluated by several radiation-based, temperature-based or pan evaporation-based methods formulated by various researchers. But, as per the recommendations of the United Nations Food and Agriculture Organization (FAO), Penman-Monteith (PM) FAO-56 method is the most accurate method for estimation of ET and is the universally acceptable method in absence of experimental data of ET by using the lysimeter.

The PM FAO-56 method is also known as the combination-based method since it merges methods applied in temperature-based and radiation-based methods for evaluation of ET. Hence in this combination-based method, radiative as well as aerodynamic factors affecting ET changes are taken into account while estimating the ET_0.

The changes in ET rates may influence water availability for various crops and also alter crop water requirements leading the need for irrigation planning, and more so in changing climate. The IPCC [6] has reported rise in global mean air temperature by 0.6°C over the last 100 years, and the general expectation is that global warming will lead to an increase in ET. However, various researchers have reported decreasing trends in ET instead of increase in evaporative demands. For example, decreasing trends in ET_0 were reported over a few sites in India [3], in the Changjiang (Yangtze River) in China [5], in the Haihe river basin of China [19], in a few sites of Iran [4], and in humid climates of northeast India [8].

The climate in arid environments has revealed well-substantiated extensive fluctuations. The spread of irrigation areas in arid regions of the northwest India along with changing land-use pattern has considerably modified the local climatic conditions over the Thar Desert region of Rajasthan state, India. Most of the recently observed temperature increases is due to the rise in greenhouse gas concentrations, and one would expect an increase in evaporation. Identifying trends in ET_0 is quite important for water resources management and suitable agriculture as this parameter plays crucial role in irrigation planning, and more so in the arid environments.

Therefore in this chapter, trends are investigated in the ET_0 using the Mann-Kendall (MK) test as there has been no previous study on the behavior of ET rates over Pilani under arid environment. This study was carried out with following objectives:

1. To estimate the ET_0 by using the PM FAO-56 method in different time scales from 1961 to 2005;
2. To investigate trends in the ET_0 through the MK test;
3. To estimate magnitudes of trends in the ET_0 by using the Theil-Sen's test; and
4. To identify the most dominating meteorological variables affecting the ET_0 process under arid-climatic conditions.

2.2 STUDY AREA AND METEOROLOGICAL DATA

The present study was carried out for Pilani, an arid site located in north-eastern part of Rajasthan, India (Figure 2.1). Pilani is chosen for studying trends in the ET_0 as water availability becomes the main constraining factor in arid sites, and thus this study may be useful for estimating principal crop water requirements accurately under changing climates in the arid region. Pilani (28.37° N, 75.6° E, elevation of 280 m amsl) is a small town situated in the Shekhawati region of Jhunjhunu district with total population of about 30, 000 [2]. Pilani is known for extreme climate and summer (winter) temperatures reach up to 44°C (1°C) in May and June (December and January) over Pilani.

The climatic data of Pilani of various meteorological parameters used in the estimation of the ET_0, (maximum and minimum temperature, maximum and minimum relative humidity, sunshine hours, wind speed, etc.) are obtained from India Meteorological Department (Pune) for 45 years from the duration of 1961 to 2005. The average annual rainfall of Pilani is 430 mm with standard deviation and coefficient of variation of annual rainfall of 182.4 mm and 42.3%, respectively. The total rainfall in winter and post-monsoon seasons is quite low, which is less than 10% of the total annual rainfall in Pilani. On the other hand, about 60% of the total annual

FIGURE 2.1   Location map of Pilani (Rajasthan), northwest India.

rainfall is received during the two months (July and August) of monsoon season over Pilani. Recurring droughts occur in Pilani as 18 years out of 45 years received total annual rainfall with deficit of 20% of average annual rainfall or higher deficit up to the tune of 62% of average annual rainfall of Pilani. Thus, Pilani has mostly suffered from meteorological droughts based on total annual rainfall once in three years. Thus, assured irrigation water is the need of hour for growing crops in rainfall-deficit places in Rajasthan similar to Pilani.

2.3 PENMAN-MONTEITH FAO-56 METHOD

The Penman-Monteith (PM) FAO-56 method was used to estimate the ET_0 over Pilani using the monthly data from 1961 to 2005. The universally acceptable PM FAO-56 method was derived by Allen et al. [1] for calculating ET_0 in various types of climates. The FAO defined the reference crop as hypothetical crop with an assumed height of 0.12 m having the surface resistance of 70 s/m and albedo of 0.23, closely resembling the evaporation of an extension surface of green grass of uniform height and actively growing and adequately watered. The PM FAO-56 method is expressed as:

$$ET_O = \frac{0.408\Delta\left(R_n - G\right) + \gamma\dfrac{900}{T+273}U_2(e_a - e_d)}{\Delta + \gamma(1 + 0.34U_2)} \qquad (1)$$

$$R_n = R_{ns} - R_{nl} \qquad (2)$$

The reference crop was assumed as green grass. A complete set of equations, proposed by Allen et al. [1] according to the available weather data and time step computation, constitutes the PM method. Since no actual solar radiation or sunshine duration data were available at Pilani, we estimated R_n from the sunshine hour data of Bikaner site situated in the Thar Desert having similar kind of environment. The assumed values for albedo equal to 0.23 and angstrom coefficients 'a' and 'b' equal to 0.25 and 0.5, respectively as per the recommendation of Shuttleworth [16], was used in this study.

2.4   TRENDS IN ET_0 AND IDENTIFICATION OF ITS KEY METEOROLOGICAL VARIABLES

The calculated evapotranspiration rates were used in the detection of trends by using the non-parametric Mann-Kendall (MK) test in monthly and annual time scales. Then the Theil-Sen's slope estimator method was applied to obtain the magnitudes of the observed trends in the ET_0. Subsequently stepwise regression analysis was performed keeping the ET_0 as dependent variable and other climatic parameters as independent variables to investigate the causal mechanism of the observed ET_0 changes under the arid climatic conditions of Pilani.

2.4.1   MANN-KENDALL TEST

The MK test is a non-parametric method to test the trend in a given series. It is most widely used method as it is impartial to distribution of data in the series. It is rank-based and is not sensitive to outliers [11, 12]. The Mann-Kendall test was performed to detect the presence of trends in the ET_0 in monthly and annual time scales. For a time series $X = \{x_1, x_2, ..., x_n\}$, the null hypothesis Ho states that there is a sample of n independent and identically distributed random variables. The alternative hypothesis H_1 of a two-sided test is that the distribution of x_k and x_j are not identical for all k, j $\leq$ n with k $\neq$ j. The statistic S is calculated as [10]:

$$S = \sum_{k=1}^{n-1}\sum_{j=k+1}^{n} \text{sign}(x_j - x_k) \qquad (3)$$

The mean and variance of the S statistic are obtained under the assumption that data are independent and identically distributed as [11]:

$$E(S) = 0 \qquad (4)$$

$$V(S) = \frac{n(n-1)(2n+5) - \sum_{i=1}^{m} t_i \times (t_i - 1) \times (2t_i + 5)}{18} \qquad (5)$$

The standard normal variable Z is computed as:

$$Z = \begin{cases} \dfrac{S-1}{\sqrt{\mathrm{Var(S)}}} & S > 0 \\[2mm] 0 & S = 0 \\[2mm] \dfrac{S+1}{\sqrt{\mathrm{Var(S)}}} & S < 0 \end{cases} \tag{6}$$

The values of test statistics are computed, and if the Z values lie within the limits −2.56 and +2.56, the null hypothesis of no trend in the series can't be rejected at the 1% level of significance.

2.4.2 SLOPE ESTIMATOR

The slope of n pairs of data points was estimated using the Theil–Sen's estimator [15, 17]. The Theil–Sen's estimator is the most popular nonparametric technique for estimating the magnitude of any trend, and is insensitive to outliers. It can be significantly more accurate than non-robust simple linear regression, and has been widely used in identifying the slope of the trend line in hydrological time series [20]. The further details of the Theil–Sen's estimator can be referred in Dinpashoh et al. [4] and Jhajharia et al. [7].

2.5 RESULTS AND DISCUSSION

For any arid site, such as, Pilani that has high evaporative demand as the advection of sensible heat from warm, dry regions provides an important additional energy source for conversion of water to vapor besides the natural heat from the sun. Thus the demand for water resource management and better irrigation planning is of foremost importance, and more so in current scenario of global warming and anthropogenic-induced climate change.

2.5.1 ESTIMATION OF ET_0

2.5.1.1 Annual ET_0

The monthly data of the four climatic parameters (temperature, wind speed, relative humidity and sunshine hours) were used in estimating ET_0 by using the

PM FAO-56 method. The estimated monthly ET_0 values were added together to obtain the total annual ET_0 over Pilani. Figure 2.2 shows the time series of total ET_0 and the trend line (broken line) in case of total ET_0 in annual time scale over Pilani from 1961 to 2005. The average of total ET_0 in annual duration is found to be about 1380.05 mm with standard deviation of 352.5 mm and coefficient of variation of 25.6% in arid environments of Pilani. Figure 2.2 shows a consistent decline in the total ET_0 during last 45 years from 1961 to 2005. Total decadal ET_0 remained in the range of 1723.9–2034.7 mm from 1961–1970 over Pilani with decadal average of 1863.7 mm during the decadal period from 1961 to 1970. Similarly, the total decadal ET_0 are found to varying from 994.8–1935.02 mm from 1971–1980 over Pilani with decadal average of 1512.8 mm during the decade. However, the following two decades, i.e., 1981–1990 and 1991–2000, witnessed the decadal average values of about 1157.6 mm and 1060.10 mm respectively, which are quite less than the average of total annual ET_0 of 1380.05 mm over Pilani.

It is worth to mention that the five lowest values of total annual ET_0 were found during the period from 1990 to 1999 over Pilani. On the other hand, the five highest values of total annual ET_0 were found from 1961 to 1972 over Pilani. Figures 2.3 and 2.4 show the comparison of the five highest and the five lowest ET_0 values with the corresponding climatic parameters in these ten years on annual basins for last 45 years over Pilani. It is clear

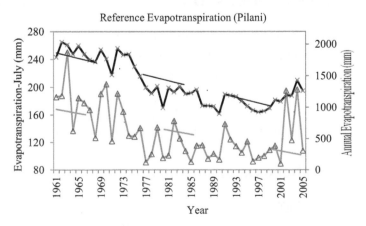

FIGURE 2.2   Time series of ET_0 and trend lines (broken) in July and in the annual time scale: Time series with triangle represents ET in the month of July.

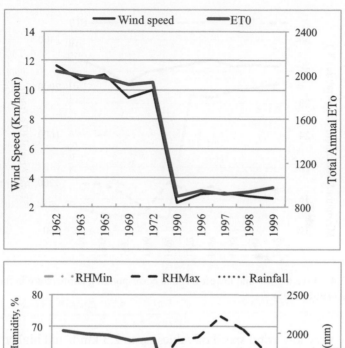

FIGURE 2.3   Five highest and lowest ET_0 and corresponding climatic parameters in these years over Pilani (1961–2005).

from Figure 2.3 that comparatively very high values of ET_0 occurred in early 1960's to 1970's due to occurrence of relatively very high wind speed of values up to 11.7 km/h and low humidity of values up to 55% (morning RH) and 36% (afternoon RH) over Pilani. On the other hand, comparatively very low values of ET_0 occurred in 1990 and in 1996–1999 due to occurrence of

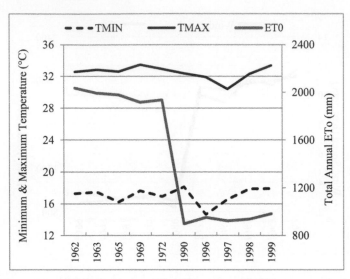

FIGURE 2.4   Five highest and lowest ET_0 and corresponding temperature in these years over Pilani (1961–2005).

relatively very low wind speed of values up to 2.3 km/h and higher humidity of values up to 73% (morning RH) and 50% (afternoon RH) over Pilani.

2.5.1.2   Monthly ET_0

The estimated monthly total ET_0 through the PM FAO-56 reached up to values of 100–130 mm (160–185 mm) during the months of November, December, January and February (in March and October) over Pilani. However, the monthly total ET_0 values in April, May and June were comparatively higher than remaining months. The monthly ET_0 values reached to peak values of 250–360 mm during these three months of pre-monsoon season. The comparatively high values of minimum temperature (up to 33 to 36°C) and maximum temperature (up to 40 to 44°C) and wind speed (up to 12 to 23 km/h) along with the low values of humidity (up to 10–13% in morning and 25–37% in evening) were responsible for the occurrence of very high ET_0 values in hot and bright sunny months of summer season over Pilani. A sample time series of monthly ET_0 for the months of May and December are shown in Figure 2.5 along with respective trend lines (denoted by dashed line).

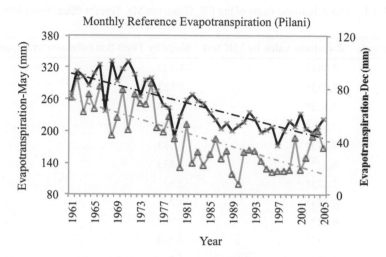

FIGURE 2.5 Time series of ET_0 and trend lines (broken) in May and in December over Pilani: The time series with triangle represents ET in the month of May.

2.5.2 INVESTIGATION OF TRENDS IN ET_0

The Mann-Kendall test was carried on the calculated ET_0 in monthly and annual time scales to investigate the presence of trends. Table 2.1 shows the trend values in the ET_0 obtained by using the MK test over Pilani for the duration of 1960–2005. On annual time scale, the test-statistics (Z) value obtained through the MK test is found to be –5.997, which indicates that the presence of statistically significant decreasing trend in the ET_0 at the 1% level of significance. Similarly, the Z values obtained through the MK test in all the twelve months are found to varying from –7.05 to –8.52, which indicate the presence of very strong decreasing trends in the ET_0 at the 1% level of significance on monthly time scale under arid environments of Pilani.

Further the Theil-Sen's test was used to estimate the magnitude of the observed trends in the ET_0 in annual and monthly time scales. The values of changes in the ET_0 (in mm/year) by using the non-parametric Theil-Sen's test are shown in Table 2.1. On monthly time scales, the magnitudes of the observed trends in the ET_0 were found to varying from –1.43 mm/year (January) to –7.83 mm/year (June). It is worthwhile to note that the

TABLE 2.1   The Z-Statistic Value of the ET_0 Using the MK Test for Pilani From 1960 to 2005

Time scale	Z-statistic value by MK test	Slope by Theil-Sen estimator (mm/year)
Jan	−7.494*	−1.433
Feb	−7.053*	−1.622
Mar	−7.522*	−3.346
Apr	−7.542*	−4.188
May	−7.855*	−5.133
Jun	−8.52*	−7.833
Jul	−7.875*	−3.65
Aug	−7.553*	−1.76
Sep	−7.21*	−1.907
Oct	−8.012*	−2.244
Nov	−7.494*	−1.7
Dec	−7.62*	−1.463
Annual	**−5.997***	**−23.0**

Note: values marked by * denotes statistically significant trends at 1% level of significance.

magnitude of trends in estimates of ET_0 are varying from −1.43 to −1.70 during the months of January, November and December, which are comparatively less than the magnitude of the observed trends in the months of summer season. The magnitudes of the observed trends in the ET_0 were found to varying from −4.2 mm/year (April) to −7.8 mm/year (June). Therefore, the sharp differences in the magnitudes of the ET_0 changes in the summer and winter seasons indicate the presence of some seasonality in the ET_0 under arid environments of Pilani. On annual time scale, strong decreases in the ET_0 at the rate of 230 mm/decade are witnessed in Pilani.

2.5.3   INFLUENCE OF METEOROLOGICAL PARAMETERS ON ET_0 CHANGES

Several researchers have reported that the ET_0 values have witnessed strong decreasing trends in-spite of reported rise in temperature in different types of environments in India and other parts of the world. Such paradoxical situation has been reported at various sites from northeast India where the ET_0 declined while temperatures have increased in humid environments [8]. Therefore, in the present study, the stepwise regression analysis was carried out in monthly and annual time scales using the SYSTAT, commercial software for statistics, for probing the variables, which has larger influence on the observed behavior of the dependent variable, i.e., the ET_0. The evaluated reference evapotranspiration rates were used as dependent variable and various meteorological parameters as independent variables. The regression analysis helps to identify the dominant causal parameters affecting the ET_0 by selecting one or a group of meteorological variables and eliminating the redundant variables during the analysis [9].

The results of the stepwise regression analysis are shown in Table 2.2. Table 2.2 shows the rank of the step at which a given climatic parameter was selected during the regression analysis, for example, variable selected at step one indicates the most dominating variable influencing the ET_0 process. It is clear from Table 2.2 that wind speed is found to be the most dominant variable influencing the ET rates in all the twelve months as well as in annual time scale in arid environments of Pilani. McVicar et al. [14] have reported statistically significant decreasing trends in terrestrial wind speed over different parts of India, including arid sites from India, in a global review paper. It has been reported that the global wind speed decreases may have occurred due to the possible role played by the increase in terrestrial surface roughness leading to the obstruction of wind flow [13, 18].

TABLE 2.2   Results of Stepwise Regression Analysis Over Pilani Site

Time scale	T_{min}	T_{max}	T_{mean}	Sunshine	Wind	RH_{min}	RH_{max}	RH_{mean}
Jan	–	III	–	IV	I	V	–	II
Feb	–	–	IV	III	I	–	–	II
Mar	IV	III	–	V	I	II	–	–
Apr	–	–	–	IV	I	II	III	–
May	–	–	–	IV	I	II	–	III
Jun	–	–	–	IV	I	–	II	III
Jul	–	III	–	–	I	II	–	–
Aug	IV	III	–	–	I	II	–	V

TABLE 2.2   Continued

Time scale	T_{min}	T_{max}	T_{mean}	Sunshine	Wind	RH_{min}	RH_{max}	RH_{mean}
Sep	III	IV	–	V	I	II	–	–
Oct	–	V	–	IV	I	II	–	III
Nov	–	V	–	IV	I	II	III	–
Dec	–	III	IV	V	I	II	–	–
Annual	–	–	**III**	**IV**	**I**	–	–	**II**

I, II and III indicate the step at which the meteorological variable was selected in the stepwise regression method (I first, i.e., the most dominant variable; and III third).

2.6 CONCLUSIONS

In the present study, the ET_0 estimates were obtained through the PM FAO-56 method from 1961 to 2005 in annual and monthly time scales at Pilani. The mean of decadal ET_0 varied from 1863.7 mm to 1060.10 mm during the 1961–1970 and 1991–2000, respectively. Comparatively higher values of ET_0 occurred in early 1960s to 1970s due to occurrence of relatively very high wind speed and low relative humidity during the same period. However, comparatively very low values of ET_0 occurred in 1990 and during 1996 to 1999 because of occurrence of relatively very low wind speed and higher humidity over Pilani in late 1990s.

The changes in ET_0 estimates obtained through the PM FAO-56 were quantified using the MK test in annual and monthly time scales at Pilani. The magnitudes of the observed trends in the ET_0 were obtained by using the Theil-Sen' test. The results indicate strong decreasing trends in the ET_0 at the 1% level of significance in arid environment of Pilani situated in northeastern region of Rajasthan in all time scales. On monthly time scales, the magnitudes of the observed trends in the ET_0 were found to varying from –1.43 mm/year (January) to –7.83 mm/year (June). On annual time scale, the decreasing trends in the ET_0 are witnessed at the rate of 230 mm/decade in Pilani. The ET_0 changes in the summer and winter seasons indicate the presence of seasonality in the ET_0 under arid environment.

Subsequently, the causal mechanism of the observed changes in ET_0 was ascertained by stepwise regression analysis. Wind speed is found to be the most dominant variable influencing the ET_0 rates in all the twelve months as well as in annual time scale. Relative humidity followed by temperature was found to be other two dominating variables influencing the ET_0 process. These results may help to explain the paradoxical situation (declining ET_0 inspite of rising temperature), which may have occurred due to significant wind speed decreases and rising trends in relative humidity under arid climatic conditions.

2.7 SUMMARY

The values of decadal ET_0 were found to be comparatively very high during the pre-1975 period in comparison to the decadal ET_0 of 1990s because of occurrence of high wind speed and low relative humidity from 1961 to 1974 at Pilani. The ET_0 values declined significantly during the early and late 1990s because of drastic decline in wind speed and increase in relative humidity in arid environments of Pilani. The decline in wind speed and rise in humidity is most likely due to the increase in surface resistance and extensive spread of irrigation facilities in the arid region, respectively. Strong decreases in ET_0 trends arc witnessed in all the time scales at Pilani. Comparatively higher magnitudes of ET_0 decreases are witnessed in pre-monsoon season comprising of months of March to June than the ET_0 decreases during the remaining seasons, which indicate the presence of seasonality in the ET_0. Wind speed followed by relative humidity and temperature are found to be three main dominating factors influencing the process of ET_0 under arid climatic conditions of Pilani.

ACKNOWLEDGMENTS

The authors gratefully acknowledge the support from the India Meteorological Department (Pune), Maharashtra (India) for providing the meteorological data of Pilani (Rajasthan). Also, the authors thank Dr. Megh R. Goyal (Senior Technical Editor-in-Chief, Agriculture and Biomedical Engineering, Apple Academic Press, and Professor in Agric. and Biomedical Eng. University of Puerto Rico – Mayaguez) for his suggestions and support.

KEYWORDS

- Aerodynamic
- Arid environment
- Changing land use pattern
- Climate change
- Combination-based ET_0 method
- Decreasing trends
- Dominating variable
- Evaporative demand
- Evapotranspiration
- Global warming
- Mann-Kendall
- Meteorological parameters
- Non-parametric
- Northwest India
- Penman-Monteith FAO-56
- Radiative factor
- Rajasthan
- Reference evapotranspiration
- Relative humidity
- Seasonality
- Stepwise regression
- Temperature
- Terrestrial surface roughness
- Theil-Sen
- Trends
- Wind speed

REFERENCES

1. Allen, R. G., Pereira, L. S., Raes, D., & Smith, M. (1998). *Crop Evapotranspiration: Guidelines for Computing Crop Requirements*. FAO Irrigation and Drainage Paper No. 56. FAO, Rome, Italy.

2. Census (2011). *Census of India 2011*. Ministry of Home Affairs, Government of India. http://www.census2011.co.in/data/town/800478-pilani-rajasthan.html. Access Date: 06ᵗʰ October 2015.

3. Chattopadhyay, N., & Hulme, M. (1997). Evaporation and potential evapotranspiration in India under conditions of recent and future climate change. *Agricultural and Forest Meteorology*, 87, 55–73.

4. Dinpashoh, Y., Jhajharia, D., Fakheri-Fard, A., Singh, V. P., & Kahya, E. (2011). Trends in reference crop evapotranspiration over Iran. *J. Hydrology*, 399, 422–433.

5. Gong, L., Xu, C., Chen, D., Halldin, S., & Chen, Y. D. (2006). Sensitivity of the Penman–Monteith reference evapotranspiration to key climatic variables in the Changjiang (Yangtze River) basin. *J. Hydrology*, 329, 620– 629.

6. IPCC (2007). Summary for policymakers. In Climate Change 2007. In: *The Physical Science Basis*, Solomon, S., Qin, D., Manning, M., Chen, Z., Marquis, M., Averyt, K. B., Tignor, M., & Miller, H. L. (Eds.). Cambridge University Press: New York, USA.

7. Jhajharia, D., Dinpashoh, Y., Kahya, E., Choudhary, R. R. and Singh, V. P. (2014). Trends in temperature over Godavari River basin in Southern Peninsular India. *Int. J. Climatology*, 34, 1369–1384.

8. Jhajharia, D., Dinpashoh, Y., Kahya, E., Singh V. P., & Fakheri-Fard, A. (2012). Trends in reference evapotranspiration in the humid region of northeast India, *Hydrological Process*, 26, 421–435.

9. Jhajharia, D., Shrivastava, S. K., Sarkar, D., & Sarkar, S. (2009). Temporal characteristics of pan evaporation trends under the humid conditions of northeast India. *Agricultural and Forest Meteorology*, 149(5), 763–770.

10. Kahya, E., & Kalayci, S. (2004). Trend analysis of streamflow in Turkey. *J. Hydrology*, 289, 128–144.

11. Kendall, M. G. (1975). *Rank Correlation Measures*. Charles Griffin, London, 202 pp.

12. Mann, H. B. (1945). Non-parametric tests against trend. *Econometrica*, 13, 245–259.

13. McVicar, T. R., & Roderick, M. L. (2010). Atmospheric science: winds of change. *Nat. Geosci.*, 3(11), 747–748.

14. McVicar, T. R., Roderick, M. L., Donohue, R. J., Li, L. T., Niel, T. G. V., Thomas, A., Grieser, J., Jhajharia, D., Himri, Y., Mahowald, N. M., Mescherskaya, A. N., Kruger, A. C., Rehman, S., & Dinpashoh, Y. (2012). Global review and synthesis of trends in observed terrestrial near-surface wind speeds: Implications for evaporation. *J. Hydrology*, 416–417, 182–205.

15. Sen, P. K. (1968). Estimates of the regression coefficients based on Kendall's tau. *Journal of the American Statistical Association*, 63, 1379–1389.

16. Shuttleworth, W. J. (1993). Evaporation. In: *Handbook of Hydrology*. Maidment, D.R. (Ed.), McGraw-Hill, New York, pp. 4.1–4.53.

17. Theil, H. (1950). A rank invariant method of linear and polynomial regression analysis, Part 3. *Netherlands Akademie van Wettenschappen Proceedings*, 53, 1397–1412.

18. Vautard, R., Cattiaux, J., Yiou, P., Thépaut, J. N., & Ciais, P. (2010). Northern hemisphere atmospheric stilling partly attributed to increased surface roughness. *Natural Geoscience*, 3(11), 756–761.

19. Wang, W., Peng, S., Yang, T., Shao, Q., Xu, J., & Xing, W. (2011). Spatial and temporal characteristics of reference evapotranspiration trends in the Haihe river basin, China. *J. Hydrologic Engineering*, 239–252.

20. Yue, S., Pilon, P., Phinney, B., & Cavadias, G. (2002). The influence of autocorrelation on the ability to detect trend in hydrological series. *Hydrological Processes*, 16, 1807–1829.

CHAPTER 3

EVAPOTRANSPIRATION OF WOODY LANDSCAPE PLANTS

RICHARD C. BEESON

Associate Professor (Plant Physiologist) University of Florida, USA; Agri Res and Education Center, APOPKA 2725 S. Binion Road, Apopka, FL 32703-8504, Phone: (407) 814.6172, Fax: (407) 814-6186; Mailing address: 2725 South Binion Rd, Apopka, Florida 32703. E-mail: rcbeeson@ufl.edu

CONTENTS

This chapter is printed from, "*Richard C. Beeson (2011). Evapotranspiration of Woody Landscape Plants. Chapter 18, pp. 347–370, In: Evapotranspiration—From Measurements to Agricultural and Environmental Applications by Giacomo Gerosa (Ed.). ISBN: 978-953-307-512-9, Open Access Article by InTech, Available from: http://www.intechopen.com/books/evapotranspiration-from-measurements-to-agricultural and-environmental-applications/evapotranspiration-of-woody-landscape-plants.*"

3.1   INTRODUCTION

Landscape trees and woody shrubs are important components of residential, commercial and municipal sites, and are nearly ubiquitous in modern non-nomadic societies. Attractive landscaping can increase property values from 6 to 20% in the USA [24, 51]. Proper selection and placement increases energy efficiency for heating and cooling of structures, provides storm water management, air and noise pollution abatement and carbon sequestration in urban and suburban areas [14, 32, 37, 45]. An advantage of urban trees is that they are located at the source of highest CO_2 and pollution concentrations, and thus can have the greatest impact, if healthy. McPherson et al. [33] calculated dollar benefits of the existing urban forest in Modesto, California compared to its annual tree budget cost. Including storm water and air pollution abatement, energy saving and other tangible cost, benefits outweighed cost by 2 to 1. Non-tangible aspects, such as esthetics, positive psychological being and wildlife habitat are improved when tree and shrub quality are maintained [25, 53]. Healthy maintained landscapes have a calming effect, especially for those who are under stress or depression [53].

Though most beneficial in populated areas, urban sites where woody plants are transplanted are often areas not conducive to normal growth. These sites often have constrained soil volumes [30, 31]. This is frequently due to underground utilities or street and building foundations [43]. Highly compacted soils are also a major problem [11]. If drainage is poor or nonexistent, anaerobic conditions can lead to tree decline and death [8, 30]. Conversely, where drainage is acceptable, soil volume and/or plant available water are often inadequate to supply more than a few days of transpirational demands [21].

In addition to soil volumes and soil water availability, urban environments are often vastly different from natural woodlands or rural settings.

Heilman et al. [26] planted *Ligustrum japonicum Thunb.* 0.5 m from exterior walls facing the cardinal directions, and they found that the actual evapotranspiration (ET_A) was increased with exposure to sun due to emission of long wave radiation from walls, which increased plant temperature during and after sun exposure. Maximum sap flow was about 30% higher at the west wall than other directions. Whitlow et al. [57] measured temperature, humidity and solar radiation at street level in a study of urban trees in Manhattan, New York City. The data were compared to concurrent measurements a few blocks away in Central Park during the summer for a three-year period. Duration of hours of direct sunlight was severely truncated due to shading by tall buildings at street level. The microclimate within this urban canyon had much higher daily vapor pressure deficits (VPD) than those measured at Central Park [57]. Higher VPD was due to both drier air and higher temperature. Despite high VPD, stomata closure was rare. In a similar study conducted in Seattle, Washington, USA, solar radiation in an urban canyon was restricted to 44% of that received at a nearby park [28]. Like canyon trees in Manhattan, trees maintained an appearance of vigor. This was proposed due to acclimation to generally deep shade, resulting in larger and more leaves. In both studies, trees in the urban canyons rarely exhibited signs of water stress and stomata closure was uncommon. The limited exposure to direct sunlight lowered daily ETA in both locations to within the volumes of water available to the trees. Additionally, reduced solar radiation would also reduce re-radiated heat from adjacent buildings and pavement. In Seattle, unlike the Manhattan canyon, VPD were similar to those at the park. In contrast, trees growing in a non-shaded paved plaza, a traffic island bordered by three major arteries in Seattle, exhibited symptoms of severe stress, such as small leaves, limited growth and low xylem water potentials. This was proposed due to chronic water stress induced by limited soil volumes and higher evaporative demand. Higher evaporative demand was principally due to two factors, both of which are common, in varying degrees, in all urban and suburban landscapes.

The first factor was wind. In the open area of the plaza surrounded by multilane highways, there were no barriers to impede natural wind. In addition, vehicles traveling at highway speeds would have augmented nature winds and created almost perpetual winds during the majority of a day. Constant winds through isolated trees would have reduced boundary layers of still air around leaves, increasing transpiration. The thinner the boundary layers, the more quickly water molecules are lost to turbulent air around a leaf [36]. Higher wind speeds would have also increased transpiration from shaded interior leaves in the canopy, by both lowering boundary layers and increasing VPD within the canopy by advection of drier air [41] arising from the heated concrete and pavement surrounding the trees.

The other environment factor contributing to severe stress was convection and re-radiation of energy absorbed by the concrete and pavement surrounding these trees. This heat energy absorbed by the leaves would have increased leaf temperature, resulting in increased density of water vapor inside a leaf, leading to a greater gradient of water vapor between inside and outside a leaf. These higher gradients would accelerate transpiration (latent heat loss). To adapt to these two forces, leaf area was substantially reduced to limit heat loading by the surroundings. Smaller leaves also have thinner boundary layers, allowing for greater conduction of heat across the boundary layer.

Whether in urban or rural areas, trees and woody shrubs usually need supplemental irrigation immediately after transplanting into a landscape to speed root growth to re-establish their natural root to leaf ratios. The duration of supplement irrigation depends on plant size and species. In central Florida (latitude 28°N), root growth can occur year round for evergreen species and 10 months for deciduous species. Here root regeneration occurs at a rate of 1–5 months per cm in trunk caliper (measured at 15 cm above the soil line) [19]. For trees with broadly spreading roots, such as *Quercus laurelifolia* and *Pinus elliotii*, establishment was 1–1.2 months per cm caliper whether trees were grown in above ground containers or in the ground. Root regeneration was slower for species with dense, slowly expanding root systems, such as *Ilex attenuata* East Palatka". Root regeneration for this species was 2.4 month per cm trunk caliper if grown in ground, or 4.8 months if transplanted from a container.

Upon analysis of existing literature, Watson [56] noted that tree establishment after transplanting was generally consistent across hardiness zones if based on trunk caliper and length of growing season when well-watered. Under-irrigated trees have sparse canopies. Insufficient leaf area restricts new root generation and elongation. Such trees do not perform their intended functions in a landscape. If they survive, it can take years to regain the vigor and health they had during nursery production. To maintain esthetically pleasing, healthy plants, supplemental irrigation is frequently required in landscapes after plant establishment. How much water woody plants use, and equally important, how much water they require, is generally unknown. Water use efficiency of most woody plants increases as soil water availability declines, up to a point. Thereafter, excessive stomata closure limits growth and eventually survival. Most research quantifying tree and woody shrub water use has been related to trees under forest conditions, or trees and shrubs during nursery production. Under these conditions, the answers have been to the question of how much water woody plants use. These answers generally establish the highest water use of a species for a given size. These values are ideal for establishing new plants into a landscape or rapidly growing plants to fulfill their place in the landscape. But once this is achieved, continuing to irrigate at production levels can result in excess pruning to maintain plant esthetics, and is generally wasteful. It is the latter question, how much do plants require that the least is known.

Going forward, the discussion will first focus on what is known of how much water plant use. Thereafter the discussion will be of how to modify known plant water use to plant water needs in landscapes and an alternative irrigation strategy that takes advantage of woody plant stomata control and perennial root systems.

3.2 WOODY SHRUBS

In terms of number of plant species, there has been more research quantifying water use of woody shrubs than quantifying that of trees. However, much of this research is based on small plants in 3.8 L containers, with nearly all related to irrigation needs during production. Knox [29] grew five species of woody shrubs in north Florida (lat. 30.31 N). As plants neared market size, daily water use during late spring to early summer ranged between 0.15 to 0.30 L/day. Burger et al. [9] grew 22 species of landscape plants in 3.8 L containers at three locations in central (38.33 N) and southern (33.56 N) California. When plants reached marketable size, daily ETA was quantified by weighing. ETA was then divided by the container upper surface area and reference evapotranspiration (ETo) at each location to calculate a coefficient (Kc). Actual volumes were not reported, but Kc's were. The Kc ranged from 1.1 to 5.1. Later, Regan [40] conducted a similar but much larger experiment, calculating Kc for 50 species of woody shrubs and trees in 3.8 L containers in northern Oregon (45.13 N). These Kc's at market size ranged from 2.3 for *Tsuga Canadensis* (l.) *Carriere* to 5.6 for *Hydrangea macrophylla Thumb*, 'Nikko Blue'. Since container diameters were approximately 15 cm in both studies, if ETo for a local condition is known, water use for these species could be estimated. However, basing Kc on a fixed container diameter is only accurate, if plant canopy sizes are similar between the reference plant and the plant ETA that should be estimated for. Larger plants in the same container would have higher a Kc since larger plants would transpire more water, but the volume would be normalized by the same upper container surface area.

Garcia-Navarro et al. [16] transplanted four plant species from 3.8 L containers into 200 L drainage lysimeters, then allowed them to establish and grow for 15 months in central California (38.33 N). Daily ETA was then quantified and averaged for three additional months. Mean ETA was 2.3 L/day for *Spiraea x vanhouttei* (Briot.) Xabel, 1.8 L/day for *Viburnum tinus* L., 1.45 L/day for *Arctostaphylos densiflora* M.S. Baker 'Howard McMinn', and 2.0 L for *Leucophyllum frutescens* (Berl.) I.M. Johnst. Beeson [3] summarized results for three woody shrub species averaging marketable size in three container sizes (Table 3.1: 28.40 N).

TABLE 3.1 Mean Canopy Dimensions and Daily Water Lost (Liters) From Container Grown Plants

Species	Container size		
	3.8 L	11.4 L	26.6 L
Viburnum odoratissimum	0.20 x 0.30 *** 0.40	0.45 x 0.60 *** 1.14	0.90 x 1.2 *** 2.53
Ligustrum japonicum	0.20 x 0.30 *** 0.34	0.45 x 0.60 *** 1.08	0.90 x 1.2 *** 2.36
Rhaphiolepis indica	0.23 x 0.15 *** 0.25	0.57 x 0.38 *** 0.66	0.90 x 0.6 *** 1.48

Legend: Mean canopy width (m) x height (m); *** Mean L/day.

Means are based on 3 replications per species and container size.

Data was collected 1 Jan 1995 to 31 Dec. 1996 from Beeson [3].

According to Beeson [3], mean daily ETA ranged from 0.255 L for *Rhaphiolepis indica* (L) Lindl. Ex Ker-Gawl grown in 3.8 L containers to 2.53 L for *Viburnum odoratissimum* Ker-Gawl grown in 26.5 L containers. Winter daily water use was generally half the average, while summer water use was 50% more. In 2004, Beeson [4] presented a graph of daily ETA of *Ligustrum japonicum Thumb.* grown in 11.4 L containers (28.40 N). At marketable size (0.6 m tall and 0.45 m width) and spaced 43 cm on center, ETA was about 0.80 L in the spring, with an average ETo of 4.6 mm/day. These ETA volumes were comparable to flow rates of the same species reported by Heilman and Ham [30.36 N; 27]. Steinberg et al. [50] reported somewhat smaller daily water use of 0.3 to 0.5 L for market size *Ligustrum japonicum* in 7.6 L containers in early June in a shaded greenhouse in eastern Texas (30.36 N).

3.3 TREES

Most studies of short-term water use of both conifer and hardwood trees have occurred in forest, and thus are not accurately related to water use in landscapes. Yet the same species are among the mainstays of most landscapes. With some adjustment, detailed later, the values derived from forest trees can be used as estimates for trees in landscapes. Wullschleger et al. [58] reviewed forest tree water use reported from 1970 through 1998. The review covered 67 species of trees gleaned from 52 studies. It focused on contributions of the different methods of quantifying tree water use related to a holistic approach for understanding water movement within trees, between canopies and the air above. Ninety percent of the tree water use surveyed ranged from 10 to 200 L/day. Selected studies included in the review have been included in the discussions in this chapter where specific values are relevant to landscape ETA.

A limited number of studies have quantified water use of small landscape trees. Costello et al. [12] reported cumulative water loss from three tree species grown in 15.6 L containers in a shaded courtyard versus an open knoll (38.33 N). The daily average over a 14-day period for the open location was 1.2, 1.18 and 0.89 L for *Liquidambar styraciflua* L., *Sequoiadendron giganteum* (lindl.) *J. Bucholz,* and *Magnolia grandiflora* L., respectively. The average reduction in ETA for trees in the courtyard was 54%. Tree size was not indicated. Similar daily ETA (1.3 L) for a *Taxodium distichum* (L.). Rich growing in a 15.6 L container in July in Texas (30.36 N) was reported by Steinberg et al. [48]. In Northwest Texas (33.2 N), ETA of *Prosopis glandulosa* Torr. with trunk diameter of 1.9 cm ranged from 0.25 to 2.5 L per day [13]. Cumulative water use of trees over a full year's growth is seldom reported. However, the annual cumulative ETA of seedlings of 10 tropical trees transplanted into 200 L drainage lysimeters in India (26.85 N) ranged from 1.358 L for *Pongamia pinnata* (L.) Pierre to 5,324 L for a *Eucalyptus* hybrid [10].

3.3.1 ISOLATED TREES

For large to mature trees, daily water use of only a small number of isolated trees have been quantified, and then principally for only a few days to weeks. Montague et al. [35] measured the ETA of five species of trees transplanted as balled and burlapped material into in-ground weighting lysimeters (41.44 N) during the summer. Mean daily ETA averaged between 3.2 L/day of *Acer platanoides* L. 'Emerald Queen' (25.2 cm^2 trunk area) to 5.1 L/day for *Fraxinus pennsylvanica* Marsh. 'Patmore' (36.3 cm^2 trunk area). Steinberg et al. [49] reported daily ETA in Texas (32.12 N) of an 11cm caliper *Carya illinoensis* (Wangenh.) K.Koch 'Wichita' during the summer to range from 100 to 150 L per day over a 12-day period in August, based on weighing lysimeter measurements. The tree was 3.9 m tall, with a *diameter at breast height* (dbh, 1.4 m) of 7.9 cm. Green [22] reported summer daily water use of a 10 year old *Juglans* L. species (walnut; 3.4 x 3.1 m canopy widths) in New Zealand (40.2 S) to range from 14 to 40 L, depending on environmental factors. In New Zealand pastures (40.08 S), where canopy coverage by *Populus deltoids* (Bart. Ex Marsh, Clone 178) was 66%, ETA determined by sap flow ranged from 162 to 417 L/day for trees with projected crown areas of 97 to 275 m^2 [23]. Part of the variability among trees was due to location on the hillside and shading by neighboring trees. Total ETA was highly correlated with sapwood area (r = 0.86) and dbh (r = 0.93). Ruiter [42] reported ETA of *Pinus radiata* with a caliper of 13 cm (at 15 cm) to average 21 L maximum daily ETA over a 4 week period (37.49 S) in the summer.

The earliest large isolated trees measured by lysimetery were reported by Edwards [15]. From these, graphs were published for the daily ETA of four species of trees, ranging from about 8.9 to 20 cm in caliper at 15 cm (40.21 S). For the deciduous species *Populus x euramericana* (Dode) Guinier cl 'Flevo', with a trunk diameter of about 11 cm, ETA from fall through early spring ranged from 0 to about 5 L/day. ETA was increased in spring, to maintain between 70 and 90 L/day during summer, dropping quickly to near zero in mid-autumn. A similar seasonal pattern was presented for *Salix matsudana Koidz*. When leafless, ETA ranged from 0 to 5 L/day. During the peak of summer, ETA ranged from 50 to 60 L/day for this 8.8 cm trunk diameter tree. An *Eucalyptus fastigata* Dean and Maid 3.3 m tall generally transpired 10 L/day during winter, but increased to a mean of 65 L/day during the summer. The other evergreen species *Pinus radiata* was about 20 cm in caliper. Winter to late spring ETA was around 40 L/day. Throughout the summer, ETA ranged from 85 to 125 L/day, decreasing to the 60 to 80 L range in the fall.

3.3.2 FOREST TREES

Within forest, daily tree water use is generally smaller than for isolated trees. In Germany (51.13 N), *Fraxinus excelsior* L. saplings 4.5 m tall and 2.84 cm dbh registered maximum transpirations of about 7 L/day in early August under optimum conditions [52]. In England (51.27 N), whole tree ETA was calculated for a *Populus trichocarpa Toor* and *A. Gray x P. tacamahaca* L. hybrid based on sap flow through branches, scaled to the whole tree level. Trees were growing on a site with a shallow water table. With trunk dbh of 4.2 to 4.6 cm and heights of 5 to 6 m, peak ETA was 23 to 28 L/day in June [59]. In a forest near Melbourne, Australia (37.34 S), the porometer method was used to measure ETA of a large *Eucalyptus regnans* F.J. Muell. [55]. For a tree 55 m tall and 0.83 m dbh, flow rates through the sapwood ranged from 61 to 323 L/day. Granier [21] found that dominant trees in a *Pseudotsuga menziesii* Carrière forest (48.44 N) averaged 22 L per day for trees about 18 cm dbh in France during the summer. Schulze et al. [46] estimated ETA of a *Picea abies* L. (28 cm dbh; 49.2 N) and a *Larix hybrid* (26 cm dbh; 49.95 N) growing in the Czech Republic and Germany, respectfully. In the summer, ETA's were 63.1 L the *Picea* and 74.4 L for the *Larix*. More recently Verbeeck et al. [54] reported maximum daily sap flows of 28 L/day in May for *Pinus sylvestris* L. growing in a forest in the Belgian Campine region (51.18 N). The tree was nearly 20 m tall with a dbh of 26.7 cm.

3.3.3 TROPICAL TREES

Daily water use of tropical trees is among the highest of species quantified to date. Andrade et al. [1] measured sap flow of five species of tropical trees in the Panama rainforest (8.58 N) during a period when soil water availability was not limited. Eighteen meter tall *Cecropia longipes* Pitt. with a dbh of 19.7 cm had a cumulative sap flow rate of 46.5 L/day. A *Spondias mombin* L. tree 23 m tall with a dbh of 33.1 cm averaged 80 L/day. *Luchea seemannii* Triana and Planch. was 129 L/day for a 29 m tall tree with a 38.2 cm dbh. A *Ficus insipida* Willd. with a 56.7 cm dbh trunk sap flow rate was 164 L/day for a 30 m tree. The highest sap flow was from an *Anacardium excelsum* (Bertero and Bald. ex Kunth) Skeels tree at 379 L/day from a 15 m tall tree with a 101.8 cm dbh. Total sap flow rate was highly correlated (r = 0.99) with trunk dbh. With the exception of isolated trees in high-density urban areas, once trees approach the small forest sizes detailed above, there is little need for supplemental irrigation. On average, if unrestricted, roots systems of landscape trees are generally three times the width of their canopy [17, 18]. At this point, unless there are substantial disruptions to the rooting volume, such as new construction, these large trees in landscapes are self-sufficient in terms of water needs.

3.3.4 DAILY TREE WATER USE

While there have been studies quantifying short term water use of landscape trees, or trees that could be used in landscapes, until recently there have been none that track tree water use for the same tree from propagated material (seedlings or rooted cuttings) to trees of landscape stature. In 2001, a project was initiated in Central Florida (latitude 28° 40') to quantify daily water use of three landscape tree species: *Acer rubrum* L.

(red maple), *Quercus virginiana* Mill. (Live oak), and Ilex x 'Nellie R Stevens' (holly). Both the oak and holly are evergreen species. In early spring 2001, propagules were transplanted in 26 L polyethylene containers and placed in suspension lysimeters [6]. Each tree was surrounded by like trees, but spaced so that canopy coverage was about 50%. Tree ETA was recorded daily for the following 5 to 6 years as trees were transplanted into ever-larger containers up to 1.4 m diameter.

3.3.4.1   Ilex

In six years, the holly tree grew from a trunk caliper of 3.7 mm to 130 mm measured 15 cm above soil line. The holly cultivar, 'Nellie R. Stevens', is slow growing, with an average maximum height of 9 m [20]. Like most hollies, it was produced with lower branches left close to the ground. Because this holly was a true evergreen species, after the first two years the tree retained two or three years' worth of leaves. Thus leaf area was relatively constant or increasing throughout the experiment. Bell-shaped variations in ETA over a year (Figures 3.1–3.5) are therefore reflective of

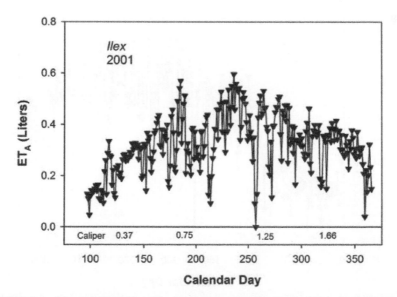

FIGURE 3.1   Mean daily ETA of the holly, Ilex x 'Nellie R. Stevens', for 2001. Trunk diameter (cm), indicated at the bottom of the graph (Caliper), measured at 15 cm inches above the substrate level.

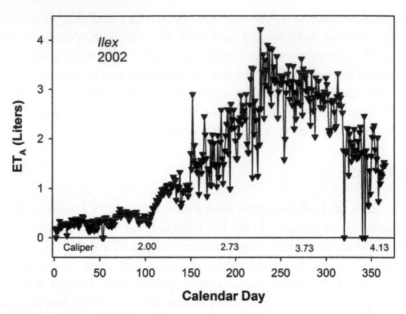

FIGURE 3.2 Mean daily ETA of the holly, Ilex x 'Nellie R. Stevens', for 2002. Trunk diameter (cm), indicated at the bottom of the graph (Caliper), measured at 15 cm inches above the substrate level.

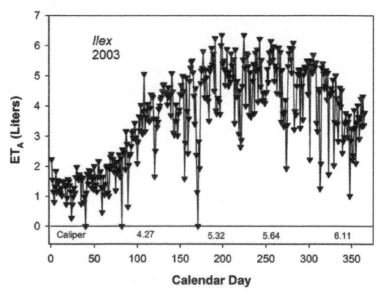

FIGURE 3.3 Mean daily ETA of the holly, Ilex x 'Nellie R. Stevens', for 2003. Trunk diameter (cm), indicated at the bottom of the graph (Caliper), measured at 15 cm inches above the substrate level.

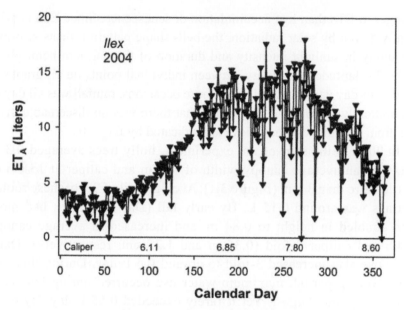

FIGURE 3.4 Mean daily ETA of the holly, Ilex x 'Nellie R. Stevens', for 2004. Trunk diameter (cm), indicated at the bottom of the graph (Caliper), measured at 15 cm inches above the substrate level.

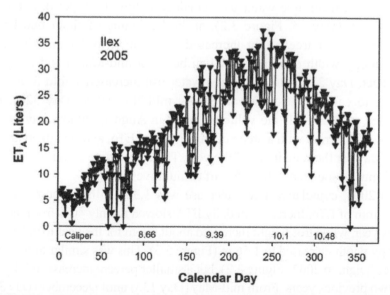

FIGURE 3.5 Mean daily ETA of the holly, Ilex x 'Nellie R. Stevens', for 2005. Trunk diameter (cm), indicated at the bottom of the graph (Caliper), measured at 15 cm inches above the substrate level.

changes in ETo than leaf accumulation or senescence. Since ETo is principally driven by solar radiation, the bell- shape pattern reflects seasonal variability in sunlight intensity and duration of the northern hemisphere at the 28° latitude. Variability between individual points on all graphs is the day-to-day variation in ETA. On rare occasions, rainfall was all day or intermittently frequents enough such that there was no discernable mass loss from a lysimeter. These days are indicated by $ET_A = 0$.

In 2001 at the start of the experiment, holly trees averaged 0.3 m tall, with an average canopy width of 12 cm and caliper at 15 cm of a little more than 3 mm (Figure 3.1). Average daily ETA of these rooted cuttings was around 0.15 L. By early fall (Day 255) trees had more than doubled in height to 0.84 m, and increased in average canopy width and caliper 3-fold (0.35 m and 1.25 cm, respectively). Daily water use also increased 3-fold to around 0.5 L/day. During this first year growing period, maximum water use occurred during late summer in July and August. Yet it rarely exceeded 0.55 L/day. By early November (Day 305) tree water use was generally less than 0.4 L/day. Over the course of the 257 days this young tree was measured in 2001, cumulative ETA was 86 L. The cumulative ETo was 1,074 mm.

In 2002, mean tree water use remained below 0.4 L per day through mid-March (Day 75, Figure 3.2), at the beginning of shoot bud break. During the year tree height increased about 60%, to 2.1 m, with average canopy width doubling to 1.1 m between mid-March and early fall (October, Day 274). However, tree water use increased 7-fold from mid-March to early August (Day 213). By mid-November (Day 320), ET_A dropped to about half the value recorded in August, without any loss of leaves, concurrent with a steep decline in daily ETo, and remained around 2.2 L/day for the remainder of the year. The cumulative ETo for 2002 was 1477 mm, associated with 566 L of cumulative ET_A.

In 2003, cumulative tree water use was 1,305 L that occurred during 1,328 mm of ETo. Increases in daily ETA slowed to only 40% from the previous year, while tree height increased about 0.6 m for the year to 2.4 m, and canopy width increased to 1.56 m (Figure 5.3). This was similar to increases in tree height in 2002 (Figure 3.2), but a smaller percent increase in ET_A than the two previous years. From mid-May (Day 135) until December (Day 335) ETA was generally consistent between 4.7 to 6.2 L/day (Figure 3.3). There was little reduction in ETA in early winter as observed in previous years.

In 2004, ETA remained at around 4.5 L/day until early April (Day 93, Figure 3.4). As in previous years, tree height increased around 0.6 m over the year. Average canopy spread increased to 2.0 m, similar to earlier years. With the increase in canopy spread, tree ETA increased 3.5 - fold compared to the winter months. Like in previous years, with exception of 2003, ETA declined by more than half (13 to 6.5 L/day) from early October (Day 275) to early December (Day 336). Although three hurricanes buffeted the research plot in 2004 (Days 220 to 255), they had little effect on ETA or tree growth the holly. Cumulative ETA, as trunk caliper increased from 5.9 to 8.0 cm was 2,976 L, with a cumulative ETo of 1,403 mm.

Tree height maintained its constant pace of a 0.6 m increase in 2005. Canopy width increased at a similar constant rate of 0.5 m, while trunk caliper increase was also on pace with previous years at 19 mm. Also consistent was peak ETA, increasing from 6 to about 34 L/day during summer (Figure 3.5). Like previous years, ETA began declining in early October (Day 275) and was down to about half of its summertime peak by December, corresponding to an approximate halving of daily ETo. Cumulative ETA for 2005 was 6,789 L with 1,363 mm of ETo.

In 2006, ETA peaked at around 50 L/day in late June (Figure 3.6). At this point the tree was 4.2 m tall with an average canopy width of 2.78 m. From the median ETA of 17 L/day in January, there was again a 3-fold increase in ETA from winter to late summer. At its peak, ETA frequently varied about 50% over periods as short as a week. Similar variability can be seen through all graphs for the holly. In 2006, mean cumulative ETA was 10,227 L from this tree in a 1.4 m diameter container. Cumulative ETo was 1451 mm. When the experiment was terminated at the end of 2006, total ETA to grow this tree from a rooted cutting to a tree with a trunk caliper of 13.0 cm, height and width of 4.2 m and 2.8 m, respectively, was 21,949 L.

3.3.4.2   Quercus

The Live oak seedling was germinated during the winter of 2000–2001. At the beginning of March 2001, the seedling was 0.38 m tall, with a stem caliper at 15 cm 3.04 mm (Figure 3.7). During the first year of growth, tree height increased 3.6-fold, to 1.4 m, with a trunk caliper increase of 4-fold to 1.27 cm. Similar growth rates of Live oak seedlings have been reported before for trees in 11.4 L containers [2]. Initial mean ETA was similar to

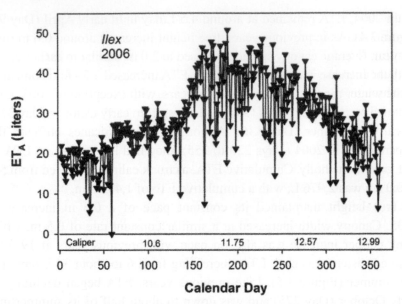

FIGURE 3.6 Mean daily ETA of the holly, Ilex x 'Nellie R. Stevens', for 2006. Trunk diameter (cm), indicated at the bottom of the graph (Caliper), measured at 15 cm inches above the substrate level.

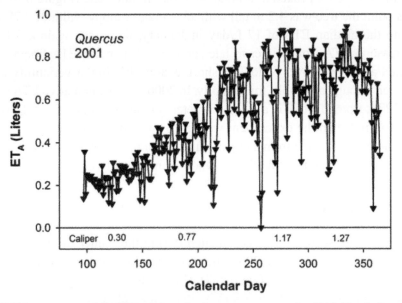

FIGURE 3.7 Mean daily ETA of the oak, *Quercus virginiana* for 2001. Trunk diameter (cm), indicated at the bottom of the graph (Caliper), measured at 15 cm inches above the substrate level.

that of the holly, but ETA of the oak increased more rapidly, obtaining a 4-fold increase in daily ETA to 0.76 L/day within 120 days after transplanting (Figure 3.7). Unlike holly, ETA did not decline as much late in the year until late December. Cumulative mean ETA for 2001 was 136 L.

Through the first part of 2002, oak ETA was around 0.5 L/day until early March (Day 60; Figure 3.8). Though not evident in the graph, there was a dip in ETA from mid-February (Day 45) until early March. This corresponded with annual late winter leaf drop and bud burst of Live oak. With leaf development, ETA increased relatively gradually from early March through mid-May (Day 140), quadrupling with increasing shoot growth. By mid-August (Day 227), mean daily ETA had increased to 9.5 L/day with a further increase in early October (Day 275) with a final cycle of shoot flush. In November, tree growth had stopped at 2.63 m in height and increased average canopy width of 2.7 m and trunk caliper of 4.82 cm. By late December, ETA had declined to 4 L/day, and remained so until bud burst the following spring. Cumulative mean ETA for 2002 was 1,527 L.

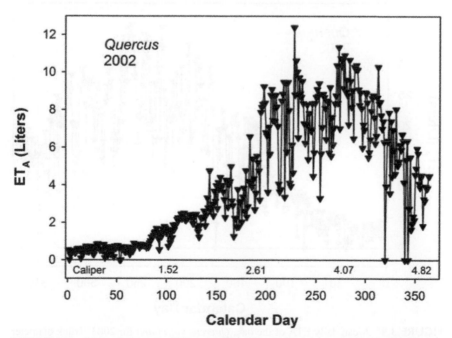

FIGURE 3.8 Mean daily ETA of the oak, *Quercus virginiana* for 2002. Trunk diameter (cm), indicated at the bottom of the graph (Caliper), measured at 15 cm inches above the substrate level.

In 2003, leaf drop in February briefly reduced daily ETA to 1 L/day (Figure 3.9). ETA peaked over 36 L/day in mid-September (Day 250) as the tree grew from a height of 2.63 to 3.92 m. Daily ETA rates between 26 and 30 L/day persisted from early July (Day 182) through mid-November (Day 315), later declining to around 16 L daily by early December. Cumulative mean ETA for this Live oak in 2003 was 5,728 L. At the end of the year, the tree was 4.12 m tall, with an average canopy width of 2.9 m and trunk diameter of 9.42 cm.

In 2004, leaf change occurred closer to mid-March (Day 75), with a substantial decline in ETA from a few weeks before (Figure 3.10). The tree began transpiring at its 2003 peak rates of 30 L/day by early May (Day 125). About 60 days later in early July (Day 185), mean daily ETA had doubled to over 65 L/day. During this time, tree height had increased by 0.47 m, with mean a mean width increase of 1.0 m. From mid-August (Day 211) through late September (Day 274), the tree was impacted by

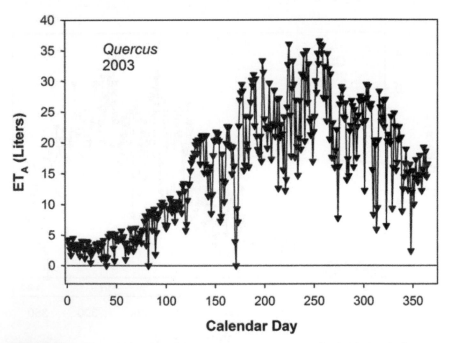

FIGURE 3.9 Mean daily ETA of the oak, *Quercus virginiana* for 2003. Trunk diameter (cm), indicated at the bottom of the graph (Caliper), measured at 15 cm inches above the substrate level.

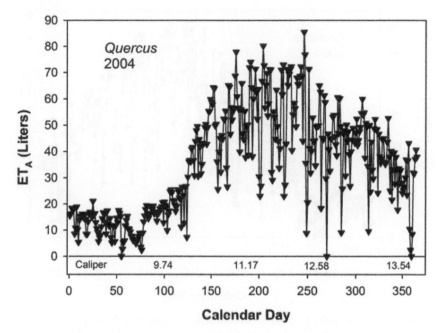

FIGURE 3.10 Mean daily ET_A of the oak, *Quercus virginiana* for 2004. Trunk diameter (cm), indicated at the bottom of the graph (Caliper), measured at 15 cm inches above the substrate level.

three hurricanes. The impact was minor with the first two, until the last hurricane, Jeanne, stripped leaves from southeastern side of the tree in late September (Day 273). Peak winds of 145 Km/h were recorded near the site. These leaves were replaced, resulting in a slight bulge in ETA in late October/early November (Day 290) but with little change in canopy size. By December, ETA declined to 30 to 35 L/day. Final dimensions of the tree for 2004 were a height of 5.28 m, mean width of 4.0 m and trunk caliper of 13.5 cm. Cumulative ETA was 12,827 L.

In 2005, the effect of leaf drop on ET_A in the spring was more evident, occurring between late February (Day 50) and late March (Day 80; Figure 3.11). By early May (Day 125), ET_A regained its summer peak of 75 L/day the year before. From late June (day 170) to mid- September (Day 255), ET_A was generally around 115 L/day or better. During this time, peak ET_A was 140 L/day with a trunk caliper of 16.7 cm. Tree height increased 0.97 m, while average tree spread increased 0.8 m during 2005. Trunk caliper had the most

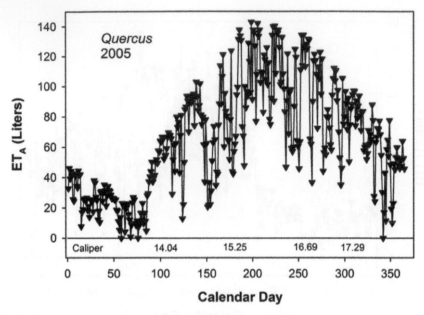

FIGURE 3.11 Mean daily ET_A of the oak, *Quercus virginiana* for 2005. Trunk diameter (cm), indicated at the bottom of the graph (Caliper), measured at 15 cm inches above the substrate level.

impressive increase, starting at 14.0 cm in spring and increasing to 16.69 cm by late September. From mid-September (Day 255) to early December, ET_A declined from 130 to 70 L/day with no loss of leaf area, due solely to decreases in ETo from an average of 5.0 mm to 1.6 mm daily. Cumulative ET_A for 2005 was 23,898 L.

In 2006, leaf drop began earlier than in 2005, starting in mid-February (Day 45) and running through early March (Day 70 Figure 3.12). During this transition, ET_A fluctuated between 30 and 60 L/day. The tree was 6.88 m tall and 5.05 m in average canopy width. Like 2005, trees obtained their previous summer's ET_A in early May (Day 125). Mean ET_A increased to 177 L/ day in late June. Shortly after the peak, the tree was blown over during a thunderstorm, destroying two of the load cells. Data collection was terminated at this time. Through the duration of data collection, tree height increased from 0.38 to 6.74 m, with a mean canopy width of 5.3 m, a trunk caliper increase from 2.3 mm to 18.73 cm. Until its demise in June, the tree transpired 13,709 L during a period of 689 mm of ETo. Mean cumulative ET_A for this live oak from seedling though mid-June 2006 was 57,825 L.

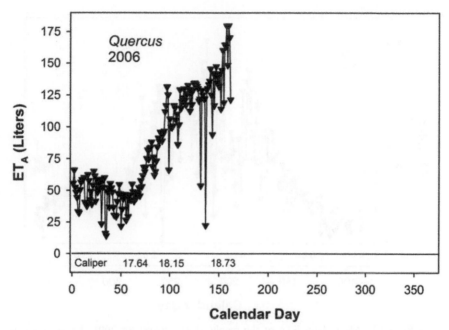

FIGURE 3.12 Mean daily ETA of the oak, *Quercus virginiana* for 2006. Trunk diameter (cm), indicated at the bottom of the graph (Caliper), measured at 15 cm inches above the substrate level. Data collection was terminated early due to storm damage.

3.3.4.3   Acer

Like the holly, the maple was a rooted cutting from a maple selection by Trail Ridge Nursery in north Florida. The maple was the most rapid growing of the species in terms of both shoots and root growth. Each year at transplanting, maples had completely extended their roots throughout the container volume. Being deciduous, fluctuations in ET_A were very dramatic between seasons, like those reported by Edwards [15]. From late December until early March, the tree was leafless. Thus the rise in ET_A in spring was quite rapid. This increase coincided with increasing day lengths and ETo. Conversely, leaf senescence in December lowered ET_A at the same time ETo was declining with shorter days and cooler temperatures. Thus differentiating between changes in leaf area and function, with those related to seasonal changes in ETo are difficult.

In 2001, ET_A of the maple was similar to other species initially, around 0.18 L/day (Figure 3.13). As the tree grew from 0.45 to 2.05 m in height the first

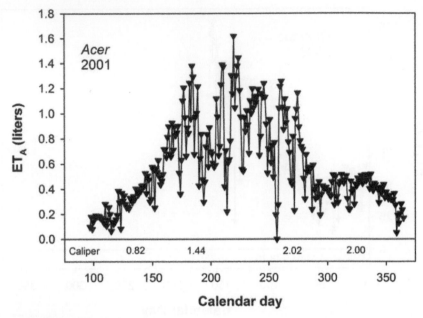

FIGURE 3.13   Mean daily ETA of the maple, *Acer ruburm* for 2001. Trunk diameter (cm), indicated at the bottom of the graph (Caliper), measured at 15 cm inches above the substrate level.

season, ET_A increased to around 1.1 L/day in mid-August. This ET_A was twice as much as measured for the oak and three times that of the holly. ET_A began declining in mid-September (Day 260) and was quite low by early October (Day 280). This corresponded with a disease on the leaves that induced early leaf senescence. By October, most leaves had fallen. In later years, protective sprays prevented early leaf senescence. When growth ceased for the year, mean canopy width had increased from 0.12 m to 0.65 m, while trunk caliper had increased from 4.9 to 20.0 mm. Cumulative mean ET_A for 2001 was 156 L.

In 2002 bud break and leaf expansion began in late March, around Day 85 (Figure 3.14). Prior to this, ETA was equivalent to that of the young tree initially, about 0.25 L/day. ETA increased from near nothing to 3.0 L/day over the next 60 days (Day 140, late May). ETA was around 11 L/day from late July (Day 205) to early October (Day 275). During this time trees increased from 2.05 to 3.54 m in height and from 0.65 to 1.8 m in canopy spread (Figure 3.14). Leaf senescence began in mid-December (Day 350) and was complete by month's end. Cumulative ETA in 2002 was 1,728 L. Over the year, trunk caliper increased from 2.0 to 5.39 cm.

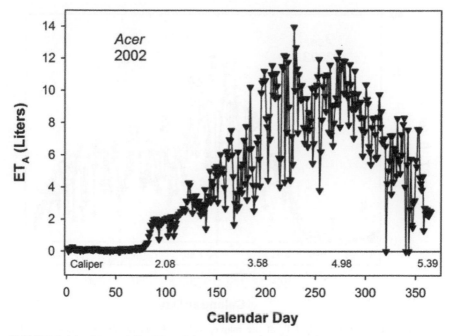

FIGURE 3.14 Mean daily ET_A of the maple, *Acer ruburm* for 2002. Trunk diameter (cm), indicated at the bottom of the graph (Caliper), measured at 15 cm inches above the substrate level.

In 2003, the tree flowered before leaf expansion (Figure 3.15). Flowering began in mid-February (Day 45), with leaf and shoot growth not beginning until late March (Day 80). Although there was likely little photosynthesis and no shoot or leaf growth, ET_A more than doubled with flowering, yet it was still less than 2 L/day. Increase in ET_A was more rapid with the larger tree in 2003 than in 2002 (Figure 3.14), with ET_A increasing from 2.0 to 23 L/day in 45 days (Figure 3.15). ET_A peaked over 30 L/day in early June (Day 155) and continued generally around 25 L/day through the first week of July. During this period, stems with red leaves, indicative of new and expanding tissue, were 30 to 45 cm in length on most major branches. The second week of July shoot elongation nearly stopped, resulting in a rapid and dramatic drop in expanding leaves. This coupled with maturation of previously expanding leaves resulted in substantial rapid decline in ET_A of 8 L/day. Shoot elongation remained limited thereafter, growing 0.58 m in height, with no measurable increase in average canopy width after the middle of July. Leaf senescence initiated in mid-December and was nearly

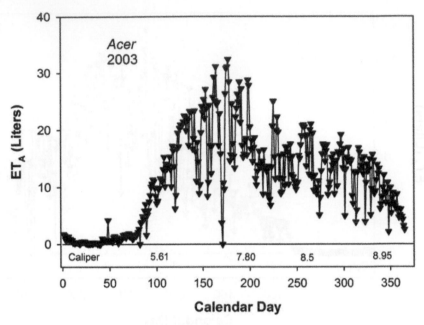

FIGURE 3.15 Mean daily ET_A of the maple, *Acer ruburm* for 2003. Trunk diameter (cm), indicated at the bottom of the graph (Caliper), measured at 15 cm inches above the substrate level.

complete by end of the year. Cumulative ET_A for 2003 was 4,121 L as the tree grew from 3.54 to 5.17 m in height and from 1.56 to 2.9 m in width, with trunk caliper increases of 3.6 cm.

Prior to flower initiation in mid-February 2004 (Day 45), ET_A was less than 4 L/day, and increased to less than 7 L/day during flowering (Figure 3.16). With shoot and leaf expansion beginning in mid-March (Day 75), ET_A increased from 7 to 43 L/day over a 45-day period. From early June until mid-July (Day 200), ET_A ranged around 55 L/day. As in 2003, shoot elongation slowed dramatically in mid-July, resulting in ET_A of 38 to 45 L/day until the mid-August leaf loss due to the first hurricane (Charlie) of that year (Day 226). For the remaining of the fall, ET_A dropped following each sequential hurricane as leaves were torn or lost entirely. The 0 ET_A on Day 270 occurred during the peak of Hurricane Jeanne. Thereafter ET_A then steadily declined until leaf senescence was completed in mid-December (Day 350). Despite fall storms, the tree increased in height from 5.17 to 6.9 m and width by 2.9 m to 4.1 m. Cumulative ET_A for 2004 was 9,459 L. Trunk caliper at the end of leaf senescence was 13.25 cm.

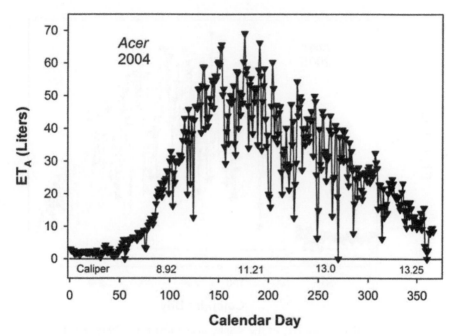

FIGURE 3.16   Mean daily ET_A of the maple, *Acer ruburm* for 2004. Trunk diameter (cm), indicated at the bottom of the graph (Caliper), measured at 15 cm inches above the substrate level.

In 2005, ET_A was around 7 L/day until flowering in mid-February (Day 45, Figure 3.17). With onset of shoot elongation (Day 90), tree water use rapidly increased from 11 to 75 L/day over a 2-week period. From first of June (Day 150) until mid-August (Day 225) ET_A was generally around 85 L/day, peaking up to 110 L/day. As in previous years, slowing of shoot elongation reduced ET_A for remainder of the fall, though not as dramatically as in 2003 (Figure 3.15). This slowdown in shoot growth occurred about a month later than in previous years. Whether this was due to being a larger tree with more branches, a residual from stress from hurricanes the year before or a new pattern of grows cannot be determined. Data collection on the maple stopped the first week of November. During this last year, the tree increased 1.1 m in height growth to final height of 8.0 m, and, increased in width by 0.45 m for an average canopy spread of 4.55 m. When terminated, tree caliper was 18.6 cm. Cumulative ET_A for 2005 was 16,491 m during a period when cumulative ETo was 1,242 mm. To grow from a maple of 0.34 m tall to one of 8 m tall required 4.75 years and 31,955 L of ET_A.

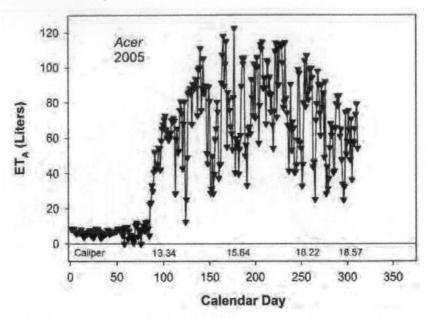

FIGURE 3.17 Mean daily ET$_A$ of the maple, *Acer ruburm* for 2005. Trunk diameter (cm), indicated at the bottom of the graph (Caliper), measured at 15 cm inches above the substrate level.

Beginning in spring 2005, data collection of three additional tree species began in similar 26 L polyethylene containers that were placed in suspension tripod lysimeters [6]. This second set consisted of *Pinus elliotii*, *Ulmus parvafolia* and *Magnolia grandiflora*. Data collection was the same as the first set of tree species, beginning in March and continuing through January 2011. Single tree examples of daily ET$_A$ for each of these species and each year can be found (http://mrec.ifas.ufl.edu/rcb/default.html).

3.4 TRANSLATING KNOWN ET$_A$ TO LANDSCAPE IRRIGATION

As noted above, ET$_A$ of trees in forest are usually lower than that measured for smaller isolated trees. This effect is due to the degree of canopy closure, measured as *projected canopy area* (PCA) density. In January 2003 (28.48 N), market-quality shrubs of *Viburnum odoratissiumum* in three sizes (3.8, 11.4 and 26.6 L containers) were used to verify a result obtained in 1997 [5]. Mean plant heights were 0.55, 0.80 and 1.23 m for the three container sizes. Plant ET_A was measured at four levels of canopy closure, ~0, 33, 67 and 100% closure with weighting lysimeters [6] concurrently for nine replications of each container size.

The response to canopy closure was identical for each plant size and that observed in 1997 (Figure 3.18). Daily ET_A normalized by ETo was the same for isolated plants as those with 33% and 67% canopy closure, i.e., where the cumulative PCA's of border and lysimeter plants was 33% and 67% of the total area on which the plants were set. At 100% canopy closure, normalized ETA declined to 60% of that of plants at 67% closure or less. The ET_A of plants at 100% canopy closure was associated with transpiration from only the upper 40% of a canopy.

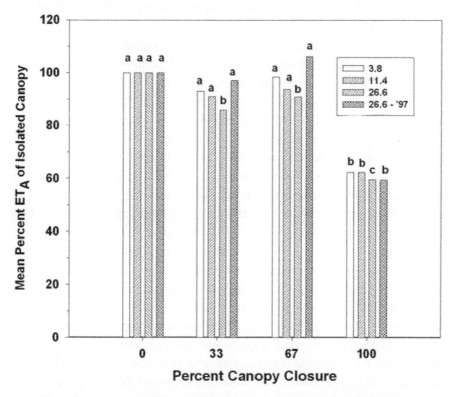

FIGURE 3.18 Normalized ETA of *Viburnum odoratissimum* of marketable size in three container sizes at four levels of canopy closures. Each bar is the mean of nine plants. Bars with different letters within a container size are significantly different (α=0.5) based on F- Protected LSD.

A complimentary opposite of this effect, increased ET_A, has been reported for trees that remained after forests had been partially harvested to reduce stand density [7, 34]. Thus in forested stands, production nurseries and landscape plantings, maximum plant water use will decrease between 70 and 100% canopy closure. This occurs due to mutual leaf shading, increases in leaf boundary layers, and development of canopy boundary layers caused by reductions in canopy roughness. Water use values derived from closed forest or plant production cited above would underestimate ET_A in landscapes with less than 70% canopy closure or for isolated trees. Conversely, as landscapes increase in canopy coverage to above 70% canopy closure, water use of the landscape in its entirety will decline. These conclusions formed the basis of the ASABE S623 Landscape Plant Water Demand standard for woody plants adopted in 2015.

After accounting for canopy closure, both in examples above and at the location where the information will be applied, utilizing this information is a two-stage process. As noted earlier, for optimum establishment, newly transplanted trees and shrubs should be irrigated at their water use rates based on well-watered conditions and size. However, once established, research indicates irrigation can be curtailed with woody plants, while still providing their intended esthetic and functional expectations.

How much irrigation can be reduced has been researched mainly in the western USA. There, research has focused on the minimum level of irrigation that maintains esthetically pleasing landscapes of established plants. Sachs et al. [44] limited irrigation frequency of established plants of eight species in central (37.2 N) and southern (33.44 N) California. All species maintained acceptable appearance with only a bimonthly fixed-volume irrigation during the summer dry season. Paine et al. [38] used a more flexible approach, basing irrigation rates on historical ETo rather than a fixed volume and frequency. They reported acceptable plant appearance of *Photinia fraseri* and *Rhamnus californica* when irrigated at 63.8% of historical ETo, without rain, nine months after transplanting (33.56 N). Pittenger et al. [39] extended the versatility of using ETo for triggering irrigations (33.56 N), by multiplying daily ETo by a fraction (deficit irrigation level, DI). Irrigation occurred when the cumulative sum exceeded 4 cm, replenishing deep soil moisture. Deficit irrigation levels were 20–50% of measured ETo. With typical fluctuations in ETo and winter rainfall with southern California's Mediterranean climate, *Baccharis*, *Drosanthemum*, and *Hedera* irrigated at 20% ETo and Vinca at 30% ETo provided acceptable or higher esthetic quality.

More recently, Shaw and Pittenger [47] applied their deficit irrigation method on 30 common woody shrub species beginning two years after transplanting (33.56 N). Deficit irrigation treatments were initially 0.36, 0.24 and 0.12 times ETo, but were reduced to 0.36. 0.18 and 0 for the last two years. Aesthetic ratings were the same as used by Pittenger et al. [39]. After three years, 11 of 30 species maintained accept esthetic quality with no (DI = 0) summer–fall irrigation. Both *Hibiscus* and *Ligustrum* exhibited continuous slow declines suggesting the 0.36 × ETo was insufficient. Three other species were not successful at the 0.36 level, three more (*Cassia*, *Leucophyllum* and *Galvezia*) were unacceptable during the last of summer through winter months. The other 22 species maintained acceptable esthetical quality at the 0.12 deficit level. These included several shrubs also used in Florida landscapes, which have the opposite climate, i.e., cool dry winters and wet hot summers. These species were *Calliandra*, *Cassia*, *Lantana*, *Leptospermum*, *Otatea*, *Pittosporum tobira*, *Prunus caroliniana*, *Pyracantha* and *Rhaphiolepis indica*.

3.5 SUMMARY

Little research of plant water use to date is directly applicable landscape environments. Most information must be extrapolated from production or ecological research, which often contains insufficient details of plant size or microclimate. Further research with deficit irrigations will increase the list of species that can be successfully infrequently irrigated. However, major advances will require quantifying the evaporative demand in urban landscapes and linking that to plant size and algorithms that predict plant water use.

KEYWORDS

- **Acer**
- **Actual evapotranspiration, ETA**

- **Air pollution**
- **Alternative irrigation**
- **Available water**
- **Carbon sequestration**
- **Crop coefficient**
- **Deciduous species**
- **Deficit irrigation**
- **Drainage**
- **Ecological research**
- **Energy efficiency**
- **Evaporative demand**
- **Evapotranspiration**
- **Evergreen species**
- **Florida**
- **Forest trees**
- **Holly tree**
- **Ilex**
- **Landscape**
- **Landscape environment**
- **Long wave radiation**
- **Microclimate**
- **Minimum irrigation**
- **Natural woodlands**
- **Plant size**
- **Plant temperature**
- **Plant water use**
- ***Quercus***
- **Root regeneration**
- **Sap flow**
- **Senescence**
- **Shrubs**
- **Solar radiation**

- **Stomata closure**
- **Storm water**
- **Supplement irrigation**
- **Transpiration**
- **Tree decline**
- **Tree trunk**
- **Tropical forest**
- **Tropical trees**
- **Trunk caliper**
- **Urban canyon**
- **Urban landscape**
- **Urban trees**
- **Vapor pressure deficit, VPD**
- **Water flux**
- **Water management**
- **Water use**
- **Water use efficiency, WUE**
- **Weighting lysimeter**
- **Wildlife habitat**
- **Woody landscape plants**
- **Woody shrubs**
- **Xylem**
- **Xylem water potential**

REFERENCES

1. Andrade, J. L., Meinzer, F. C., Goldstein, G., Holbook, H. M., Calvelier, J., Jackson, P., and Silvera, K. (1998). Regulation of water flux through trunks, branches and leaves in trees of a lowland tropical forest. *Oecologia*, 115, 463–471.
2. Beeson, Jr., R. C., and Haydu, J. J. (1995). Economic feasibility of micro-irrigating container-grown landscape plants. *Journal of Environmental Horticulture*, 15, 23–29.
3. Beeson, Jr., R. C. (2001). Water use of marketable-size shrubs in containers: A 2-year average. *Proceeding of the Southern Nurseryman Association Research Conference*, 46, 592–594.

4. Beeson, Jr., R. C. (2004). Modeling actual evapotranspiration of *Ligustrum japonicum* from rooted cuttings to commercially marketable plants in 12 liter black polyethylene containers. *Acta Horticulturae*, 664, 71–77.

5. Beeson, Jr., R. C. (2010). Response of evapotranspiration of *Viburnum odoratissimum* to canopy closure and the implications for water conservation during production and in landscapes. *HortScience*, 45, 359–364.

6. Beeson, Jr., R. C. (2011). Weighing lysimeter systems for quantifying water use and studies of controlled water stress for crops grown in low bulk density substrates. *Agricultural Water Management*, 98, 967–976.

7. Bréda, N., Granier, A., and Aussenac, G. (1995). Effects of thinning on soil and tree water relations, transpiration and growth in an oak forest (*Quercus petraea* (Matt.) Liebl.). *Tree Physiology*, 15, 295–306.

8. Berrang, P., Karnosky, D. F., and Stanton, B. J. (1985). Environmental factors affecting tree health in New York City. *Journal of Arboriculture*, 11, 185–189.

9. Burger, D. W., Hartin, J. S., Hodel, D. R., Lukaszewski, T. A., Tjosvoid, S. A., and Wagner, S. A. (1987). Water use in California's ornamental nurseries. *California Agriculture*, 41(9/10), 7–8.

10. Chaturvedi, A. N., Sharma, S. C., and Srivastava, R. (1988). Water consumption and biomass production of some forest tree species. *International Tree Crops Journal*, 5, 71–76.

11. Chi, Y. J. (1993). Soil compaction as a constraint to tree growth in tropical and subtropical urban habitats. *Environmental Conservation*, 20, 35–49.

12. Costello, L. R., Thomas, D., and DeVries, J. (1996). Plant water loss in a shaded environment: a pilot study. *Journal of Arboriculture*, 22, 106–108.

13. Dugas, W. A., Heuer. M. L., and Mayeux. H. (1992). Diurnal measurements of honey mesquite transpiration using stem flow gauges. *Journal of Range Management*, 45, 99–102.

14. Dwyer, J. F., McPherson, E. G., Schroeder, H. W., and Rowntree, R. A. (1992). Assessing the benefits and cost of the urban forest. *Journal of Arboriculture*, 18, 227–234.

15. Edwards, W. R. N. (1986). Precision weighing lysimetry for trees, using a simplified tared- balance design. *Tree Physiology*, 1, 127–144.

16. Garcia-Navarro, M. C., Evans, R. Y., and Montserrat, R. S. (2004). Estimation of relative water use among ornamental landscape species. *Scientica Horticulturae*, 99, 63–174.

17. Gilman, E. F. (1988a). Predicting root spread from trunk diameter and branch spread. *Journal of Arboriculture*, 14, 85–88.

18. Gilman, E. F. (1988b). Tree root spread in relation to branch dripline and harvestable root ball. *HortScience*, 23, 351–353.

19. Gilman, E. F., Beeson, Jr., R. C. (1996). Production method affects tree establishment in the landscape. *Journal of Environmental Horticulture*, 14, 81–87.

20. Gilman, E. F. and Watson, D. G. (1993). *Ilex x 'Nellie R. Stevens': 'Nellie R. Stevens' Holly*. Environmental Horticulture Department, Florida Cooperative Extension Service, Institute of Food and Agricultural Sciences, University of Florida. EDIS EHN472.

21. Granier, A. (1987). Evaluation of transpiration in a douglas-fir stand by means of sap flow measurements. *Tree Physiolology*, 3, 309–320.

22. Green, S. R. (1993). Radiation balance, transpiration and photosynthesis of an isolated tree. *Agricultural Forest Meteorology*, 64, 201–221.
23. Guevara-Escobar, A., Edwards, W. R. N., Morton, R. H., Kemp, P. D. and Mackay, A. D. (2000). Tree water use and rainfall partitioning in a mature popular-pasture system. *Tree Physiology*, 20, 97–106.
24. Hardy, J., Behe, B. K., Barton, S. S., Page, T. J., Schutzki, R. E., Muzii, K., Fernandez, R. T., Haque, M. T., Booker, J., Hall, C. R., Hinson, R., Knight, P., McNeil, R., Rowe, D. B. and Safley, C. (2000). Consumers' preferences for plant size, type of plant material and design sophistication in residential landscaping. *Journal of Environmental Horticulture*, 18, 224–230.
25. Hartig, T., van den Berg, A. E., Hagerhall, C. M., Tomalak, M., Bauer, N., Hansmann, R., Ojala, A., Syngollitou, E. and Carrus, G. (2010). Health benefits of nature experience: psychological, social and cultural processes. In: *Forest, Trees and Human Health*. (Nilsson, K., de Vries, S., Sangster, M., Seeland, K., Gallis, C., Schipperijn, J. and Hartig, T., Eds.). Springer Science Business Media B. V., doi: 10.1007/978-90-481, 9806–9815.
26. Heilman, J. L., Brittin, C. L., and Zajicek, J. M. (1989). Water use by shrubs as affected by energy exchange with building walls. *Agricultural and Forest Meteorology*, 48, 345–357.
27. Heilman, J. L., andHam, J. M. (1990). Measurement of mass flow rate of sap in *Ligustrum japonicum. HortScience*, 25, 465–467.
28. Kjelgren, R., Rupp, L., and Kilgren, D. (2000). Water conservation in urban landscapes. *HortScience*, 35, 1037–1040.
29. Knox, G. W. (1989). Water use and average growth index of five species of container grown woody landscape plants. *Journal of Environmental Horticulture*, 7, 136–139.
30. Krizek, D. T., and Dubik, S. P. (1987). Influence of water stress and restricted root volume on growth and development of urban trees. *Journal of Arboriculture*, 13, 47–55.
31. Lindsey, P., and Bassuk, N. (1991). Specifying soil volumes to meet the water needs of mature urban street trees and trees in containers. *Journal of Arboriculture*, 17, 141- 149.
32. McPherson, E. G., Nowak, D., Heisler, G., Grimmond, S., Souch. C., Grant, R., and Rowntree, R. (1997). Quantifying urban forest structure, function, and value: the Chicago urban forest climate project. *Urban Ecosystems*, 1, 49–61.
33. McPherson, E. G., Simpson, J. R., Peper, P. J., and Xiao, Q. (1999). Benefit-cost analysis of Modesto's municipal urban forest. *Journal of Arboriculture*, 25, 235–248.
34. Medhurst, J. L., Battaglia, M., and Beadle, C. L. (2002). Measured and predicted changes in tree and stand water use following high-intensity thinning of an 8-year-old *Eucalyptus nitens* plantation. *Tree Physiology*, 22, 775–784.
35. Montague, T., Kjelgren, R., Allen, R., and Wester, D. (2004). Water loss estimates for five recently transplanted landscape tree species in a semi-arid climate. *Journal of Environmental Horticulture*, 22, 189–196.
36. Nobel, P. S. (2009). *Physiochemical and Environmental Plant Physiology*. 4th Ed. Academic Press. ISBN: 978-0-12-374143-1.
37. Nowak, D. J., Hoehn III, R. E., Crane, D. E., Stevens, J. C. and Walton, J. T. (2006). Assessing urban forest effects and values, Washington, D.C.'s urban forest. Resource Bulletin NRS-1. Newtown Square, PA: U.S. Department of Agriculture, Forest Service, Northern Research Station. 24 pages.

55. Vertessy, R. A., Hatton, T.J., Reece, P., O'Sullivan, S. K., and Benyon, R. G. (1997), Estimating stand water use of large mountain ash trees and validation of the sap flow measurement technique, Tree Physiology 17, 747-756.

56. Watson, W. T. (2005) Influence of tree size on transplant establishment and growth, HortTechnology 15, 118-122.

57. Whitlow, T. H., Bassuk, N. L., and Reichert, D. L. (1992), A 3-year study of water relations of urban street trees, Journal of Applied Ecology 29, 436-450.

58. Wullschleger, S. D., Meinzer, F. C., and Vertessy, R. A. (1998), A review of whole plant water use studies in trees, Tree Physiology 18, 499-512.

59. Zhang, H., Morison, J. I. L., and Simmonds, L. P. (1999), Transpiration and water relations of poplar trees growing close to the water table, Tree Physiology 19, 563-573.

CHAPTER 4

EVAPORATION AND SEEPAGE LOSSES FROM DUG-OUT TYPE FARM PONDS

YOGESH MAHALLE

Former Student at Soil Water Conservation Engg. Agril College, Dr.PDKV, Akola; Postal Address: A T Mahendra Colony, Near Sharda Agency, P.O. VMV Amravati – 444604 (MS), India. Mobile: +91 9326279798 or 9420124699; E-mail: yrmahalle1@gmail.com

CONTENTS

4.1 INTRODUCTION

Water is one of the most vital requirements for economic and social development. Human population of Indian subcontinent is ever increasing thereby increasing demand for water for domestic, agricultural and industrial use. To satisfy this demand, fresh resources of water supply are being topped and serious thought is being given towards adoption of different methods of conservation of water. In the arid and semi-arid region of the world, where the lack of sufficient water resources may prove the limiting factor in economic development, increased attention is being focused on the search for area specific methods to conserve existing water supplies.

In any region where water supply from natural resource is not uniformly abundant, its maximum use can be secured by storing water in times of surplus for use in times of storage. Rainwater harvesting is the gathering, accumulating and storing of rainwater [7]. Rainwater harvesting has been used to provide drinking water, water for livestock and water for irrigation or to refill aquifers. The surface runoff water harvesting can be achieved through check dams, farm ponds, irrigation tanks, dugout ponds, diversion ponds, roof top rainwater harvesting etc. Ground water recharge is the process by which water percolates down the soil and reaches the water table, either by natural or artificial methods.

India has ancient history of tank technology. Runoff collection ponds exist in nearly every district in India. Runoff collection irrigation tanks are more concentrated in coastal districts of Tamil Nadu and Andhra Pradesh (Telangana), south central Karnataka, and east Vidarbha in Maharashtra state. *"Tank Irrigation Authority"* was created to manage all aspects of tank irrigation by Government of India. Irrigation has great importance in agriculture. In India total cultivable land is 139.36 M ha, out of this 37.64 M ha is under irrigation, which comes to 27%. In Maharashtra, the total cultivable land is 18.5 M ha and gross cropped area is 22.66 M ha out of which 4.04 M ha is the gross irrigated area and the percentage of gross irrigated area to gross cropped area comes as 17.8 [1], which is quite below the national average of 27%.

The Purna Valley of Vidarbha region is an east-west elongated basin with slight convexity to the south occupying part of Amravati, Akola and Bhuldhana district of Vidarbha and extends from $20^0 45'$ N latitude and $75^0 15'$ to $77^0 45'$ E longitude with east-west length of about 100–150 km having width of about 10 to 60 km covering an area of about 2.74×10^5 ha in 547 villages [21]. The unique feature of salt affected soils of Purna Valley is that though the salinity and sodicity is widely reported in this tract, the use of well water, which is poor quality, makes the situation more problematic.

The farming practices adopted in saline tract of western Vidarbha is totally rainfed since ground water resources are not suitable for irrigation. The productivity in this area is totally dependent on amount and distribution of monsoon rainfall. Harvesting of rainwater and reusing it for providing life saving irrigation to crops has thus become an urgent need of the hour for stabilizing and further improving the production potential of dryland farming in saline tract for the benefit of farmers under ongoing changing rainfall situation and recurring droughts.

To over come the drought prone situation in drought affected districts of Vidarbha, constructions of number of dug-out type ponds (water storage structures) for harvesting of excess rainwater on farmers field is the best option available and is being implemented since last three years. Water stored in the ponds or reservoirs is subjected to loss by seepage and evaporation. The loss due to seepage is governed by the properties of soil forming the floor and the banks of the storage while evaporation loss is a function of the climate characteristics of the location of the storage. The data on seepage and evaporation losses from dug-out type ponds (small water bodies) in saline tract areas is not available. Seepage and evaporation loss data needs to be quantified in saline tract of Vidarbha region so that maximum amount of harvested water can be made available for protective irrigation.

The harvested water in farm ponds is being used for providing life saving irrigation to dryland crops by lifting and applying to the fields. In the semi-arid region like Vidarbha, the evaporation rates from water storage structures are generally high due to high temperature, low relative humidity, and high wind speeds, To minimize the evaporation loss from such small water harvesting structures [34] in arid and semi-arid regions, scientist world over, have tried several types of anti-evaporants such as floating sheet cover of plastic membrane, polystyrene sheet, foamed wax blocks, plant residue, oil mulches, polyethylene oxides, gum mixtures and fatty alcohols. There was not much success with these attempts. As yet, an efficient, economical and durable evaporation suppressant has not emerged from the research, which can be used widely.

Large numbers of dug-out type farm ponds (120) are being constructed in Ghusar village of Akola taluka by State Agriculture Department which varies in design, dimensions and sizes. Both lined and unlined type dug-out farm ponds are available in the study area and there is a need to quantify storage losses from farm ponds in saline tract region [3]. Hence, the present study was planned and undertaken with following objectives on farmer's field.

This chapter presents tank technology with the objectives: 1. To quantify evaporation and seepage losses from dug-out type farm ponds in saline tract; To study utility of harvested water on farmers field.

4.2 REVIEW OF LITERATURE

4.2.1 IMPORTANCE OF FARM POND

Howard [13] reported on need of farm pond in general and reported that farm pond is chief source of water supply for livestock, irrigation and farmstead uses and fish production. Pond properly constructed, fertilized, stoked and managed will produce as much as 500 pounds of fish annually for each acre of surface area. The author also reported on planning of farm pond right from selection of site, its design and use.

Sastry et al. [32] studied farm ponds for assured protective irrigation of rabi crops in doon valley. The study conducted at the Central Soil and Water Conservation Research and Training Institute, Dehradun indicated that the harvested runoff water can be recycled which forms an integral part of successful crop management program. The limited water available in the farm ponds should be utilized at the earliest opportunity, i.e., at pre-sowing stage covering larger area for a given depth of water for optimum yields. Authors stated that farm ponds serve dual purpose of storing water for crops life saving irrigation and at the same time minimize the flood hazard in the downstream.

At Central Arid Zone Research Institute (CAZRI) Jodhpur, Man and Rao [19] stated that the utilization of rainwater harvesting and recycling increases the efficiency of available land and water resources. The potential productivity of treated regions appeared to be two to three times higher than that was attained by the traditional system of production. The authors reported positive results on rainwater harvesting and management and the quantity of possible increase in productivity. The limitations of the existing technique of rainwater harvesting have been received an indicating further research need.

Gajri et al. [10] reported on rainwater harvesting and its recycling for maximization of crop production. Water harvesting system, which consists of collection and storage of water in suitable reservoir, the excess runoff from the catchment and its use as crop life saving irrigation would help increase and stabilized yield in dry land areas. The author described the water harvesting technology pertaining to the designs of tank, methods for decreasing seepage and evaporation losses. Responses to small irrigation, obtained in the research experiments conducted under the All India Coordinated Research Project on Dryland Agriculture are summarized.

Sastry et al. [30, 31] studied farm ponds and their influence on flood retardance. The farm ponds are effective in retarding flow peaks as well as flood volume considerably. During the month of July, August and September, which are responsible for serious flood problems. The average flood peak retardance is 82, 90 and 99%, respectively. The watershed treated with graded terraces generates low runoff in the land phase as compared to untreated watersheds. The size of the farm pond can be increased to affect a further flood retardance value for the pond. The scope of further improving the performance of farm pond through more rational schedule of using available water for crops such as paddy could be further explored.

Verma et al. [39] revealed from the studies on feasibility of storage of runoff in dug-out ponds and its use for supplemental irrigation in Punjab that dug-out ponds which hold greater promise for storing runoff and using it for supplemental irrigation. Stored water may be applied as a presowing irrigation, if seed zone is dry, or as an early irrigation to remote growth.

Phadnvis et al. [26] studied the Impact of water harvesting structures on ground water recharge on shallow black soil in semi arid region in Maharashtra. Authors carried out experiment at Padalsing watershed located in Beed district of Maharashtra and found that the water table in the wells located on the down streamside of the water harvesting structure increased considerably. This was related to the water harvesting structure making more water available for irrigation through the open wells. The shrinkage percentage of nala bund during first 2 to 3 years was found to be 9.17%, which was within safe limits.

Sahu et al. [28] reported that lack of water at times of need is the dominant constraint to increasing productivity in most of the rainfed area. Small farm reservoirs (SFR) can play important role in alleviating drought in this area. Research on this storage systems have been conducted in Bangladesh, India, Philippines and Indonesia to determine their value in conserving rainwater, improving land productivity and the impact on farmer's income. An SFR generally has individual ownership and cover a small part of the total area. Therefore, SFR can be managed easily by the farm owner with very little social problem .It has multiple potential uses such as reducing drainage congestion, irrigation and aquaculture with improved cropping system using SFR productivity and income can be increased.

4.2.2   DESIGN OF FARM POND

Carreker [5] suggested that the first consideration in the pond construction is its location and a narrow depression between two opposites slope with a wide flat area slope generally makes the best side. To prevent seepage from the pond, a core wall should extend down under each dam to impervious materials free of roots and other organic matter.

Potter and Krimgold [27] studied on area relationships that simplify the hydrologic design of small farm ponds. The great majority of farm ponds are small. The spillways are usually excavated channels protected by vegetation. Hydrologic consideration indicates that to keep the flow over the spillway to a minimum the depth of water and therefore the size of the surface area must fluctuate fairly widely. The two major losses components are seepage and evaporation. Seepage losses can be reduced by selecting a proper shape and that gives the ratio of water stored to the wetted surface area value should be higher. Circular and square shape would give higher than rectangular, semicircular and horse-shoe shapes minimum surface area for the same capacity of tank is important factor for minimizing evaporation losses.

Chittaranjan [6] conducted studies on rainwater harvesting and recycling and stated that water is normally stored in tanks, dug-out s or impounded or a combination of both. In design of storage capacity the following points needs consideration:

- **Catchments and water yield**: There should be a proper relationship between catchment and size of storage based on average excepted amount of runoff. It is estimated that water yield works out to 0.5 ha-m. While selecting the optimum size of storage, allowance has to be made for storage losses such as evaporation and seepage, which are inevitable. To avoid silting of pond the catchment area should have high percentage of vegetation as also adoption of recommended soil and water conservation measures. In addition to the above, a silt trap may have to be provided a few feet above the point of entry of water into the pond.

- **Economic storage**: Best location is the area that permits the maximum storage with least amount of earthwork i.e., we should aim at greater storage – excavation ratio.
- **Dimensions of pond**.

4.2.3 STORAGE LOSSES

Cluff (1977) suggested compartalization of SFR to reduce seepage and evaporation losses by reducing wetted area. Seepage losses in light textured substrata of SFR were reduces from 71.3 to 24.1% by mixing the clay soil in SFR bed followed by trampling by animals and puddling: compacting by hand compactor and roller. Reduction in seepage losses has been reported from 37.8 to 15.2% by compaction alone.

Grewal et al. [12] concluded that the effect of some soil and site properties on seepage losses from three small storage reservoirs developed in the Kalka area of Siwalik region for rainwater harvesting and supplemental irrigation was evaluated. It was observed that on average 16–42% of the total stored monsoon rainwater was lost through evaporation and seepage before it is utilized for 'Rabi' crop irrigation. The seepage loss on an average varied from 6.2 to 20.0, 0.4 to 1.2 and 2.0 to 10.1 mm/day in Sukhomadry reservoir numbers SM-I, II and III with the corresponding heads varying from 2.0 to 3.4, 3.4 to 3.3 and 2.5 to 4.1 meters, respectively.

Sastry et al. [33] reported structural measures for efficient control of seepage from dug-out farm ponds at the Dhod Kot Block of the Research farm, Dehra Dun in 1976. High storage losses (10 cm per day) were observed during the first year of storage of runoff in the unlined farm pond. The storage efficiency of farm ponds depends mainly on storage losses (seepage, evaporation and evapotranspiration and storing or excavation ration. The seepage losses alone are enormous in freshly dug-out farm ponds in the region. It is observed that seepage losses tend to stabilize after about 8 to 10 years of construction of farm ponds. The sides of the farm pond were lined with brick mortar. The brick work was plastered with 12 mm thick cement mortar (1: 5) in compartments of 10 m. By a layer of 7.5 cm thick cement concrete (1:5:10) over the compacted surface of 15 cm broken stone of 3–4 cm size along with course ranged. The cement concrete was then covered with approximately 12 mm thick cement plaster (1:4) over the compacted surface, a layer of 7.5 cm thick cement concrete (1:5:10). The cement concrete was then covered with approximately 12 mm thick cement plaster (1:4).

Kale et al. [17] has studied the effect of various sealent materials on seepage losses in tanks in lateritic soil in Konkan region of Maharashtra. The maximum seepage loss was observed in the control lank followed by the lank treated with cow dught paddy husk + soil (91.49%), the bentonite clay plastering (65.03%), cement plaster at bottom (69.2%) and paddy pusa ash (62.50%) behaved more or more in less in similar way in reducing the seepage losses as compared to the control. The seepage losses in tanks lined with polythene, cement + soil plasper (2:10) and cement + soil plaster (3:10) were 1.72, 4.56 and 4.04%, respectively. These three materials could be considered for lining purpose in lateritic soil.

Singh and Reddy [36] stressed the need of resource management in rainfed watershed. The author stated that farm ponds in watershed should be so designer and located that they can irrigate fields lying below by gravity food. In dug-out ponds, the sides as well as bottom have to be made impermeable. Suitable sealent in light soils for minimizing seepage looses at different centers were evaluated. Materials like soil-cement, asphalt, alkali soil, etc., have been found to perform better than others.

Sastry and Mittal [29] studied on water harvesting and its recycling in doon valley. They observed that lining of dug-out ponds with brick and cement mortar was effective in reducing seepage losses about 20% but seepage was dependent on head of water and varied from 0.28 cm/day to 10.4 cm/day as the head increased from 0.5 m to 2.0 m.

Kale and Deshmukh [16] used black LDPE polythene (1000 gauge) as lining materials under lateritic soil condition. Lining of only beds of big dug-out farm ponds was found inadequate in significantly reducing the seepage losses. The average specific percolation loss of 5.14 L hr-m^2 under unlined condition was reduce to only 3.88 L/hr-m^2 i.e., by 24.51% after lining the bed of farm ponds. The farm pond with total storage capacity of 8315 m^3 would be emptied within a period 32 days at the average percolation rate of 3.88 L/hr-m^2 and therefore, very purpose of providing irrigation to rabi crops would not be achieved.

Khan [18] conducted a study on influence of climatic parameters on rate of evaporation from free water surface. The results revealed that air temperature was the major factor affecting evaporation, water temperature influence evaporation indirectly by conveying radiation energy and was in itself affected by evaporation. The effect of relative humidity of the air seemed to be far out weighted by other climatic factors. At tanks wind had a marked effect on evaporation. Climatic factor under the conditions prevailing at the test site complemented each other in increasing or decreasing evaporation.

Wagh et al. [40] observed that there were fluctuations in the water levels of the pond during rainy season. This was due to dry and wet spells. The pond becomes totally empty after monsoon season. The average seepage rate of this pond was between 1.8 to 4.4 L/hr/m². The average evaporation loss during crop growing period (from 1st November to 15th January) is about 73 m³ (17.8 %) farm pond of 4 × 10⁵ liter capacity.

Codie and Webster [8] studied that evaporation from fetch-limited water bodies has been investigated for the first time using a coupled atmospheric boundary layer-water body model. The model incorporates a simplified atmospheric boundary layer in which heat and moisture are advected horizontally and diffused vertically. The wind field evolves over the water body through the formation of an internal boundary layer, which is initiated by the change in roughness from the land to the water surface. The wind also responds to local stability through the inclusion of Monin-Obukhow similarity functions. This system is coupled to a dynamically active water body based on primitive equations with full thermodynamics. This is achieved through continuity of stress and heat flux through the air-water interface. The model results reveals that along-wind gradients in wind stress, humidity, and temperature can all significantly influence evaporation. The most important effects are growing wind stress and increasing humidity as we move downward across the water body. However, there is a tendency for these effects to cancel, so that the behavior can range from averaged evaporation weakly decreasing with fetch for very smooth land surfaces to weakly increasing with fetch relatively rough land terrain.

Codie and Webster [8] showed that evaporation increases substantially with speed (e.g., Increasing wind speed from 8 km/hr to 18 km/hr produces twice the evaporation), the depth of water storage had almost no effect on the evaporation per unit area increasing storage depth from 1 m to 6 m and maintaining the same area evaporation per unit area by less than 1%.

4.2.4 POND UTILITY

Arnold and Stockle [2] stated that on-farm runoff collection and supplemental irrigation systems can increase and stabilize crop production in many regions of the world. The simulation model is linked to determine optimum pond size with respect to net profits from crop production under supplemental irrigation systems. A comprehensive basin-scale soil and water-resources model is modified to simulate crop yield, supplemental irrigation, furrow disking, and economics. The model produces monetary-return frequency distributions for various pond sizes and is capable of determining the effects of various management strategies on crop production and water and sediment yields. The frequency distributions give decision makers a risk assessment tool for designing the pond. Simulated hydrologic components and crop yields are compared with measured values, showing the model performs well under various climate, soil, crop and management conditions. Finally, the model was applied to two basins in Texas.

Sharada (1994) reported that the water harvesting technique being adopted under different situations in Northern hilly regions with special on their design criteria, rainfall runoff relationship, catchment area storage capacity ratios and methods to contain storage losses. Studies revealed that properly designed dug-out cum embankment type ponds/reservoir when used for providing supplemental irrigation can help improving the crop yields by 2 to 3 tons.

Gorantiwar [11] conducted a study on effect of percolation tank in augmenting the ground water recharge at Singhway watershed in Ahmadnagar district. The water level in the selected wells was recorded periodically for the period from 1993 to 2003. The study revealed that water in the well decreased with its distance from percolation tank. It was also observed that, water level in the percolation tank increases from month of June to December and again drops from December onwards. This indicates that, water level in the wells was influenced by the storage of water in the tank. The average recharge observed for the period 1993 to 2003 was almost 87.08% of the total inflow in to the tank.

Panigrahi et al. [24] used daily water balance simulation in rainfed rice lands and estimated the probable supplemental irrigation (SI) requirements to meet the water deficit during the reproductive stage of rice and the surface runoff (SR) generated that can be harvested in on farm reservoir (OFR). Value of SI of rice during reproductive stage at 25% probability of exceedence (PE) was found to be 144 mm, neglecting distribution and application losses. Water harvesting potential of study area indicated that at 50% PE, 85% of SI can be met from the SR generated from the rice lands and stored in OFR. Rest amount of SI can be met from the direct conservation of rainfall in lined OFR of 2 m depth with 1:1 side slopes occupying 9rice lands. Economic analysis of OFR irrigation system reveals that the OFR of 9% rice lands gives net profit (NP) of Indian Rs. 13,445/- (US $ 295.49) for 1 ha sown with dry seeded rainfed upland rice with benefit-cost ratio (BCR) 1.25. Values of NP and BCR indicate that investment in OFR irrigation system is profitable in the study region.

Srivastava, et al. [38] stressed the need of proper design of rainwater harvesting in high rainfall area based on spatial and temporal behavior of rainfall, water requirement of the crops, in addition to catchment characteristics. A water harvesting system can facilitate timely transplantation of rice and proper utilization of the rainwater in addition to the prospect of second crop in the post-monsoon season. He developed a simulation model to design a system for determining catchment/command area ratio, size of tank, desirable command area of a single tank and the feasibility/ economics of lining of tank. The methodology facilitates the design of the system using the catchment area characteristics, command area, and efficiency of conveyance system. Tank size of 1750 m³/ha command area and catchment/command ratio of 3.0 was found suitable for a rice-based cropping system in eastern India, as it facilitates desirable moisture regime for rice and two irrigations to succeeding crop. The command area should not be extended beyond a limit to have a conveyance efficiency of 80% or more. Using traditional methodology i.e., by total irrigation requirement method, the tank capacity requirement will be 4500–6000 m³/ha command area. The sensitivity analysis showed that this holds good for almost all types of land use in this particular region.

Panigrahi and Panda [23] developed a computer simulation model for prediction of optimal size of an OFR so as to provide supplemental irrigation to rice in monsoon season and pre-sowing irrigation to mustard in winter for a rainfed farming system of eastern India. Daily simulations of water balance parameters of both the cropped field and the OFR, economic analysis, and irrigation management practices were inputs to the model. The study predicted an OFR of depth 2 m requiring 12% of the 800 m^2 farm area with a volume of 61 m^3 to be optimum. The above-mentioned optimal size of the OFR gave a BCR of 1.22, internal rate of return (IRR) of 15%, and payback period (PBP) of 15 years. Simulated results were verified by conducting three years of field experiments to justify the investment in the OFR irrigation system. The observed BCR, IRR, and PBP from the experimental study were 1.17, 14.8%, and 16 years, respectively. There was an increase of 39 and 15% in the yield of rice grain and mustard seed over rainfed conditions because of application of 84 and 45 mm of supplemental irrigation, respectively.

Pandey et al. [22] stressed the need of optimum size of the rainwater storage and recycling processes for mitigating the uncertainty of rainfall, thereby developing rainfed-farming systems on sustainable basis. Authors developed a model to predict the optimum size of the on-farm reservoir so as to provide supplemental irrigation to rice in monsoon season, pre-sowing irrigation to mustard in winter and to support fish cultivation in rainfed farming systems in Eastern India. The model is based on water balance approach for both the crop field and the reservoir, where hydro-meteorological parameters and the irrigation management practices are input to the model. The model has predicted the reservoir sizes in terms of the percentage of the farm area under unlined and plastic lined conditions for two side slopes (horizontal:vertical) of 1.5:1 and 1:1 and two reservoir depths of 2 and 2.5 m. The OFR sizes are in the range 10.7 to 17.3% for lined and 17.5 to 21.5% for unlined ponds

Dhanapal et al. [9] reported that arable lands in the micro-watersheds in *Alfisols* are more prone to runoff and nutrient losses lead to degradation of natural resources. A long-term research cum demonstration is being conducted with regard to soil and water conservation and utilizing the harvested water for protective irrigation and Pisciculture. During the cropping season, maximum runoff collection per hectare was recorded in cultivated catchments than forest catchments. In the cultivated catchments area, maximum runoff and soil loss was observed in groundnut and lowest runoff and soil loss was observed in maize and finger millet + red gram inter cropping system respectively. It also indicated that *Nase* grass as live barrier is effective to reduce run-off and soil loss. The runoff was collected and stored in farm ponds and the same was used for protective irrigation in double cropping system. Highest forage yield was produced by giant bajra followed by South African maize grown during the early *Kharif.* Significantly higher French bean green pod yield was recorded in early grown sweet sorghum plots followed by giant bajra; Suvidha recorded significantly higher green pod yield as compared to Anoop. However, French bean receiving 100% recommended dose of fertilizers recorded higher green pod yield as compared to 75% of recommended dose of fertilizers. Growth and development of all the four breeds of fishes was normal and each weighed 450–500 g after complete maturity. By adopting scientific method of fish production about 50–60 kg fully matured fishes (4–6 months) could be harvested in farm pond (180 m^3) depending upon the maintenance of pond water. So that, on an average Rs. 2100/- would be the additional income from fish production. In nutshell, crop soil and water productivity is enhanced by good rainwater management in *Alfisols.*

Srinivasulu et al. [37] reported that South Coastal Andhra Pradesh [20] is prone to both droughts and floods and receives rainfall from southwest and northeast Monsoon. The study was initiated to overcome prolong dry spells by providing one lifesaving irrigation to improve the yield of flue cured Virginia (FCV) tobacco. Nine farm ponds were dug for rainwater harvesting and recycling. The yield of cured leaf improved in all the six years of study period and ranged between 12 and 31% with one life saving irrigation over control (rainfed). This translated to an additional net income ranging from Rs. 2255.00 to Rs. 17,049.00 per hectare. The BC ratio for lined pond was worked out to be 1.34, while net present value (NPV) and IRR was estimated as Rs. 21,938.00 and 1.6%, respectively. The payback was estimated as 12 years for lined pond after discounting the cost and returns.

Bhandarkar [4] observed that for sustainable agriculture in rainfed vertisols water harvesting and recycling is atmost important. Water harvesting pond is constructed in 10–12% of watershed area. Minimum depth of pond should be 3 m .The runoff received is 300 mm (3000 m^3/ha) to fill up pond every year. About 60–70% of stored water can be utilized for irrigated crop. Entire kharif and 50% of rabi crop can be irrigated twice with two fold increase in yield. Additional benefit obtained through irrigation net benefit-cost ratio was work out to be 1.13. The benefit-cost ratio works out of to be 2.03 considering total benefit–cost ratio under irrigated condition. Increase in yields was 6 q/ha (Rs. 1000/q), 15.9 q/ha (Rs. 700/q), 15.6 q/ha (Rs. 800/q), and 13.9 q/ha (Rs. 1200/q) in case of soybean, rice, wheat and chickpea respectively [Note: one q = 100 kg].

4.3 MATERIAL AND METHODS

4.3.1 STUDY AREA

The Ghussar village in Akola taluka was selected as an experimental site for the present study, since, the large number of dug-out farm ponds are constructed during 2008 summer under Prime Minister package program for drought affected districts in Vidarbha region and few of them which were constructed under National Horticulture Mission had polyethylene linings, so that the objectives of this study can be evaluated. It is located at 20° 7' N latitude, 77⁰ 07' E longitude and at an altitude of 282 m above mean sea level. This Ghussar village comes under saline tract area of western Vidarbha. The soils are mainly black cotton soils with depth varying from 4 m to 10 m having clayey texture. They are poorly drained and have moderately high runoff potential.

The climate is characterized by relatively hot summer, and cold winter. In Akola taluka average annual temperature ranges from a high of 48°C to a low of 10°C. The overall climate can be classified as semi-arid tropical. The total average annual rainfall in Akola taluka is 751.52 mm. The village under the study received 655.10 mm annual rainfall during 2009 in 34 rainy days, out of which monsoon season (June to September) contributed 87%. The village under the study received 1041.5 mm annual rainfall during 2010 in 38 rainy days, out of which monsoon season (June to September) contributed 89%. The Ghusar village, agro-climatically falls under assured rainfall zone of Vidarbha region of Maharashtra.

4.3.2 DATA ACQUISITION

There are about 120 dug-out farm ponds constructed under the Prime minister Package and National Horticulture Mission Program in Ghusar village of Akola taluka. During 2008–09, only 27 ponds were constructed on farmers field while 93 farm ponds were constructed during 2009–10 under both the programs as stated above. The dug-out farm ponds constructed in the village under study are either constructed on-stream or off-stream and few of them are lined.

The farm ponds constructed during 2008–09 and 2009–2010 were only considered for this study; and from which 11 dug-out type ponds were selected for the study. Three ponds out of this 11 are with polyethylene sheets (400 micron). The details of dug-out type farm ponds, the field number (Gat number) and sizes of field along with owner name are given in Table 4.1a.

4.3.2.1 Climatic Data

The climatic weather data on minimum and maximum temperature, morning and evening relative humidity, bright sunshine house, wind speed, pan evaporation and rainfall collected at Meteorological observatory of Agronomy Department, Dr. Panjabrao Deshmukh Krishi Vidyapeeth, Akola and was considered for the study, Since the experimental site is 12 km away from the location of the observatory. The average daily weather data on weekly basis during the rainy season 2009 and 2010 of the study period is given in (Appendix IA and Appendix IB).

TABLE 4.1A. Details of Selected Dug-Out Farm Ponds in Saline Tract of Ghussar Village

Name of Farmers	Pond Number	Location of Farm Pond	Size of Farm Pond (m)	Lined/ Unlined
Damodhar Khadase	11	877	82 × 26 × 3	Unlined
Dipak Prakash Raut	7	137	30 × 30 × 3	Unlined
Gunwant B. Wakode	4	107	82 × 26 × 3	Lined
Laxman W. Behare	9	820	30 × 30 × 3	Unlined
Nilesh Raut	6	138	30 × 30 × 3	Unlined
Rajesh S. Laharia	3	37	30 × 30 × 3	Unlined
Rameshwar J. Pagrut	5	107	82 × 26 × 3	Lined
Santosh S. Pagrut	10	847	82 × 26 × 3	Lined
Sardar S. Laharia	2	33	30 × 30 × 3	Unlined
Shivprasad S. Laharia	1	32	30 × 30 × 3	Unlined
Shrikrushna W. Behare	8	898	30 × 30 × 3	Unlined

4.3.2.2   Hydraulic Conductivity and pH of Soil and Water Samples

The hydraulic conductivity and pH of the soil samples in few selected farmers field before starting of rainy season of 2010 after the end of kharif season of 2010 and lastly after the end of rabi season are taken and analyzed in the Department of Soil Science and Agricultural Chemistry, Dr. Panjabrao Deshmukh Krishi Vidyapeeth, Akola, according to Israelsen [14] and Jackson [15] methods, respectively.

Similarly hydraulic conductivity and pH of the runoff water samples collected in dug-out farm ponds during 2009 and 2010 were also determined according to Jackson [15] method.

4.3.2.3   Saturated Hydraulic Conductivity

The saturated hydraulic conductivity of the soil samples at different layers in 3 m deep dug-out farm ponds is determined by the adopting the producer of Israelsen [14] under disturbed soil condition in laboratory under constant head method. The saturated hydraulic conductivity of soil samples was determined in the ranges of 40 cm to 320 cm depth from the surface to bottom of the dug-out type farm pond.

4.3.2.4   Seepage and Evaporation Loss Measurement

The water storage losses from dug-out farm ponds were quantified by monitoring water level fluctuations in the ponds periodically weekly after the end of rainy season, when ponds are in mostly filled condition. The water storage losses so monitored in consist of seepage losses from sides and bottom of unlined ponds and evaporation from surface area. To separate out from seepage losses and evaporation losses components from water losses recorded in the unlined dug-out farm pond, water storage losses due to only surface evaporation from black polyethylene (400 micron size), lined dug-out pond of same size was monitored.

Water loss from unlined dug-out type farm pond was monitored in four ponds of same size (30 m × 30 m × 3 m). Similarly for monitoring only evaporation loss components from dug-out farm pond, three polyethylene lines ponds (pond number 4, 5 and 10) were monitored during 2009–10 and 2010–11.

4.3.2.5 Utility of Harvested Water

Utility of harvested rainwater in dug-out type ponds was recorded for the rabi season at 2009–10 and 2010–11. The number of protective irrigation given by different farmers under study, area covered with available water resources, type of crops for which protective irrigation given and productivity obtained was recorded. The productivity of dry land crops taken by adjoining farmers without protective irrigation was also recorded.

4.4 RESULTS AND DISCUSSION

This chapter presents the result of *assessment of storage losses from dug-out type farm pond in saline tract*. The field data collected on the storage losses from the experimental farm ponds and utility of harvested water were analyzed losses.

4.4.1 STORAGE LOSSES FROM DUG-OUT TYPE FARM PONDS

Water storage losses from dug-out type farm ponds in saline tract area of Ghussar village were studied by monitoring stored water levels in lined and unlined ponds constructed on the field during 2009 by the State Agriculture Department under different development programs. Water level observations were recorded on weekly basis in four unlined ponds of 30 m × 30 m × 3 m size, one unlined pond of 82 m × 26 m × 3 m size and three lined ponds of 82 m × 26 m × 3 m size during 2009 rabi season. Water level observations in above farm ponds were also monitored in the next rainy season of 2010 except the unlined pond of 82 m × 26 m × 3 m size (Pond number 11) which was damaged during the rainy season of 2010. Since, the pond number 11 was constructed on stream due to large volume of stream flow rates due to in cessant rains of 2010, the pond was completely damaged and water could not be collected in it that season.

4.4.1.1 Evaporation Losses From Lined Dug-Out Farm Pond

The weekly evaporation losses recorded from three lined ponds of 82 m × 26 m × 3 m size during 2009 and 2010 are given in Table 4.1b. It indicates that the weekly evaporation during 2009 and 2010 in general varies from 3.65 cm to 5.75 cm and 3.50 cm to 7.00 cm, respectively. The total evaporation during three month period are in general higher during 2010 than that of 2009 except from pond number 5, were reverse trend is recorded. This reduction of evaporation loss during 2010 in only 5 number pond might be due to change of crop from Gram (2009) to Cotton (2010) around the pond of Shri. R. J. Pagrut. The total evaporation loss during 2009 and 2010 ranges from 58.23 cm. to 62.61 cm and 56.00 cm to 67.00 cm respectively during October to December.

Monthly evaporation from black polyethylene lined dug-out type ponds (82 m × 26 m × 3 m) in Table 4.2 indicates that higher evaporation occurs

TABLE 4.1B. Weekly Evaporation Losses From Lined Type of Farm Pond During 2009 and 2010

Met Week	Evaporation losses (cm)				Evaporation losses (cm)			
	Pond Number			Average, cm	Pond Number			Average, cm
	4	5	10		4	5	10	
		2009				2010		
41	5.08	5.37	5.17	5.21	5.00	5.00	5.00	5.00
42	5.75	5.43	4.91	5.23	6.00	5.00	6.00	5.67
43	5.75	5.75	5.67	5.02	7.00	4.00	6.00	5.67
44	5.50	4.92	4.33	4.92	7.00	4.00	5.00	5.33
45	5.00	4.60	4.85	4.82	6.00	5.00	6.00	5.67
46	5.20	4.90	5.30	5.13	5.00	4.00	6.00	5.00
47	6.00	5.80	5.90	5.40	6.00	5.00	5.00	5.33
48	5.80	6.33	4.42	5.00	6.00	6.00	6.00	6.00
49	5.00	4.90	5.10	5.00	3.67	6.00	4.34	4.67
50	4.60	4.70	4.85	4.72	4.66	5.33	3.50	4.50
51	3.70	3.65	3.65	3.67	5.67	3.67	3.17	4.17
52	5.23	4.83	4.08	4.71	5.00	3.00	4.49	4.16
Total	**62.61**	**61.18**	**58.23**	**58.82**	**67.00**	**56.00**	**60.50**	**61.17**

TABLE 4.2 Monthly Evaporation Losses from Lined Type of Farm Pond During 2009 and 2010

Month	Evaporation losses (cm)			Avg cm	Evaporation losses (cm)			Avg cm
	Pond Number				Pond Number			
	4	5	10		4	5	10	
	During 2009				During 2010			
October	22.08	21.47	20.08	20.37	25.00	18.00	22.00	21.67
November	22.00	21.63	20.47	20.35	23.00	20.00	23.00	22.00
December	18.53	18.08	17.68	18.10	19.00	23.00	15.50	17.50
Total	**62.61**	**61.18**	**58.23**	**58.82**	**67.00**	**56.00**	**60.50**	**61.17**

in the month of October followed by November and December, which may be due to higher temperature prevailing in the month of October and reducing light hours in winter months (December). The Multiple regression equation was fitted between average daily evaporation recorded during different meteorological weeks (41 to 52) and minimum and maximum temperature, morning and evening relative humidity, bright sunshine hours and wind speed during corresponding meteorological weeks of 2009 and 2010 (Appendix IA and Appendix IB). The results of regression analysis are given Table 4.3.

4.4.1.2 Water Storage Losses From Unlined Dug-Out Farm Pond

The water storage loss from unlined dug-out type farm ponds 30 m × 30 m × 3 m size recorded at four different locations during 2009 and 2010 are given in Tables 4.4 and 4.5, respectively.

The water weekly storage loss from four unlined ponds varies from 5.89 to 8.02 cm (Table 4.4) and 2.34 to 6.00 cm (Table 4.5) during 2009 and 2010, respectively. The total storage loss from October to December during 2009 and 2010 varies from 77.50 cm to 91.82 cm and 41.66 cm to 63.66 cm respectively. The above results of water storage losses during 2009 and 2010, indicates that during 1st year of construction of ponds (2009), water storage losses are in general more than that of the 2nd year after construction (2010). The average water storage losses during October to December 2009 come out to be 82.10 cm and the same value

TABLE 4.3   Results of Regression Analysis

Variable	Value
Multiple regression	0.913
R square	0.833
Adjust R Square	0.775
S.E.	0.358
Observations	24

TABLE 4.4   Weekly Storage Losses From Unlined Type of Farm Pond During 2009

Meteorological Weeks	Evaporation losses (cm) during 2009				
	Pond Number				Average cm
	2	6	8	9	
41	7.08	8.02	6.71	6.14	6.98
42	6.89	7.72	5.92	5.89	6.60
43	7.10	7.35	5.95	7.20	6.90
44	6.17	7.65	6.66	6.50	6.74
Total (Oct)	27.23	30.73	25.24	25.72	27.23
45	6.12	7.54	6.72	6.64	6.74
46	6.06	7.62	6.75	6.58	6.75
47	6.08	7.70	6.76	6.70	6.81
48	6.11	7.73	6.79	6.74	6.84
Total (Nov)	24.37	30.59	27.02	26.66	27.10
49	6.20	7.68	6.85	6.75	6.87
50	6.41	7.65	6.90	6.78	6.94
51	6.52	7.60	6.93	6.50	6.89
52	6.77	7.57	6.99	6.77	7.02
Total (Dec)	25.90	30.50	27.67	26.80	27.27
Total	**77.50**	**91.82**	**79.92**	**79.18**	**82.10**

during 2010 comes out to be 53.19 cm. The water storage losses from unlined dug-out farm pond consist of both surface evaporation and seepage components. The considerable difference in water storage loss (28.91 cm) during first and second year of construction clearly indicates that the seepage component in first year of construction (2009) is quite high which subsequently reduces in 2nd year of the study period (2010).

TABLE 4.5   Weekly Storage Losses From Unlined Type of Farm Pond During 2010

Meteorological Weeks	Evaporation losses (cm) during 2009				Average cm
	Pond Number				
	2	6	8	9	
40	6.00	5.50	5.00	4.50	5.25
41	6.00	5.50	5.00	4.50	5.25
42	6.00	5.90	6.00	3.75	5.41
43	6.00	5.85	4.28	4.00	5.10
Total (Oct)	24.00	22.75	20.28	17.00	21.01
44	6.00	4.85	4.10	3.25	4.55
45	5.50	4.95	4.50	4.00	4.74
46	5.00	5.00	4.00	4.00	4.50
47	5.16	5.00	4.20	3.67	4.46
Total (Nov)	21.66	19.80	16.60	14.92	18.25
48	5.66	4.29	4.00	2.34	4.07
49	4.67	4.00	3.08	2.51	3.56
50	4.33	4.17	2.67	2.42	3.40
51	3.34	3.33	2.50	2.47	2.91
Total (Dec)	18.00	15.79	12.24	9.74	13.94
Total	**63.66**	**58.34**	**49.12**	**41.66**	**53.19**

The Physical analysis (saturated hydraulic conductivity) of soil samples were taken at different depths up to the pond bottom in different unlined ponds (Appendix II) and the overall range of saturated hydraulic conductivity in each pond are given in Table 4.6. The results of soil analysis at different depths in different ponds indicate that the saturated hydraulic conductivity of soil samples reduces with increasing depth (Appendix II). From Table 4.6, it is observed that a saturated hydraulic conductivity of soil sample, which was determined under disturbed condition (Laboratory method), varies in the range of 0.60 to 0.16 cm/hrs. The overall range of hydraulic conductivity of soils at different depth in all the unlined ponds under study is near about in same range. This indicates that seepage losses in all the unlined ponds are taking place at near about same rate. Hence, the seepage loss from unlined ponds can be determined by deducting

TABLE 4.6   Saturated Hydraulic Conductivity of Soil Samples at Different Depths in Dug-Out Type Farm Pond

Farm pond number	Total depth of soil samples taken (m)	Range of hydraulic conductivity (cm/hr)
2	0.60 to 2.90	0.54 to 0.24
6	0.50 to 3.20	0.60 to 0.20
8	0.40 to 3.00	0.51 to 0.16
11	0.60 to 3.00	0.60 to 0.16

evaporation losses recorded in lined ponds from the overall storage losses recorded in unlined ponds.

4.4.1.3   Evaporation and Seepage Losses From Dug-Out Type Farm Pond

The quantification evaporation and seepage losses from dug-out type farm pond in Ghussar village was done by monitoring the water level fluctuation in unlined and lined farm ponds of similar size (82 m × 26 m × 3 m) on weekly basis starting from October 2009 till December 2009. Weekly water loss observations from both three lined and one unlined farm ponds are given Appendix-III. The water loss in three lined ponds is due to only surface evaporation, where as the water loss from unlined pond is due to both surface evaporation and seepage. The average evaporation loss from lined ponds and the water storage loss from unlined pond during 41 to 52 meteorological weeks are given in Table 4.7.

The difference between these two types of losses in different meteorological weeks indicate that seepage losses that occurred in unlined pond. The result of quantified evaporation and seepage losses during 2009, indicate that weekly values of evaporation and seepage loss ranges from 3.67 cm and 5.23 cm and 0.83 cm to 1.21 cm, respectively in different meteorological weeks. The monthly evaporation loss during October to December ranges from 18.10 cm to 20.37 cm and seepage losses ranges from 3.60 cm to 4.70 cm.

The total evaporation and seepage loss from dug-out type farm pond was found to be 58.82 cm and 12.60 cm, respectively. The percentage of seepage losses was found to be 21% of evaporation during the three months storage period.

TABLE 4.7　Weekly Evaporation and Seepage Losses (cm) From Dug-Out Farm Ponds During 2009

S. No.	Meteorological Weeks	Evaporation losses (cm) lined ponds	Storage losses (cm) unlined pond	Seepage loss (cm)
1	41	5.21	6.42	1.21
2	42	5.23	6.41	1.18
3	43	5.02	6.17	1.15
4	44	4.92	6.07	1.15
5	Total-Oct	**20.37**	**25.07**	**4.70**
6	45	4.82	6.00	1.18
7	46	5.13	6.25	1.12
8	47	5.40	6.40	1.00
9	48	5.00	6.00	1.00
10	Total-Nov	**20.35**	**24.65**	**4.30**
11	49	5.00	5.90	0.90
12	50	4.72	5.70	0.98
13	51	3.67	4.50	0.83
14	52	4.71	5.60	0.89
15	Total-Dec	**18.10**	**21.70**	**3.60**
Grand Total		**58.82**	**71.42**	**12.60**

4.4.1.4　Water Losses From Lined and Unlined Farm Ponds 2010

The volume of water lost due to evaporation in lined ponds and due to both evaporation and seepage in unlined ponds during the study period is given in Table 4.8. The water storage capacity of dug-out type farm ponds was calculated in Table 4.8, by using the Prismodial formula [35].

$$\text{Water storage capacity (V)} = [A + 4B + C]\,[D]/\,6 \tag{1}$$

where, V = volume of water in farm ponds (m³); A = top surface area of farm ponds in (m²); B = middle surface area of farm ponds in (m²); C = bottom surface area of farm ponds in (m²); D = depth of farm ponds in (m).

The percent of volume of water lost from the farm ponds during October to December, over the total storage capacity, in unlined and lined ponds varies in the range of 14.93 to 22.21% and 19.51 to 24.96%, respectively

(Table 4.8). This indicates that the average volume of water lost during three months storage period (October to December) in unlined and lined farm ponds comes out to be 18.73 and 22.70%, respectively. There it can be concluded that in saline tract area or deep clay soils, on an average 21% of stored volume of harvested water is lost till its utilization as protective irrigation.

4.4.2 UTILITY OF HARVESTED POND WATER

The utility of runoff water harvested in dug-out type farm ponds at different farmer's field was monitored during 2009 and 2010 and results are discussed in following head.

4.4.2.1 Quality of Harvested Water

The quality of harvested runoff water in dug-out farm ponds during the rainy season of 2009 and 2010 was tested by analyzing the water samples for pH values and electrical conductivity. The results of chemical analysis of water samples are given in Table 4.9. This indicates that the pH of water samples varies from 7.42 to 8.50 and 7.95 to 8.34 during 2009 and 2010 respectively. The pH values of harvested water (Table 4.9) in dug-out farm ponds is slightly alkaline in nature, which can be used for protective

TABLE 4.8 Water Losses From Lined and Unlined Dug-Out Farm Ponds in Three Month Storage Period (October to December)

Farm Pond Number	Size of Farm pond (m)	Total Storage Volume (m³)	Volume of water loss (m³)	Percent Storage losses
2 (Unlined)	30 × 30 × 3	1971.00	384.84	19.53
6 (Unlined)	30 × 30 × 3	1971.00	352.21	22.21
8 (Unlined)	30 × 30 × 3	1971.00	295.81	18.27
9 (Unlined)	30 × 30 × 3	1971.00	250.07	14.93
Average water loss in Unlined farm pond (m³)				**18.73**
4 (Lined)	82 × 26 × 3	5019.00	979.03	19.51
5 (Lined))	82 × 26 × 3	5019.00	955.18	23.64
10 (Lined))	82 × 26 × 3	5019.00	1014.51	24.96
Average water loss in Lined farm pond (m³)				**22.70**

irrigation since, the pH values of ground water is quite high (11.50) and is not suitable for irrigation. The electrical conductivity of water samples during 2009 and 2010 varies from the ranges of 0.28 to 0.42 dsm^{-1} to 0.31 to 0.55 dsm^{-1}, respectively. The electrical conductivity of harvested water runoff water is on slightly higher side of the good quality water (0.25 dsm^{-1}), but the water with an electrical conductivity in the ranges of 0.25 to 0.75 dsm^{-1} can also be used for irrigation with precautionary measures or with limited used [25].

The electrical conductivity of harvested water during 2010 is a higher side than that of 2009 which may be due to higher rainfall of 2010 with continuous long wet spells producing more number on runoff events.

4.4.2.2   Utility of Harvested Water

The rainwater harvested in dug-out farm ponds was used by the farmers in Ghusssar village during rabi season of 2009 and 2010. The type of power sources used for lifting pond water and method of irrigation used for protective irrigation by different farmers is given in Table 4.10. One farmer has used electrical pump set while other farmers have used tractor engine as the power sources for running pump sets. Almost all the farmers have used sprinkler irrigation system for applying protective irrigation. The farmers in Ghussar village could not able to give protective irrigation to their kharif crops during the dry spell that occurred in August 2009, since, almost all the ponds were dry at that time because no runoff event occurred in that year up to the occurrence of longest dry spell of 2009.

TABLE 4.9   Chemical Analyzes of Pond Water Samples

Farm Pond Number	2009		2010	
	pH	EC, (dsm^{-1})	pH	EC, (dsm^{-1})
3	8.50	0.36	8.33	0.31
4	8.32	0.31	7.95	0.40
6	8.38	0.31	8.15	0.37
8	7.63	0.42	8.34	0.55
11	7.42	0.48	8.21	0.46

TABLE 4.10   Type of Power Sources for Lifting Water and Mode of Applying Protective Irrigation

Sr. No.	Name of farmer with farm pond number	Water lifting power source		Mode of irrigation
		Electric pump (HP)	Tractor engine (HP)	
1	Gunvant B. Wakode (4)	5	-	Sprinkler
2	Nilesh Raut (6)	-	35	Sprinkler
3	Dipak P. Raut (7)	-	35	Sprinkler
4	Anil Pralhad Raut (8)	-	39	Sprinkler
5	Srikruna W. Behare (9)	-	39	Sprinkler
6	Laxman W. Behare (10)	-	35	Sprinkler

The farm ponds got filled up in the month of September 2009 and hence farmers could able to used harvested water as a protective irrigation to the rabi crops/ long duration kharif crops (Cotton). The effect of protective irrigation on the productivity of dryland crops during 2009 as given in Table 4.11, indicates that the yield of different cotton varieties under dry land condition increased from 20 to 30 q/ha., to 35 to 37.5 q/ha showing 17 to 87% increased. Similarly the productivity of gram crop increased from 17.5 to 27.5 q/ha., to 18.75 to 30 q/ha., indicating 57 to 87% increase in productivity due to one protective irrigation.

During 2010 due to good distribution of rainfall the ponds were fully filled, but there was no need of protective irrigation in that kharif season. The farmers have used harvested rainwater for protective irrigation to cotton and gram crop during the rabi season of 2010 and the resulting productivity is indicated in Table 4.11. The one protective irrigation indicates a percent increase in productivity of different varieties of cotton crop which ranges from 25 to 114%. Similarly, the productivity of Gram varieties shows on increase of 42 to 45% to one protective irrigation.

The above results of effect of one protective irrigation on productivity of dry land crops during 2009 and 2010 indicate an increase in yield of cotton crop to the tune of 47.3 to 55.3% on an average and that of gram crop indicates the increase of 58.5 to 43.5% on an average.

TABLE 4.11 Effect of Protective Irrigation on Yield of Different Dryland Crops During 2009 and 2010

S. No.	Dry land crops (Varieties)	Crop Yield (q/ha)					
		2009			2010		
		Dry land	One Irrigation	Percent of increase in productivity	Dry land	One Irrigation	Percent of increase in productivity
1	Cotton (Rajat)	30	35	17	30	37.5	25
2	Cotton (Akka)	25	35	40	27.5	35.0	27
3	Cotton (Dr Brand)	20	37.5	87	17.5	37.5	114
		Average percent Productivity		47.3	Average percent Productivity		55.3
4	Gram (Vijaya)	17.5	27.5	57	15.0	21.2	42
5	Gram (Jacky)	17.7	30.0	60	13.7	20.0	45
		Average percent Productivity		58.5	Average percent Productivity		43.5

During 2010, two farmers of the village under study had given more than one protective irrigation to cotton crop and the response of number of protective irrigations on productivity of dry land crop is given in Table 4.12. The effect of protective irrigation (Table 4.12) on cotton crop resulted to increase in productivity from 114% with a one protective irrigation, 142% with two protective irrigations and 214% with three protective irrigations. The above results of the number of protective irrigations, clearly indicates an increase of 13.3% and 46.7% in the productivity of cotton crop due additional one or two protective irrigations.

4.4.2.3 Effect of Protective Irrigation of Soil

To study the effect of protective irrigation using harvested runoff water on saline sodic soil of Purna valley, the soil sample from five farmers were taken after the harvest of rabi crop during 2009 (month of February 2010), after the end of rainy season of 2010 (November, 2010), and after harvested of rabi crop of 2010 (March, 2011). The results of chemical analysis of soil samples as given in Table 4.13, shows that the high PH and electrical conductivity of soil sample after harvest of the rabi crop reduces to normal level

TABLE 4.12   Effect of Protective/Supplemental Irrigation On Dry Land Crops

Sr. No.	Dry land crops (Variety)	Crop yield according to number of irrigation applied (100 kg/ha) 2010			
		Dry land	Number of Irrigation		
			1	2	3
1	Cotton (Dr Brand)	07	15	17	22
2	Percent increases over Dry land	-	114	142 t	214
3	Percent increases over one protective irrigation	-	-	13.3	46.7

in the month of November due to leaching of salts in rainy season. The result in Table 4.13 shows that salt concentration in crop root zone layer increases after protective irrigation due to upward movement of salts, which reduces after leaching after the rainy season, and hence it can be concluded that the practice of protective irrigation is not harmful in saline tract area.

4.5   SUMMARY

Maharashtra is prominently a rainfed farming state, subjected to the varies of monsoon in recent years, which greatly affects the agriculture production for sustainable and assured agriculture production irrigation facilities are necessary which can be Increased by constructing dug-out type farm ponds. The importance of dug-out type farm pond as a rainwater harvesting structure has greatly increased in recent years to present crop failures and increase the crop yields under the condition of moisture stress at the later stages of crop growth. Storage losses reduce the quantity of stored water in small reservoirs. Therefore, assessment of storage losses from dug-out type farm pond in saline tract was conducted. Following conclusions are presented:

- The evaporation seepage loss component from unlined dug-out type farm pond was 58.82 and 12.60 cm respectively during three months of water storage period (October to December).
- Evaporation losses from lined dug-out type farm pond were in the range of 58.23 to 61.18 cm during October to December. The evaporation component in lined black polyethylene was in the range of 19.51 to 24.60% of storage period.

TABLE 4.13 Effect of Protective Irrigation on Soil Health

S. No.	Farm pond number	February 2010		November 2010		March 2011	
		pH	EC, (dsm⁻¹)	pH	EC, (dsm⁻¹)	pH	EC, (dsm⁻¹)
1	3	8.64	0.30	8.25	0.22	8.67	0.28
2	4	8.43	0.26	8.28	0.20	8.50	0.22
3	6	8.79	0.32	8.06	0.22	8.85	0.30
4	8	8.43	0.30	8.12	0.21	8.51	0.23
5	11	8.65	0.28	8.02	0.22	8.52	0.25

- The water storage losses from unlined dug-out farm ponds were reduced from 1st year construction to the 2nd year. The average water storage loss during 2nd year of construction was 14.93 to 22.21% of storage capacity.
- The one protective irrigation using harvested in runoff water in dug-out farm ponds resulted in increase of 47 to 55% and 43 to 58% in the dry land productivity of cotton and gram crops, respectively.
- The more number of protective irrigations (two to three) resulted in increasing the productivity of cotton crop to the tune of 13 to 46% other than one protective irrigation.

KEYWORDS

- **Black polyethylene**
- **Cotton**
- **Dry land productivity**
- **Dug-out farm pond**
- **Evaporation**
- **Evaporation loss**
- **Gram**
- **Lined dug-out farm pond**
- **Maharashtra**
- **Protective irrigation**
- **Rainwater harvesting**

- **Runoff**
- **Simulation**
- **Storage capacity**
- **Supplemental irrigation**
- **Unlined dug-out farm pond**
- **Vertisol**
- **Water storage loss**

REFERENCES

1. Anonymous, (2009). *Economic survey of Maharashtra (2009–10).* Directorate of Economics and Statistics, Planning Department, Gov. of Maharashtra, Mumbai. Annual report: 3.
2. Arnold, J. G., & Stockle, C. O. (1991). Simulation of supplemental irrigation on-farm ponds. *J. of Irri. and Drain. Engg.,* 117(3), 408–424.
3. Beasley, R. P. (1952). Determining the effect of topography and design on the characteristics of farm ponds. *J. Agril. Engg.,* Nov, 702–704.
4. Bhandarkar, D. M. (2010). Rainwater harvesting and recycling technology for sustainable agriculture in Vertisols of high rainfall. National seminar on Engineering Agriculture for Evergreen Revolution, ISAE-AP Chapter. September 24–25.
5. Carreker, J. R. (1945). Construction and management of farm ponds, *J. Agril. Engg.,* Feb., 63–65.
6. Chittaranjan, S. (1982). Rainwater harvesting and recycling. *Indian J. Soil Coserv.* 9(2–3), 100–106.
7. Chuff, C. B. (1978). The use of the compartmented reservoir in water harvesting agri.-systems. *Production In Arid Land Conference on Plant Resources.* Texas. Jan., 56–59.
8. Codie, S. A., & Webster, I. T. (1997). The influence of wind stress, temperature and humidity gradients on evaporation from reservoirs, *Water Resources,* 83, 2813–2822
9. Dhanapal, G. N., Harsha, K. N., Manjunatha, M. H., & Ramachandrappa, B. K. (2010). Rainwater management for maximization of farm productivity and conservation of natural resources in Alfisols of Karnataka. *All India Coordinated Research Project on Dryland Agriculture,* Vol. 1, 340–346
10. Gajri, P. R., Verma, H. N., & Prihar, S. S. (1982). Rainwater harvesting and its recycling for maximization of crop production. *Indian J. Soil Cons.,* 9(2–3), 69–75.
11. Gorantiwar, S. D., Dhotre, R. S., & Dahiwalkar, S. D. (2008). Effect of percolation tank in augmenting the ground water recharge souvenir. National seminar on *Soil, Water Conservation and Crop Management Technology Under Rainfed Agriculture,* ZARS, Solapur, pages 10.

12. Grewal, S. S., Luneja, M. L., Debey, L. N., & Chandrapa, M. (1982). Effect of some soil and site properties on seepage losses from harvested runoff water in small storage reservoirs. *J. Agril. Engg.*, 19(4), 63–70.

13. Howard, Matson (1943). More farm ponds needed. *J. Agril. Engg.*, 24, 380–384.

14. Israelsen, O. W., & Hansen, V. E. (1962). *Irrigation Principle and Practices*. John Wiley and Sons, Inc. N.Y.

15. Jackson, M. L. (1973). *Soil Chemical Analysis*. Prentice Hall of India Pvt. Ltd., New Delhi.

16. Kale, S. R., & Deshmukh, M. T. (1986). Effect of polythene lining to the bed of dug-out type farm pond in reducing the percolation losses. *Indian J. Soil Cons.*, 18(1–2), 65–70.

17. Kale, S. R., Ramteke, J. R., Badrekar, S. B., & Chopra, P. S. (1986). Effect of various sealent material on seepage losses in tanks in lateritic soil. *J. Agric. Eng.*, 14(2), 58–59.

18. Khan, M. A. (1992). Evaporation of water from free water surface climatic influence. *Indian J. Soil. Conservation*, 20, 26–27.

19. Man, H. S., & Rao, B. V. R. (1981). Rainwater harvesting, management and its implications. *Indian J. Soil Cons*, 9(2–3), 77.

20. Narsimulu, B. (2010). Rainwater harvesting for drought proofing and productivity enhancement of FCV tobacco in South Coastal Andhra Pradesh. *All India Coordinated Research Project on Dryland Agriculture*, 1, 349–352.

21. Pal, D. K. (2004). Genesis of sodic black soils (Vertisols) of central India. Souvenir state level seminar on "*Development of soil Science 2004*," Akola chapter of Indian Society of Soil Science. Dept. of Agril. Chemistry and Soil Science, Dr. P.D.K.V, Akola.

22. Pandey, P. K., Panda, S. N., & Panigrahi, B. (2006). Sizing on-farm reservoirs for crop-fish integration in rainfed farming systems in Eastern India. *Biosyst. Eng.*, 93(4), 475–489.

23. Panigrahi, B., & Panda, S. N. (2003). Optimal sizing of on-farm reservoirs for Supplemental irrigation. *J. Irrigation and Drainage Engineering ASCE*, 129(2), 117–128.

24. Panigrahi, B., Panda, S. N., & Mull, R. (2001). Simulation of water harvesting potential in rainfed rice lands using water balance model. *Agric. System*, 69, 165–182.

25. Patil, V. D., & Mali, C. V. (2000). *Fundamentals of Soil Science*. Phoenix publications, Parbhani, pages 160.

26. Phadnvis, A. N., Mangre, M. L., Kide, D. S., & Mulewar, G. U. (1998). Impact of water harvesting structure on ground water recharge in semi arid Maharashtra. *Indian J. Soil Cons.*, 26(1), 44–47.

27. Potter, W. D., & Krimgold, D. B. (1946). Area relationship that simplify the hydrologic design of small farm pond. *Agricultural Engineering*, 27, 229–246.

28. Sahu, R. K. (2010). Rainwater harvesting and recycling for sustainable Integrated agriculture in micro-watersheds; National seminars on Engineering Agriculture for Evergreen Revolution, ISAE-AP Chapter. September 24–25.

29. Sastry G., & Mittal, S. P. (1987). Water harvesting and recycling Engineering aspects. *Indian Journal of Soil Conservation*, 15(3), 17–22.

30. Sastry, G., & Gurumel Singh (1993). Water harvesting and recycling Engineering aspect. *Indian J. Soil Conservation*, 21(2), 40–45.

31. Sastry, G. V. Hussenappa and R. C. Bansal (1983). Farm ponds and their influence on flood retardance. *Indian J. Soil Cons.*, 11(1–3), 54–57.

32. Sastry, G. V. Hussenappa, & Bansal, R. C. (1980). Farm ponds for assured protective irrigation for rabi crops in Doon valley. *Indian J. Soil Cons.,* 8(11), 35–36.

33. Sastry, G., Joshi, B. P., & Gurmel Shing (1982). Structural measures for efficient control of seepage from dug-out farm ponds. *Indian Journal of Soil Conservation,* 10(2–3), 120–123.

34. Sharda, V. N., & Shrimali, S. S. (1994). Water harvesting and recycling in Northern hilly area. *Indian J. Soil Conservation,* 22(1–2), 84–93.

35. Singh, Gurmel (2004). *Manual of Soil and Water Conservation Practices.* Oxford & IBH Publishers, New Delhi, pages 144.

36. Singh, R. P and G. Subha Reddy (1987). Resource management in a rainfed watershed. *Indian J. Soil Cons.,* 14(3), 52–58.

37. Srinivasulu, R., Osman, M., Murthy, V. K., & Rao, K. V. (2010). Impact of water harvesting structure on ground water recharge in semi arid Maharashtra. *Indian J. Soil Cons.,* 26 (1), 44–47.

38. Srivastava, R. C. (2001). Methodology for design of water harvesting system for high rainfall areas. *Agric. Water Manag.,* 47, 37–53.

39. Verma, H. N., Parihar, S. S., & Ranjodh Singh (1984). Feasibility of storage of runoff in dug-out ponds and its use for sulli-mental irrigation in Punjab. *Indian J. Soil Conservation,* 12(1), 31–36.

40. Wagh, R. G., Wankhade D. M., & Sawake, D. P. (1996). Role of engineering techniques in increasing agricultural productivity. *Farm Pond Technology for Sustainable Watershed Management in Kokam Region,* Dec (5–6), 1–6.

APPENDIX IA

Weekly Weather data for the year 2009 recorded at Meteorological Observatory Department of Agronomy Dr. PDKV, Akola

Week No.	Dates	T_{max} (°C)	T_{min} (°C)	BSH (hrs)	WS (km/hr)	RH I (%)	RH II (%)	E_{vap} (mm)
22	28–3 Jun	41.7	28.6	6.3	15.3	51	25	17.8
23	4–10	39.8	28.3	4.6	16.8	56	31	17.2
24	11–17	41.6	28.9	7.2	15.5	55	27	17.6
25	18–24	40.8	27.0	5.3	13.3	60	28	14.4
26	25–1 Jul	33.6	24.5	3.6	9.2	82	59	7.8
27	2–8	32.8	24.5	3.7	4.1	90	65	4.7
28	9–15	30.2	24.1	1.4	7.5	89	71	3.4
29	16–22	28.6	23.6	1.2	4.9	86	78	3.1
30	23–29	30.3	23.6	3.1	4.3	85	60	4.0
31	30–5 Aug	32.2	24.0	3.4	7.5	81	52	5.9
32	6–12	32.1	24.0	2.3	7.2	84	56	4.8

Continued

Week No.	Dates	T_{max} (°C)	T_{min} (°C)	BSH (hrs)	WS (km/hr)	RH I (%)	RH II (%)	E_{vap} (mm)
33	13–19	32.8	24.9	3.7	7.7	77	49	6.3
34	20–26	31.1	22.8	2.5	4.2	94	73	3.8
35	27–2 Sep	30.2	23.5	2.8	6.5	91	68	3.5
36	3–9	30.3	23.3	3.2	10.1	93	67	4.0
37	10–16	33.2	22.9	5.4	6.0	85	43	5.2
38	17–23	35.2	22.8	7.1	2.7	81	34	5.7
39	24–30	34.3	24.5	5.6	5.5	83	49	6.5
40	1–7 Oct	32.2	23.7	3.8	4.3	90	60	4.3
41	8–14	32.9	19.7	8.1	4.4	90	33	5.1
42	15–21	34.9	17.8	7.8	1.5	81	23	5.7
43	22–28	33.8	14.7	6.7	1.6	74	19	5.6
44	29–4 Nov	34.4	14.3	6.8	1.6	70	17	6.4
45	5–11	31.0	19.1	4.1	2.4	81	50	4.9
46	12–18	29.3	21.7	3.6	3.3	92	58	3.5
47	19–25	27.5	13.1	7.9	1.1	89	35	3.8
48	26–2 Dec	28.8	12.2	7.4	1.3	82	29	4.0
49	3–9	29.6	13.5	5.4	1.0	85	33	3.5
50	10–16	30.5	15.1	5.1	0.8	88	34	3.5
51	17–23	29.0	14.7	3.3	2.1	86	36	3.3
52	24–31	27.4	12.2	3.8	1.9	76	34	3.8

APPENDIX IB

Weekly Weather data for the year 2010 recorded at Meteorological Observatory Department of Agronomy Dr. PDKV, Akola

Weeks	Dates	T max (°C)	Tmin (°C)	BSH (hrs)	WS (km/hr)	RH I (%)	RH II (%)	Evap (mm)
22	28–3 Jun	34.9	24.6	5.0	6.4	85	57	5.7
23	4–10	30.8	24.0	2.9	5.0	88	67	3.6
24	11–17	33.0	24.3	4.3	9.1	87	59	5.3
25	18–24	33.0	24.8	3.9	3.7	90	63	4.7
26	25–1 Jul	25.3	19.9	1.3	5.4	78	56	3.1
27	2–8	29.0	23.0	1.5	5.9	93	74	3.1

Continued

Weeks	Dates	T max (°C)	Tmin (°C)	BSH (hrs)	WS (km/hr)	RH I (%)	RH II (%)	Evap (mm)
28	9–15	29.1	23.4	1.9	5.3	90	67	3.5
29	16–22	30.2	23.6	3.5	2.6	94	69	3.8
30	23–29	31.6	23.5	6.3	3.1	89	61	4.3
31	30–5 Aug	26.4	20.2	4.0	3.4	77	56	3.3
32	6–12	28.8	23.2	2.3	6.8	91	76	3.3
33	13–19	31.9	23.4	6.6	4.0	86	49	5.5
34	20–26	33.8	23.1	6.3	2.9	83	42	5.2
35	27–2 Sep	33.7	22.1	7.3	1.9	85	40	5.4
36	3–9	34.8	21.4	6.0	1.2	86	38	5.1
37	10–16	34.9	22.6	5.7	2.8	77	34	5.5
38	17–23	30.7	23.2	2.9	3.2	96	70	3.4
39	24–30	32.7	19.1	5.9	1.3	82	33	5.1
40	1–7 Oct	31.3	18.5	6.0	2.1	87	43	5.5
41	8–14	31.8	20.7	4.7	3.5	88	49	6.0
42	15–21	30.4	21.1	4.5	1.6	94	56	3.7
43	22–28	31.5	19.2	7.0	2.1	91	48	4.6
44	29–4 Nov	30.8	18.0	4.8	1.6	89	41	3.7
45	5–11	27.9	15.7	3.4	2.0	84	44	3.6
46	12–18	26.9	9.8	6.6	1.3	84	28	3.9
47	19–25	28.1	8.2	7.8	0.8	84	22	3.6
48	26–2 Dec	21.1	8.1	4.1	1.3	57	21	3.5
49	3–9	20.9	9.1	3.5	1.2	56.9	24.0	3.5
50	10–16	21.0	10.4	3.9	1.5	58.4	29.0	3.5
51	17–23	20.4	9.8	3.1	1.3	54.0	27.3	3.1
52	24–31	20.4	10.0	3.2	1.3	54.7	27.7	3.1

APPENDIX II

Soil profile	Depth (m)	Hydraulic conductivity (cm/hr)
Farm pond-2 (1)	0.15–0.60	0.54
Farm pond-2 (2)	0.60–1.20	0.54
Farm pond-2 (3)	1.20–1.80	0.38

Continued

Soil profile	Depth (m)	Hydraulic conductivity (cm/hr)
Farm pond-2 (4)	1.80–2.4	0.37
Farm pond-2 (5)	2.4–3.0	0.24
Farm pond-2 (6)	2.4–30	0.24
Farm pond-6 (1)	0.00–0.50	0..60
Farm pond-6 (2)	0.50–1.00	0.44
Farm pond-6 (3)	1.00–1.50	0.43
Farm pond-6 (4)	1.50–2.20	0.40
Farm pond-6 (5)	2.20–3.00	0.36
Farm pond-6 (6)	2.20–3.00	0.16
Farm pond-8 (1)	0.00–0.40	0.51
Farm pond-8 (2)	0.04–1.00	0.23
Farm pond-8 (3)	1.00–2.50	0.17
Farm pond-8 (4)	2.50–3.20	0.16
Farm pond-8 (5)	2.50–3.20	0.16
Farm pond-11 (1)	0.15–0.60	0.60
Farm pond-11 (2)	0.60–1.20	0.52
Farm pond-11 (3)	1.20–1.80	0.52
Farm pond-11 (4)	1.80–2.40	0.32
Farm pond-11 (5)	2.40–3.00	0.16
Farm pond-11 (6)	2.40–3.00	0.16

APPENDIX III

S. No.	Met week	Lined ponds (82 m × 26 m × 3 m)				Unlined Pond
		4	5	10	Average	
1	41	5.08	5.37	5.17	5.21	6.42
2	42	5.75	5.43	4.91	5.23	6.41
3	43	5.75	5.75	5.67	5.02	6.17
4	44	5.50	4.92	4.33	4.92	6.07
5	45	5.00	4.60	4.85	4.82	6.00
6	46	5.20	4.90	5.30	5.13	6.25
7	47	6.00	5.80	5.90	5.40	6.40

Continued

S. No.	Met week	Lined ponds (82 m × 26 m × 3 m)				Unlined Pond
		4	5	10	Average	
8	48	5.80	6.33	4.42	5.00	6.00
9	49	5.00	4.90	5.10	5.00	5.90
10	50	4.60	4.70	4.85	4.72	5.70
11	51	3.70	3.65	3.65	3.67	4.50
12	52	5.23	4.83	4.08	4.71	5.60
	Total	**62.61**	**61.18**	**58.23**	**58.82**	**71.42**

PART II

ESTIMATIONS OF SOIL PROPERTIES

CHAPTER 5

SEVERAL DIELECTRIC MIXING MODELS FOR ESTIMATING SOIL MOISTURE CONTENT

ERIC W. HARMSEN,[1] HAMED PARSIANI,[2] and MARITZA TORRES[3]

[1]Professor, Department of Agricultural and Biosystems Engineering, University of Puerto Rico - Mayaguez Campus, Mayaguez, Puerto Rico 00681 USA. Tel: 787 955 5102; E-mail: eric.harmsen@upr.edu, harmsen1000@gmail.com

[2]Professor, Director & PI of NOAA-CREST Center at UPRM, University of Puerto Rico at Mayaguez (UPRM), Department of Electrical & Computer Engineering, Call Box 9000, Mayaguez, PR 00681-9000. E-mail: parsiani@ece.uprm.edu

[3]Former Student, Electrical and Computer Engineering Department University of Puerto Rico at Mayagüez, Call Box 9000, Mayaguez, PR 00681-9000

CONTENTS

5.1   INTRODUCTION

The practitioner is faced with following two significant dilemmas when estimating soil moisture content with time domain reflectometry (TDR) and ground penetrating radar (GPR) instruments:

a. The various equipment available frequently operate at different frequencies and require instrument-specific procedures for deriving the dielectric constant from the reflected signal; and

b. After the dielectric constant has been obtained, it is necessary to calculate the volumetric soil moisture content using one of the numerous mixing model equations available in the literature, or perform a soil-specific calibration. Some commercially available equipment (e.g., Theta Probe) performs the dielectric determination and conversion to moisture content internally making it invisible to the user.

Factors affecting dielectric constant of a soil include: moisture content, soil texture, specific surface area, bulk density and instrument frequency. The most important influence on the dielectric constant of a soil is its water content. The dielectric constant for dry soil is approximately 4, whereas that of pure water is 81 [10]. As soil texture changes from sand (coarse grain) to clay (fine grain), or soil organic matter increases, the soil's specific surface area increases, resulting in a greater percentage of bound water. The dielectric constant associated with this water decreases relative to free water.

Wang and Schmugge [10] assigned a dielectric value to the bound water phase equal to that of ice (dielectric constant equal to 3.2 as compared to 81 for free water). Bulk density has been shown to have a strong influence on dielectric constant. Dirksen and Dasberg [3] developed a family of dielectric/moisture content curves for bulk densities ranging from 0.6 to 1.55 g/cm^3. The effect of the bulk density has to do with the increased influence of the dielectric constant of the solid phase. Increasing bulk density resulted in increased dielectric constants. The transmitted electromagnetic wave frequency of the instrument has been shown to influence the magnitude of the dielectric constant for a given value of the moisture content [2]. The dielectric constant of water decreases with increasing frequency above 1 GHz [4]. Other factors, which can influence the dielectric constant include: geometric properties [2], temperature [8], and electrochemical interactions [2].

The purpose of this study was to review several of the available mixing models and to compare estimates of soil moisture content derived from dielectric constants obtained from several instruments (GPR, TDR and Theta Probe) and soil types (sand, loam and clay).

5.2 DIELECTRIC MIXING MODELS FOR THIS STUDY

Numerous dielectric mixing models have been developed during the last twenty-five years. In this study, the authors used six dielectric mixing models for relating dielectric constant to moisture content, including: Topp et al. [9]; Alharthi and Lange [1]; Miller and Gaskin [6]; Benedetto and Benedetto [2]; Hallikainen et al. [4]; and Wang and Schmugge [10].

The equation given by Topp et al. [9], relating the apparent dielectric constant (ε) to the volumetric water content (θ_v) is given as:

$$\theta_v(\varepsilon) = -5.3 \times 10^{-2} + 2.92 \times 10^{-2}\,\varepsilon - 5.5 \times 10^{-4}\,\varepsilon^2 + 4.3 \times 10^{-6}\,\varepsilon^3 \quad (1)$$

The equation by Topp [9] is empirical and does not account specifically for the soil properties, instrument frequency, or dielectric constants of the soil constituents (i.e., water, air and solids). The apparent dielectric constant can be calculated from knowledge of the dielectric constants of the individual soil components (water, air and solids) using the following equation [1]:

$$\varepsilon(\theta_v) = \left[\phi_1\left(1 - s(\theta_v)\right)\sqrt{\varepsilon_a} + \phi_1\,s(\theta_v)\sqrt{\varepsilon_w} + (1 - \phi_1)\sqrt{\varepsilon_s} \right]^2 \quad (2)$$

where s is the degree of saturation equal to zero when θ_v equals zero and 1 when θ_v equals the soil porosity; w, a and s refer to water, air and solids, respectively.

The mixing model presented by Miller and Gaskin [6] is used to convert the dielectric constant measured by the Theta Probe to volumetric moisture content, and has the following form:

$$\theta_v = \left(\sqrt{\varepsilon} - a_o\right) \Big/ a_1 \tag{3}$$

where, a_o and a_1 are constants. Miller and Gaskin [6] provide values of a_o and a_1 for "mineral soil" as 1.6 and 8.4, respectively; and the values for "organic soil" as 1.3 and 7.7, respectively.

Benedetto and Benedetto [2] presented dielectric/moisture content data for sand and pozzolana at 0.6 and 1.6 GHz. They only investigated moisture contents in the range of 0 to 20%. By interpolating the 0.6 and 1.6 GHz curves, it was possible to derive a curve for 1.5 GHz. It should be noted that it is not advisable to use the following equation for soils [2] in which the moisture content is above around 20%.

$$\varepsilon\,(\theta_v) = 91.589 \cdot \theta_v^{\,2} + 17.007 \cdot \theta_v + 3.0547 \tag{4}$$

The empirical model of Hallikainen et al. [4] accounts for frequency and soil texture. The following equation is applicable to dielectric data collected at frequencies equal to 1.4, 6, 8, 10, 12, 14, 16 and 18 GHz, depending on the values of the parameters used.

$$\varepsilon\,(\theta_v) = (a_0 + a_1 \cdot S + a_2 \cdot C) + (b_0 + b_2 \cdot S + b_2 \cdot C) \cdot \theta_v$$
$$+ (c_0 + c_1 \cdot S + c_2 \cdot C) \cdot \theta_v^{\,2} \tag{5}$$

where, a_0, a_1, a_2, b_0, b_1, b_2, c_0, c_1 and c_2 are constants, S is percent of sand and C is percent of clay. For 1.4 GHz: the values of a_0, a_1, a_2, b_0, b_1, b_2, c_0, c_1 and c_2 were 2.862, –0.012, 0.001, 3.803, 0.462, –0.341, 119.006, –0.500, 0.633, respectively. (Note that a_0 and a_1 in the above equation are not the same parameters used in the model of Miller and Gaskin [6]).

Wang and Schmugge [10] presented a set of equations that accounts for soil texture, bulk and particle density, and wilting point. In this model, the parameter W_t, is defined as the transition moisture content at which the dielectric constant increases steeply with increasing moisture content. Consequently, following two equations ($\varepsilon_1(\theta_v)$, and $\varepsilon_2(\theta_v)$) are necessary to define the moisture content/dielectric relationship within the moisture content range of 0 to 0.5:

$$\varepsilon_1(\theta_v) = \theta_v \cdot \varepsilon_{x1}(\theta_v) + (\phi - \theta_v) \cdot \varepsilon_a + (1 - \phi) \cdot \varepsilon_s \text{ for } \theta_v \leq W_t \quad (6a)$$

with

$$\varepsilon_{x1}(\theta_v) = \varepsilon_i + (\varepsilon_w - \varepsilon_i) \cdot \frac{\theta_v}{W_t} \cdot \gamma$$

$$\gamma = 0.57 \cdot WP + 0.481$$

$$WP = 0.06774 - 0.00064S + 0.0047C$$

$$W_t = 0.49 + WP + 0.165$$

$$\varepsilon_2(\theta_v) = W_t \cdot \varepsilon_{x2} + (\theta_v - W_t) \cdot \varepsilon_w + (\phi - \theta_v) \cdot \varepsilon_a + (1 - \phi) \cdot \varepsilon_s \text{ for } \theta_v > W_t \quad (6b)$$

with

$$\varepsilon_{x2} = \varepsilon_i + (\varepsilon_w - \varepsilon_i) \cdot \gamma$$

where, θ_v = volumetric moisture content, ε_1 = apparent dielectric constant for moisture content less than or equal to W_t, ε_2 = apparent dielectric constant for moisture content greater than W_t, ε_a = dielectric constant of air (1), ε_i = dielectric constant of ice (3.2), ε_w = dielectric constant of pure water (81), ϕ = porosity, WP = moisture content at the wilting point (pore water pressure = 15 bars), S = sand content in percent of dry soil, C = clay content in percent of dry soil, W_t = transition moisture content at which the dielectric constant increases steeply with increasing moisture content, and γ = fitting parameter which is related to WP.

5.3 METHODS AND MATERIALS

Several "sandbox" experiments were conducted in the Soil and Water Laboratory of the Department of Agricultural and Biosystems Engineering, University of Puerto Rico-Mayagüez Campus [5, 7]. Air dried soil was wetted 24 hours before the experiment so that the moisture content was two to three times greater than the air-dried soil. The wet and dry soil samples were placed in layers approximately 20 cm thick within the sand box. Metal rods

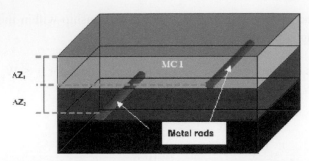

FIGURE 5.1 Schematic drawing of the "sandbox" configuration: MC1 and MC2 are moisture contents for layers 1 and 2, respectively. $\Delta Z1$ and $\Delta Z2$ are thicknesses of layers 1 and 2, respectively [7].

(5/8 inch diameter) were placed at the bottom of each soil layer to serve as a reflector of the GPR signal. Figure 5.1 shows the "sandbox" setup.

The GPR device for this research was a GSSI SIR-20 with a bow-tie transmit/receive antenna operating at 1.5 GHz. The depth resolution in sand is about 25.4 cm. Reflections are obtained at the boundary of two media of sufficiently different dielectric constants. Reflectors were laid at different depths to receive sharper reflections and be able to measure relative velocity without the aid of depth parameters (Figure 5.1). The reflections of the GPR off of the reflectors were in the form of a hyperbola (Figure 5.2).

A Tektronix 1502 TDR was used to estimate the soil dielectric constant for comparison with estimates of the GPR. The instrument is able to precisely locate and analyze discontinuities in metallic cabling. However, in this research, measuring the dielectric constant of soil is accomplished by use of a wave guide, of length 20 cm, which is pushed into the soil.

Dielectric constants were also determined using a Theta Probe [6]. With this instrument, a standing wave measurement is made to determine

FIGURE 5.2   GPR image of the sand with metal reflectors.

the impedance of a sensing rod array from which the dielectric constant can be determined. Volumetric moisture content is then calculated using equation 3. Soil moisture was measured directly with the Theta Probe using the "mineral soil" setting, after which the dielectric constant was calculated by rearranging Eq. (3). The resulting dielectric constant was then used with the other mixing models.

The soils were analyzed for their chemical and physical properties by Soilcon Laboratories, Ltd. of British Columbia, Canada. The sand was construction sand (no name), the loam soil was San Antón Loam from Juana Diaz, PR, and the clay soil was Daguey Clay from the Finca Alzamora at the UPR Campus in Mayagüez, PR.

5.4   RESULTS AND DISCUSSION

A comparison of the dielectric mixing models is presented in this section. Table 5.1 lists the inputs required by each of the six models evaluated. Three of the models solve for dielectric constant as a function of the moisture content and three solve for the moisture content as a function of the dielectric constant. Figure 5.3 compares the six mixing models graphically. For those models that required soil information, the data for sand was used. Table 5.2 lists the physical and chemical properties for the three soils used in this study. Figure 5.3 also provides the dielectric/moisture content data for sand obtained by the TDR, in which the moisture content was obtained by the gravimetric method [5]. For about 15% moisture content, all of the models compare well with the measured data, except the Hallikainen model [4]. At high moisture contents, four methods seriously under predicted the dielectric constant, while the Wang and Schmugge [10] and the Hallikanainen models performed well. It is interesting to note that only these two models require information regarding soil texture (Table 5.1).

Figure 5.4 compares the two models that account for soil texture (i.e., the Wang and Schmugge; and Hallikainen models) for sand, loam and clay. Soils data were obtained from Table 5.2. Although the Wang and Schmugge model includes a method of estimating the wilting point from the percent sand and clay, yet it did not work well. Therefore, the authors used the laboratory-measured value of the wilting point directly. For clay at low moisture contents, the Hallikaninen model actually decreases slightly, which appears to be physically unrealistic.

TABLE 5.1 Dielectric Mixing Models Compared in This Study and Inputs For These Models

Method	ε	θ_v	f	ρ_b	ρ_s	ϕ^1	%Clay	%Sand	WP	ε_i	ε_w	ε_a	ε_s
Topp [9]	IV	DV											
Alharthi and Lange [1]	DV	IV		✓	✓	✓					✓	✓	✓
Miller and Gaskin [6]	IV	DV						✓					
Benedetto and Benedetto [2]	IV	DV	✓										
Hallikainen et al. [4]	DV	IV	✓				✓						
Wang and Schugge [10]	DV	IV		✓	✓	✓			✓	✓	✓	✓	✓

Definitions: IV is independent variable, DV is dependent variable, ε is apparent dielectric constant, θ_v is the volumetric moisture content, f is frequency, ρ_b is bulk density, ρ_s is particle density, ϕ is porosity, % Clay is the percent of clay content, % Sand is the percent of sand content, WP is the wilting point, and ε_i, ε_w, ε_a and ε_s are the dielectric constants of ice, water, air and solids, respectively.

[1] If ρ_b and ρ_s are available, then ϕ can be calculated from the relation: $\phi = (1 - \rho_b/\rho_s)$.

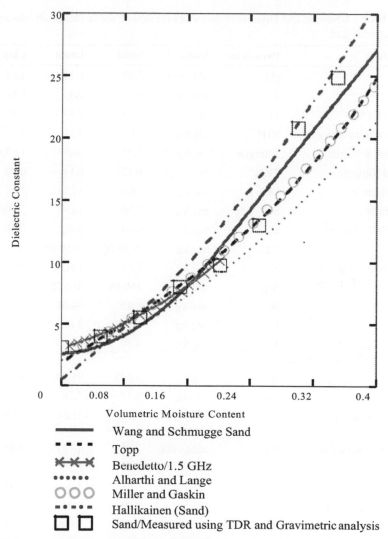

FIGURE 5.3 Comparison of the six mixing models which relate dielectric constant to the volumetric moisture content. Measured data is also provided for sand obtained using a TDR and gravimetric analysis.

Tables 5.3–5.5 indicate the volumetric moisture content estimates for the sand, loam and clay soils. Dielectric constants for wet and dry soil were obtained by the GPR, TDR and Theta Probe, except in the case of the sand, in which the Theta Probe was not available during the time of the experiment. Table 5.6 summarizes the "best" and "worst" performing mixing models

TABLE 5.2 Chemical and Physical Properties of the Soils Used in This Study (Soilcon Laboratories, Ltd.)

Analysis	Parameter	Units	Sand	Loam	Clay
	pH	pH units	7.27	6.44	4.65
	EC	(dS/m)	0.17	0.39	0.18
Total Carbon	TC	%	0.22	1.59	1.75
Ammonium Nitrogen	NH4	mg/kg	<1	5.31	2.31
Nitrate & Nitrite	Nitrogen	mg/kg	1.35	28.98	12.73
Total Nitrogen	TKN	%	0.12	0.14	0.30
Available P	Bray P	mg/kg	<0.6	11.73	7.02
Sulphate	S	mg/kg	7.90	18.00	21.00
Available Boron	B	mg/kg	<0.1	0.40	0.30
	Ca	mg/kg	2930.00	3150.00	869.57
Available Nutrients	K	mg/kg	<2	134.42	150.31
(NH_4OAc Extractable)	Mg	mg/kg	160.64	450.72	263.58
	Na	mg/kg	8.85	64.64	15.71
	Cu	mg/kg	<0.03	3.61	3.42
Available Metals	Fe	mg/kg	4.60	58.06	30.21
(0.1N HCl Extractable)					
	Mn	mg/kg	23.79	76.59	19.88
	Zn	mg/kg	0.11	2.78	1.20
Cation Exchange Capacity		meq/100g	4.15	12.61	12.84
Exchangeable Cations	Ca	meq/100g	2.80	13.37	3.58
	Mg	meq/100g	0.50	3.55	2.55
	K	meq/100g	0.16	1.04	0.59
Total Organic Carbon	TOC	%	0.05	1.54	1.67
Texture	Sand	%	96.00	35.99	3.61
	Silt	%	1.56	39.53	29.06
	Clay	%	1.62	23.87	67.27
USDA Classification			Sand	Loam	Clay
Soil Characteristic Data (Pressure vs. % Vol.)	5 kPa	% by vol	9.31	46.94	47.83
	10 kPa	% by vol	6.86	39.52	46.12
	33 kPa	% by vol	5.32	29.68	43.23
	70 kPa	% by vol	4.77	26.41	41.56

TABLE 5.2   Continued

Analysis	Parameter	Units	Sand	Loam	Clay
	100 kPa	% by vol	4.54	24.91	40.60
	300 kPa	% by vol	3.95	20.92	37.20
	500 kPa	% by vol	3.60	19.27	35.65
	800 kPa	% by vol	3.27	17.90	34.14
	1200 kPa	% by vol	2.93	16.73	32.52
	1500 kPa	% by vol	2.71	15.80	31.43
Bulk Density		kg/m3	1387.80	1350.97	1092.38
Particle Density		kg/m3	2673.17	2533.14	2537.36
Total Porosity		% vol	48.08	46.66	56.95
Air Entry Tension		kPa	0.00	3.87	0.72
Saturated Hydraulic Conductivity		cm/hr	65.70	0.24	2.77
Aeration Porosity	5 kPa	% by vol	38.77	-0.27	9.12
Aeration Porosity	10 kPa	% by vol	41.23	7.15	10.83
Available H_2O Storage Capacity		% by vol	2.61	13.88	11.79

relative to the volumetric moisture content as determined by gravimetric analysis. There was no clear "best" mixing model. However, the Miller and Gaskin method [6] appeared the most number of times in the "best" column. Furthermore it appeared only one time in the "worst" column, and in that case the error was under 5% (4.7%). Although this model ranked highest, its performance for the loam soil was disappointing. In the "worst" column, the model of Hallikainen appears eight out of sixteen times. This method only appeared once in the "best" column with an error of 7.5%.

In general, all models performed poorly for the loam soil (except estimates based on the dielectric constant from the Theta Probe/wet soil). Soil chemical properties can help explain unusual values of dielectric constant, however, in this case the loam had relatively low electrical conductivity (0.39 dS/m), and its total organic carbon (TOC) and cation exchange capacity (CEC) were roughly equivalent to the clay (Table 5.2). Additional investigation is needed to help explain the poor results for the loam soil. From this study, unfortunately, it is not possible to specify the best mixing model to use under the wide set of soil conditions that may be encountered when employing a satellite based sensors.

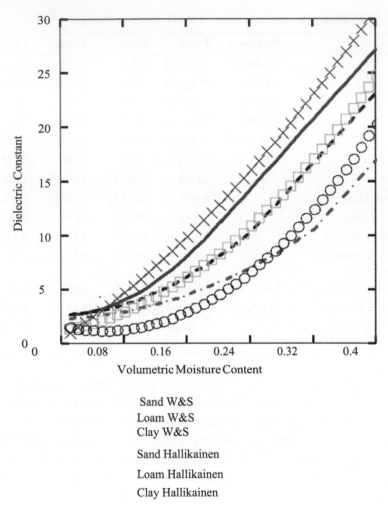

FIGURE 5.4 Comparison of the Wang and Schmugge (W&S) and Hallikainen dielectric mixing models for sand, loam and clay soils.

5.5 CONCLUSIONS

This chapter presented a comparison of six dielectric mixing models for use with time- domain dielectric data. Dielectric data were obtained from "sandbox" studies using a GPR, TDR and Theta Probe were obtained for wet and dry sand, loam and clay. Physical and chemical properties for the three soils are presented. As compared to the volumetric moisture content

TABLE 5.3 Volumetric Moisture Content for Sand Estimated from GPR and TDR Dielectric Constants Using Six Mixing Model Methods

SANDY SOIL

				Dielectric Mixing Model			
	Measured Dielectric Constant	Topp	Alharthi and Lange	Miller and Gaskin	Benedetto and Benedetto	Hallikainen et al.	Wang and Schmugge	Gravi-metric
				Moisture Content (% by volume)				
Dry	2.4/GPR	2.5	2.0	0.0	BC	4.6	7.0	7.0
Wet	9.0/GPR	15.5	19.7	16.7	15.0	13.7	27.2	21.0
Dry	3.1/TDR	6.1	4.3	1.9	7.0	5.0	14.8	7.0
Wet	6.7/TDR	19.3	14.7	11.8	20.7	11.3	31.4	21.0

Volumetric moisture content derived from gravimetric data is also shown.

TABLE 5.4 Volumetric Moisture Content for Loam Soil Estimated from GPR, TDR and Theta Probe™ Dielectric Constants Using Wang and Schmugge (W&S), Topp and Benedetto Mixing Models

LOAM SOIL	Measured Dielectric Constant	Topp	Dielectric Mixing Model					
			Alharthi and Lange	Miller and Gaskin	Benedetto and Benedetto	Hallikainen et al.	Wang and Schmugge	Gravimetric
			Moisture Content (% by volume)					
Dry	4.6/GPR	6.7	9.0	6.5	7.7	12.0	11.4	2.4
Wet	6.6/GPR	13.8	14.4	11.5	11.1	17.1	16.8	24.6
Dry	4.6/TDR	6.9	9.0	6.5	7.5	12.0	11.6	2.4
Wet	7.5/TDR	13.7	16.4	13.6	13.0	18.6	19.0	24.6
Dry	5.6/TP	9.4	11.7	9.1	9.5	14.8	14.6	2.4
Wet	10.7/TP	20.2	23.0	19.9	22.0	23.8	24.5	24.6

TP is Theta Probe

*Volumetric moisture content derived from gravimetric data is also shown.

TABLE 5.5 Volumetric Moisture Content for Clay Soil Estimated from GPR, TDR and Theta Probe™ (TP) Dielectric Constants Using Wang and Schmugge (W&S), Topp and Benedetto Mixing Models

Clay Soil				Dielectric Mixing Model				
	Measured Dielectric Constant	Topp	Alharthi and Lange	Miller and Gaskin	Benedetto and Benedetto	Hallikainen et al.	Wang and Schmugge	Gravimetric
				Moisture Content (% by volume)				
Dry	2.8/GPR	2.5	3.0	1.0	BC	15.5	7.0	7.0
Wet	8.3/GPR	15.5	18.3	15.3	15.0	26.7	27.2	21.0
Dry	4.2/TDR	6.1	8.0	5.4	7.0	19.2	14.8	7.0
Wet	10.2/TDR	19.3	22.1	19.1	20.7	29.3	31.4	21.0
Dry	5.6/TP	9.4	11.7	9.1	9.2	22.2	19.6	7.0
Wet	11.3/TP	27.6	24.2	21.0	NA	30.7	32.8	21.0

TD is Theta Probe.

* Volumetric moisture content derived from gravimetric data is also shown.

TABLE 5.6 Best and Worst Performance by Mixing Models as Compared to the Volumetric Moisture Content Determined by the Gravimetric Method

Soil/Instrument/Wetness	Best	Gravimetric minus Best	Worst	Gravimetric minus Worst
SAND/GPR/DRY	Wang and Schmugge	0.0	Miller and Gaskin	7.0
SAND/GPR/WET	Alharthi and Lang	1.3	Hallikainen	7.3
LOAM/GPR/DRY	Miller and Gaskin	4.1	Hallikainen	–9.6
LOAM/GPR/WET	Hallikainen	7.5	Benedetto	13.5
CLAY/GPR/DRY	Wang and Schmugge	0.0	Hallikainen	–9.6
CLAY/GPR/WET	Alharthi and Lang	2.7	Benedetto	11.6
SAND/TDR/DRY	Benedetto	0.0	Wang and Schmugge	–7.8
SAND/TDR/WET	Benedetto	0.3	Wang and Schmugge	–10.4
LOAM/TDR/DRY	Miller and Gaskin	–4.1	Hallikainen	–9.6
LOAM/TDR/WET	Wang and Schmugge	5.6	Benedetto	11.6
CLAY/TDR/DRY	Benedetto	0.0	Hallikainen	–12.2
CLAY/TDR/WET	Benedetto	0.3	Wang and Schmugge	–10.4
LOAM/Theta Probe/DRY	Miller and Gaskin	6.7	Hallikainen	–12.4
LOAM/Theta Probe/WET	Wang and Schmugge	0.1	Miller and Gaskin	4.7
CLAY/Theta Probe/DRY	Miller and Gaskin	–2.1	Hallikainen	–15.2
CLAY/Theta Probe/WET	Miller and Gaskin	0.0	Hallikainen	–9.7

determined by the gravimetric method, the Miller and Gaskin model (a_o and a_1 parameters for "mineral" soil) performed best. Although, this model ranked highest, its performance for the loam soil was disappointing, as was the case for all other models. Additional study is needed in the proper usage of the models as applied to the loam soil.

5.6   SUMMARY

As part of a NOAA-funded project, studies were conducted at the University of Puerto Rico-Mayagüez Campus using surface-based ground penetrating radar (GPR) to measure soil moisture content. The GPR will eventually be used to verify values of soil moisture at several locations in Puerto Rico using active radar and passive satellite-based sensors. As a part of the estimation process, it is necessary to relate moisture content to the GPR-measured dielectric constant. The motivation for this study was the need to select an appropriate dielectric mixing model for the wide range of soils being considered in the study. An important requirement of the dielectric mixing model was that it works well with input data available from NRCS Soil Survey Reports (e.g., soil texture, available water holding capacity, etc.). The advantage of using this type of data is that it can be readily incorporated into a geographic information system (GIS) to be used with the geo-referenced dielectric data of the surface and satellite-based sensors.

This chapter provides a review of several dielectric-mixing models, and compares moisture content estimates for sand, loam and clay soils, based on dielectric data obtained from a GPR, TDR and Theta Probe™. These results are also compared to soil moisture contents obtained from gravimetric data. Soils were characterized in terms of their chemical and physical properties: information needed by several of the dielectric mixing models. In some cases, especially with the loam soil, wide variations in the dielectric constants and moisture contents were observed.

ACKNOWLEDGMENTS

We are grateful to NOAA-CREST (NA17AE1625) for its financial support of this project. Thanks also to Antonio Gonzalez for his assistance in data collection and analysis.

KEYWORDS

- **Dielectric constant**
- **GPR**
- **Mixing model**
- **Moisture content**
- **Soil texture**
- **TDR**
- **Theta Probe**
- **USDA**

REFERENCES

1. Alharthi, A., & Lange, J. (1987). Soil water saturation: dielectric determination. *Water Resources Research*, 23(4), 591–595.
2. Benedetto, A., & Benedetto, F. (2002). GPR experimental evaluation of subgrade soil characteristics for rehabilitation of roads. Ninth International Conference on Ground Penetrating Radar, Steven K. Koppenjan, Hua Lee, Editors. *Proceedings of SPIE*, 4758, 605–607 and 708–714.
3. Dirksen, C., & Dasberg, S. (1993). Improved calibration of time domain reflectometry soil water content measurements. *Soil Science Society of America Journal*, 57, 660–667.
4. Hallikainen, M. T., Ulaby, F. T., Dobson, M. C., El-Reyes, M. A., & Wu, L. K. (1985). Microwave dielectric behavior of wet soil, Part I: Empirical models and experimental observations. *IEEE Transactions on Geoscience and Remote Sensing*, GE-23(1), 25–34.
5. Harmsen, E. W., & Parsiani, H. (2003). Inverse procedure for estimating vertically distributed soil hydraulic parameters using GPR. NOAA-CREST/NASA-EPSCoR Joint Symposium for Climate Studies University of Puerto Rico – Mayagüez Campus. January 10–11.
6. Miller, J. D., & Gaskin, G. J. (1999). *Theta Probe ML2x Principles of Operation and Application*. MLURI Technical Note (2nd edition). Delta-T Devices (Cambridge) and MaCaulay Land Use Research Institute (Aberdeen).
7. Parsiani, H., Harmsen, E., Rodriguez, D., & Diaz, R. (2003). Validation of an inverse procedure for estimating soil moisture content using GPR. *Proceedings of the NOAA-CREST Remote Sensing Conference*, Tallahassee, Florida, March 30–April 1.
8. Rassam, D. W., & Williams, D. J. (2000). A dynamic method for determining the soil water characteristic curve for coarse-grained soils, *Geotechnical Testing Journal*, 23(1), 67–71.
9. Topp, G. C., Davis J. L., & Annan, A. P. (1980). Electromagnetic determination of soil water content. *Water Resour. Res*, 16(3), 574–583.
10. Wang, J. R., & Schmugge, T. J. (1980). An empirical model for the complex dielectric permittivity of soils as a function of water content. *IEEE Transactions on Geoscience and Remote Sensing*, GE-18(4), 288–295.

CHAPTER 6

INVERSE PROCEDURE FOR ESTIMATING VERTICALLY DISTRIBUTED SOIL HYDRAULIC PARAMETERS USING GPR

ERIC W. HARMSEN[1] and HAMED PARSIANI[2]

[1]*Professor, Department of Agricultural and Biosystems Engineering, University of Puerto Rico - Mayaguez Campus, Mayaguez, Puerto Rico 00681 USA. Phone: 787 955 5102 E-mail: eric.harmsen@upr.edu, harmsen1000@gmail.com corresponding author*

[2]*Professor, Director & PI of NOAA-CREST Center at UPRM, University of Puerto Rico at Mayaguez (UPRM), Department of Electrical & Computer Engineering, Call Box 9000, Mayaguez, PR 00681-9000. E-mail: parsiani@ece.uprm.edu*

CONTENTS

This chapter was originally prepared by the authors for paper presentation at the NOAA-CREST/ NASA-EPSCoR Joint Symposium for Climate Studies at University of Puerto Rico – Mayaguez Campus, January 10–11.

6.1   INTRODUCTION

Knowledge of the vertical distribution of the volumetric soil moisture content is important for estimating the vertical water flux in soil. Ground Penetrating Radar (GPR) has the potential for supplying this type of detailed information. However, most investigations have been limited to estimating the bulk average moisture content (via the bulk average dielectric constant) of the soil [2, 3, 5, 6, 13]. Several studies have estimated vertically distributed moisture content using the cross–borehole GPR method [4, 7]. However, a disadvantage of the method is that boreholes need to be installed, within which one need to install unscreened tubing made of a special material.

     To implement measurements using surface GPR, a procedure is needed to extract depth-specific values of the dielectric constant from bulk average dielectric constant. This research study presents the inverse procedure to estimate vertically distributed soil hydraulic parameters using GPR.

6.2   METHODOLOGY

This section describes an inverse procedure for estimating the depth-specific dielectric constant derived from several bulk average dielectric constant values measured by surface GPR. With a minimum of soil information and the bulk average dielectric constant values, the vertical distribution of moisture content, pore water pressure, hydraulic conductivity and the Darcy flux can calculated. One only needs information on soil type (percentage distribution of sand, silt and clay) and bulk density, from which other required soil parameters (porosity, residual moisture content, and water retention parameters) are estimated using the Rosetta neural network [10]. The vertically distributed dielectric constant can be obtained from the following equations:

$$\xi_1 = \xi'_1 \tag{1a}$$

$$\xi_n = \frac{\xi'_n z_n - \sum_{i=1}^{n-1} \xi_i \Delta z_i}{\Delta z_n} \quad n \geq 2 \tag{1b}$$

where, ξ_n = square root of the dielectric constant for the nth vertical interval; ξ'_n = square root of the average dielectric constant between the soil surface and the bottom of the nth vertical interval, obtained by GPR; z_n = depth of soil between surface and the bottom of the nth vertical interval; and Δz_n = the thickness of the nth vertical interval.

Equation (1b) was obtained by using differential signal delay between the vertical intervals. The differential delay was converted to the square root of the dielectric constant. Figure 6.1 illustrates how equation 1 can be applied in practice. The example given in Figure 6.1 shows a case for: n = 5 and $Z_1 = 0.5$ (Z_2), and $\Delta Z_2 = \Delta Z_3 = \Delta Z_4 = \Delta Z_5$. GPR measurements are associated with reflections from points (represented by circles) of contrasting dielectric constant. The reflections can be produced by inserting metal rods (e.g., re-bar [6]) in the soil at intervals of Δz to a total depth Z. The maximum depth of the reflectors depends on the frequency of the GPR.

It is recommended that a trench should be made in the soil to a depth somewhat greater than Z [6], and the rods be inserted laterally through the side wall of the trench into the footprint area of the GPR. By inserting the metal rods in this way, the soil will remain relatively undisturbed. Each GPR measurement provides a value of the bulk average dielectric constant between the soil surface and the depth z_n. The closer Δz_n are to one another, the more accurately is the media represented in equation 1. Because equation 1b depends on the value of ξ_1 (estimated from the square root of the bulk average dielectric constant ξ'_1), it is recommended that z_1 should be small (e.g., 10 cm). Figure 6.2 shows the soil cross-section parallel to the direction of the reflectors, and the trench cross section.

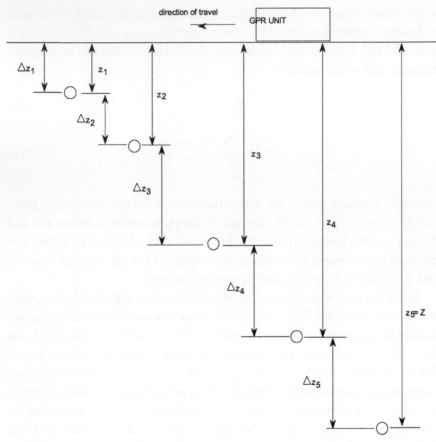

FIGURE 6.1 Cross section of soil and reflectors (circles) at five different depths below the surface.

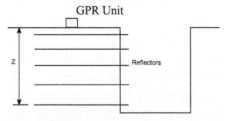

FIGURE 6.2 Cross section of reflectors (e.g., re-bars) inserted through the wall of the excavation. GPR unit is moved in a direction normal to the long dimension of the reflectors.

Several dielectric mixing models (Dobson et al. [3]; Wang and Schmugge [13]; Martinez and Byrnes [8]; Rial and Han [9]; Alharthi and Lang [1]; Shutko and Reutov [12]) for relating dielectric constant to moisture content were evaluated by Harmsen et al. [6]. The semi-empirical model of Dobson et al. [3] accounts for frequency. However, it did not provide accurate estimates of moisture content from the dielectric constant [6]. The Wang and Schmugge [13] method performed best and was selected for use in this chapter.

Wang and Schmugge [13] method consists of a set of equations that accounts for soil texture, bulk and particle density, and wilting point. In this model, the parameter, W_t, is defined as the transition moisture content at which the dielectric constant increases steeply with increasing moisture content. Consequently, following two equations ($\varepsilon_1(\theta v)$, and $\varepsilon_2(\theta v)$) are necessary to define the relationship between moisture and dielectric constants, within the moisture content range of 0 to 0.5:

$$\varepsilon_1(\theta_v) = \theta_v \cdot \varepsilon_{x1}(\theta_v) + (\phi - \theta_v) \cdot \varepsilon_a + (1 - \phi) \cdot \varepsilon_s \text{ for } \theta_v \leq W_t \quad (2)$$

$$\varepsilon_{x1}(\theta v) = \varepsilon_i + (\varepsilon_w - \varepsilon_i) \cdot [\theta v / W_t] \cdot \gamma \quad (3)$$

with

$$\gamma = -0.57 \cdot WP + 0.481 \quad (4)$$

$$WP = 0.06774 - 0.00064(S) + 0.0047(C) \quad (5)$$

$$\varepsilon_2(\theta_v) = W_t \cdot \varepsilon_{x2} + (\theta_v - W_t) \cdot \varepsilon_w + (\phi - \theta_v) \cdot \varepsilon_a$$

$$+ (1 - \phi) \cdot \varepsilon_s \text{ for } \theta_v > W_t \quad (6)$$

with

$$W_t = 0.49 (WP) + 0.165 \quad (7)$$

$$\varepsilon_{x2} = \varepsilon_i + (\varepsilon_w - \varepsilon_i) \cdot \gamma \quad (8)$$

where, θv = volumetric moisture content; ε_1 = apparent dielectric constant for moisture content less than or equal to W_t; ε_2 = apparent dielectric constant for moisture content greater than W_t; ε_a = dielectric constant of air

(= 1); ε_i = dielectric constant of ice (= 3.2), ε_w = dielectric constant of pure water (= 81); φ = porosity; WP = moisture content at the wilting point (pore water pressure = 15 bars); S = sand content in percent of dry soil; C = clay content in percent of dry soil; W_t = transition moisture content at which the dielectric constant increases steeply with increasing moisture content; and γ = fitting parameter which is related to WP.

Figure 6.3 shows a comparison of Eqs. (2) and (6) with measured data for Construction Sand and a Miller Clay [6]. The construction sand data were obtained as a part of this study. Dielectric constant values were measured using a Model 1502 Tektronix Time Domain Reflectometer (TDR) with a 5.1 cm long wave guide. Soil moisture content values were obtained by the gravimetric method.

The vertical distribution of the pore water pressure and hydraulic conductivity are estimated using the well-known van Genuchten equations:

$$h(\theta_v) = \frac{\left[\left[\frac{(\phi - \theta_r)}{(\theta_v - \theta_r)}\right]^{1/m} - 1\right]^{1/n}}{\alpha} \tag{9}$$

$$K(\theta_v) = K_s \left(\frac{\theta_v - \theta_r}{\phi - \theta_r}\right)^{0.5} \left[1 - \left[\left(\frac{\phi - \theta_r}{\theta_v - \theta_r}\right)^{1/m}\right]^m\right]^2 \tag{10}$$

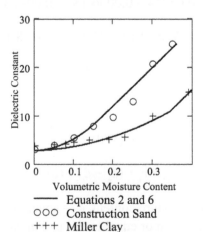

FIGURE 6.3 Comparison of measured and calculated dielectric constant for differing volumetric soil moisture content for a construction sand and miller clay.

where, $h(\theta_v)$ = pore water pressure; $K(\theta_v)$ = variably saturated hydraulic conductivity; Ks = saturated hydraulic conductivity; θr = residual moisture content. Water, held at temperatures up to 105°C, is referred to as hygroscopic water (soil tension of 31 bars) and is virtually part of the mineral structure of the soil [13]; α = inverse of the air-entry value (or bubbling pressure); and n = pore size distribution index $m = 1 - n^{-1}$ for $n > 1$.

The Darcy flux is calculated with the Darcy Buckingham equation [11]:

$$q(z) = K_v(\theta_v)\frac{dh(\theta_v)}{dx} \tag{11}$$

Equation (11) can be used to assess the instantaneous flux of water toward the soil surface, or towards the phreatic surface.

6.3 RESULTS AND DISCUSSION

6.3.1 INVERSION PROCEDURE TO OBTAIN THE DIELECTRIC CONSTANT WITH DEPTH

The inversion procedure is illustrated using hypothetical values of the bulk average dielectric constant, typical values that might be measured using surface GPR. Based on the geometrical configuration in Figure 6.1, we assume that $z_1 = 10$ cm, and $\Delta z_2 = \Delta z_3 = \Delta z_4 = \Delta z_5 = 20$ cm for the total depth Z = 80 cm. Figure 6.4 shows the hypothetical vertical distribution of the bulk average dielectric constant and the depth-specific dielectric constant calculated using equation 1. All calculations in this section were made using the software Mathcad. Figure 6.1 shows that the bulk average value of the dielectric constant (O) has been placed at the depth z/2, where z is the associated reflector depth. The depth-specific values of the dielectric constant are placed at the midpoints of their respective Δz.

6.3.2 APPLICATION OF THE DIELECTRIC MIXING MODEL

An experimental example is given here for a soil consisting of 20% sand, 30% silt, and 50% clay, and having a bulk density of 1.3 gm/cm³. The soil is classified as clay and is typical of the soils found in Puerto Rico. Using the textural information, bulk density and the calculated wilting point

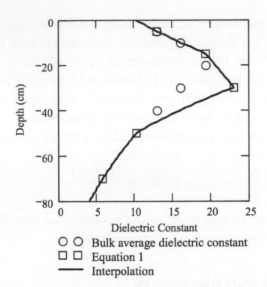

FIGURE 6.4 GPR-measured (hypothetical) bulk average dielectric constant (O), depth specific dielectric constant from equation 1, and the interpolated values of dielectric constant (——), used to estimate moisture content, pore water pressure, hydraulic conductivity and Darcy flux.

(from equation 5, WP = 0.31), the Rosetta neural network gave the following estimates of θr, φ, α, n and Ks, respectively: 0.1 cm³/cm³, 0.5 cm³/cm³, 0.0168 cm⁻¹, 1.326 and 18.63 cm/day.

Figure 6.5 shows the vertical distribution of moisture content with depth obtained from Eqs. (2) and (6). To solve for the moisture content as a function of the dielectric constant, a root search algorithm was used in Mathcad. The distribution of moisture content (Figure 6.5) could be produced, for example, by a subsurface drip irrigation line located 30 cm below the surface. Note that at this depth, the moisture content is equal to the porosity (0.5) and is therefore completely saturated.

6.3.3 ESTIMATION OF HYDRAULIC PARAMETERS WITH DEPTH

The pore water pressure, hydraulic conductivity and the Darcy velocity were calculated based on the vertical moisture content distribution and the same soil characteristics as given previously. Figures 6.6–6.8 show

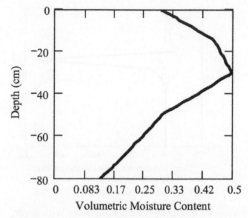

FIGURE 6.5   Volumetric moisture content from equations 2 and 6 versus depth.

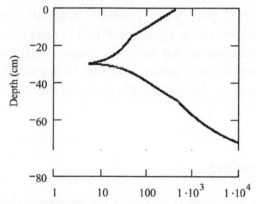

FIGURE 6.6   Pore water pressure.

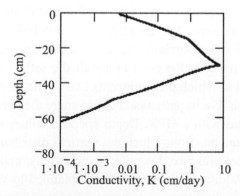

FIGURE 6.7   Hydraulic conductivity with depth.

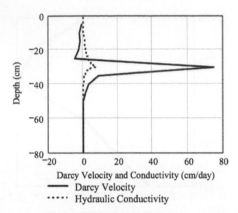

FIGURE 6.8   Darcy velocity and hydraulic conductivity with depth.

vertically distributed pore water pressure, hydraulic conductivity and Darcy flux calculated using equations 9 to 11, respectively. Figure 6.7 also shows distribution of the hydraulic conductivity for comparison. Note that between the soil surface and the 30 cm depth, the flux was upward (i.e., negative). Below the 30 cm depth, the flux was downward (i.e., positive).

6.4   CONCLUSIONS

In this chapter, a procedure is presented for estimating the depth-specific dielectric constant from several GPR - measured bulk average dielectric constants. This approach has an advantage over the borehole GPR approach, which requires the installation of boreholes and the use of special material for the borehole tubing. A disadvantage of the method presented in this paper is the need to install the reflectors, which limits the vertical extent to which measurements can be made. An example was presented in which five hypothetical bulk average dielectric constant values were measured with a GPR. Depth specific values of the dielectric constant were obtained, from which the vertical distribution of moisture content, pore water pressure, hydraulic conductivity, and the Darcy flux were estimated. Additional experiments are planned to verify the procedure under field conditions.

6.5   SUMMARY

Knowledge of the vertical distribution of the volumetric soil moisture content is essential for understanding the movement of water and its availability for plant uptake, or loss by deep percolation. A procedure is presented for estimating the depth-specific dielectric constant from bulk average dielectric constants obtained with surface GPR. An example is presented in which five hypothetical bulk average dielectric constant values were measured with a GPR. Depth specific values of the dielectric constant were obtained, from which the vertical distribution of moisture content, pore water pressure, hydraulic conductivity, and Darcy velocity were estimated.

ACKNOWLEDGMENTS

Authors are grateful to NOAA-CREST for its financial support of this project.

KEYWORDS

- Darcy flux
- Darcy velocity
- Dielectric constant
- Ground penetrating radar, GPR
- Hydraulic conductivity
- Mixing model
- Moisture content
- Pore water pressure
- Puerto Rico
- Soil texture
- TDR NOAA-CREST
- Theta Probe
- USDA
- Volumetric soil moisture content

REFERENCES

1. Alharthi, A., & Lange, J. (1987). Soil water saturation: Dielectric determination. *Water Resources Research*, 23(4), 591–595.
2. Benedetto, A., & Benedetto, F. (2002). GPR experimental evaluation of subgrade soil characteristics for rehabilitation of roads. In: *Ninth International Conference on Ground Penetrating Radar*, Steven K. Koppenjan and Hua Lee, Editors. *Proceedings of SPIE*, 4758, 605–607 and 708–714.
3. Dobson, M. C., Ulaby, F. T., Hallikainen, M. T., & El-Rayes, M. A. (1985). Microwave dielectric behavior of wet soil-Part II: Dielectric mixing models. *IEEE Transactions on Geoscience and Remote Sensing*, GE-23(1), 35–46.
4. Epptein, M. J., & Dougherty, D. E. (1998). Efficient three-dimensional data inversion: Soil characterization and moisture monitoring from cross-well ground-penetrating radar at a Vermont test site. *Water Resources Research*, 34, 1889–1900.
5. Hallikainen, M. T., Ulaby, F. T., Dobson, M. C., El-Rayes, M. A., & Wu, L. (1985). Microwave dielectric behavior of wet soil-Part I: Empirical models and experimental observations. *IEEE Transactions on Geoscience and Remote Sensing*, GE-23(1), 24–46.
6. Harmsen, E. W., Hamed Parsiani, & Maritza Torres (2003). Evaluation of several dielectric mixing models for estimating soil moisture content in sand, loam and clay soils. Written for presentation at the ASAE Annual International Meeting, Las Vegas, Nevada, USA, July 26–30.
7. Kowalsky, M. B., & Yoram Rubin (2002). Suitability of GPR for characterizing variably saturated sediments during transient flow. In: *Ninth International Conference on Ground Penetrating Radar*, Steven K. Koppenjan, Hua Lee, Editors. *Proceedings of SPIE*, 4758, 605–608.
8. Martinez, A., & Byrnes, A. P. (2001). Modeling dielectric-constant values of geologic materials: An aid to ground-penetrating radar data collection and interpretation. *Current Research in Earth Sciences*, Bulletin 247, part 1: (http://www.kgs.ukans.edu/Current/2001/martinez/martinez1.html).
9. Rial, W. S., & Han, Y. J. (2000). Assessing soil water content using complex permittivity. *Transactions of the American Society of Agricultural Engineers*, 43(6), 1979–1985.
10. Schaap, M. G., & van Genuchten, M. Th. (1998). Neural network analysis for hierarchical prediction of soil water retention and saturated hydraulic conductivity. *Soil Sci. Soc. Am. J.*, 62, 847–855.
11. Selker, J. S., Keller, C. K., & McCord, J. T. (1999). *Vadose Zone Processes*. Lewis Publishers, Boca Raton, USA.
12. Shutko, A., M., & Reutov, E. M. (1982). Mixture formulas applied in estimation of dielectric and radiative characteristics of soils and grounds at microwave frequencies, *IEEE Transactions on Geoscience and Remote Sensing*, GE-20(1), 29–32.
13. Wang, J. R., & Schmugge, T. J. (1980). An empirical model for the complex dielectric permittivity of soils as a function of water content. *IEEE Transactions on Geoscience and Remote Sensing*, GE-18(4), 288–295.

PART III

MANAGEMENT OF
MICRO IRRIGATION SYSTEMS

CHAPTER 7

THE APPROACH FOR TRANSLATING METEOROLOGICAL PATTERNS INTO INFRA-RED SIGNALING FOR VARIABLE DISPENSATION IN CROP IRRIGATION SYSTEMS

KEVIN RODRÍGUEZ

Former Technical Assistant, School of Business and Computer Science, 49 Alenore Gardens Phase-2, Off De Gannes Street, Arima, Trinidad and Tobago. Mobile: 868-372-3448 / 868-667-5457; E-mail: starkisuke7@gmail.com

CONTENTS

Edited version of, "*Kevin Rodriguez, 2015. Investigating the approach for translating meteorological patterns into infra-red signaling for variable dispensation in crop irrigation systems. Unpublished B. Sc. (Computing and Information Systems) under the Supervision of Anjan Lakhan, School of Business and Computer Science, Trinidad, Pages 137.*"

7.1   INTRODUCTION

This chapter is an edited version of an academic research project that focuses on the many changes and improvements in crop irrigation in the Caribbean including Trinidad. The methods of crop irrigation were through ponds, streams, rivers and rainfall. A few years later new innovations were introduced such as drip lines used to conserve water; sprinklers used to control humidity levels and irrigate a larger area of crops and the gravity flow of water. These were operated with the aid of a water pump and a large source of water supply. Later solar energy was implemented to power water pump used in drip lines and boom sprayer that unleashed a powerful mist on crops. These methods of crop irrigation were labor intensive.

Infrared (IR) technology can be considered as a feasible option, mainly infrared sensors/detectors. Infrared detectors were mainly used in the military in late 1990s. In 2000, there were further production and development of third generation Infrared systems. These developments are used for monitoring global environment pollution and climate changes, chemical process monitoring, Fourier transform IR spectroscopes, IR astronomy, car driving, IR imaging in medical diagnosis long time prognoses in agriculture crop yield and others. Therefore, it begs the question, whether IR (Infrared) Sensors can be used to read meteorological weather patterns in the aid to automatically trigger variable water dispensation and conservation for crop irrigation for local farmers.

7.1.1 SCOPE

This project was an initiative to determine, discover, and design a feasible solution to crop irrigation thus reducing labor input, crop water stress and soil moisture stress to optimize yield of lettuce crop.

- Java programming and respectable classes were used to build a Graphical User Interface to simulate and demonstrate the use of infrared technologies and a data logger to record environmental weather patterns and its effect on lettuce crop.
- To display crops, placement of technologies, workstations and crop irrigation system and their works.
- To perform calculations using appropriate formulas where data gathered were inputted into the formula to trigger the crop irrigation system.
- Further, the chapter presents an analysis of existing crop irrigation mechanisms and identifies the farmers' perspective on the role of technology in supplementing their tacit knowledge of weather patterns and other environmental conditions versus that of the IR approach.

To demonstrate suitable infrared sensor/detector technologies to gauge water dispensation and conservation based on climatic readings and evapotranspiration to facilitate crop water consumption, crop water stress and water use efficiency in land cultivated areas where lettuce is grown, following objectives were considered:

- To research case studies and literature on infrared sensor technology that have been used and implemented in various fields.
- To research similar white papers on infrared sensor technology that have been used in many organizations or government bodies.
- To interview engineers and pioneers in the field of infrared technology and agriculture via emails or telephone. Identify locally the most

common and efficiently used irrigation systems to water lettuce via interviews with farmers or agriculture bodies.

- To identify whether software applications and hardware devices are needed or has been used to improve crop production.
- To identify formulas to calculate evapotranspiration, crop water stress index and soil moisture stress in order for efficient use of water dispensation to replenish crops.
- To design a UML state activity diagram to show components behavior in terms of activities and their precedence constraints. In addition to a layout diagram of an implemented auto irrigation system.
- To build a simulation prototype of the automated irrigation system.

Following information was generated:

- Literature review of published papers and field trials.
- Analysis of lettuce production using conventional methods.
- Analysis of software applications used.
- Analysis of technologies used.
- A proof of concept based on a computer-based simulation:

  ❖ IR droplet and fog detector/sensor to switch off irrigation cycles at the presence of rain
  ❖ Infrared technology to measure ambient and object temperature, over an optimal field view
  ❖ Recording crop temperature for calculations.
  ❖ Calculating evapotranspiration and crop water stress index.

- A UML activity diagram of software and hardware components.
- A layout diagram, table of parameters and GUI interface of the software used.

7.2 LITERATURE REVIEW

Agricultural regions in the Caribbean which is regarded as the archipelago of islands stretches from Cuba in the north to Trinidad in the south. This region uses mostly drip and mini irrigation systems for fruit trees and vegetation production mostly for vegetation. The meteorology patterns of these regions are mostly hot and dry summers with average temperatures ranging from 20 to 30°C year round. Annual rain takes place during July to December months. This varies both between and within countries.

Therefore, large volumes of water are produced during wet season whereas dry seasons results to soil water deficit. As a result crop production tends to drop during dry season and irrigation systems such as drip lines and sprinklers are crucial for horticultural crops. It ensures consistent water supply for commercial fruit and vegetation farming.

Latin America and the Caribbean accounts for 67% of water consumption. Therefore, irrigation provides crops with adequate amount of water to reduce stress. The repetition of irrigation and quantity of water depends on local climate conditions, crops species, stage of plant growth and soil plant moisture characteristics. Crop irrigation can be determined based on visual observation and not necessary on acquiring data on rate of evapotranspiration such as plant color and leaf wilting. On the other hand, studies showed that this visual observation of information arises too late to prevent reduction of crop yield and quantity yield.

7.2.1 METHODOLOGIES

Researched methods of irrigation scheduling include determination of plant water stress, soil moisture status and soil water potential [10]. Estimating crop water requirements are accomplished by using evapotranspiration along with soil characteristics is useful to determine not only when to irrigate but the quantity of water needed. Enough water supplies for a growing season are crucial for the optimum production of crops. Crop water requirements are often provided by both rainfall and irrigation with the understanding that in places where sufficient rainfall occurs throughout the growing period, irrigation is minimum. The use of infrared thermometry can be used as a tool for irrigation scheduling.

7.2.2 INFRARED SENSOR TECHNOLOGY

Objects on the earth's surface emit radiation in the thermal infrared region of spectrum. This energy emitted from the earth's surface is proportional to the absolute surface temperature of an object. This result has shown its usefulness in assessing crop water stress due to temperature of most plant leaves that are mediated strongly by soil water availability and its effect on crop evapotranspiration. Furthermore plant temperature often differs substantially from air temperature. Studies have revealed that many physical and biological stresses that intrude with transpiration causes an elevation in plant temperature and are correlated with plant water status and reduce potential yield [14].

Infrared thermometry or radiometer instruments are used in crop irrigation scheduling to determine the surface temperature of an object without physical contact. Infrared thermometers were used in the agriculture sector in early 1960s as experiments to detect crop water stress remotely. Hand held infrared thermometers (commonly known as IR guns) were initially used to measure crop canopy temperatures, which are mostly reported in terms of crop water stress index. The measurements made with these portable devices were point measurements of canopy temperatures. Where canopy temperature measurements were most commonly reported as crop water stress index and where canopy temperatures comprises of a number of factors such as radiation, air temperature, humidity, evaporating cooling and wind speed.

The IR guns were basically the first invention used for agricultural crop to measure temperature and were mostly used during peak daily radiation between the hours of 12 noon and 2 PM when the crop is under maximum evaporative demands. The measurement of vapor pressure deficit, which comprise of air temperature, humidity and radiation were plugged in the crop water stress index formula. On the other hand, handheld infrared thermometers have limitations to their uses such that it does not produce measurements of crop water status of the entire field and it will be laborious to make daily measurements. In addition, errors can occur if the IR gun measures an area greater than the target object. This is due to the background soil which is inadvertently included in the sampling area when acquiring canopy measurements. To correct this, one must minimize the distance between the instrument and the crop canopy. Also temperature measurements of leaf /canopy were taken in different directions to ensure the difference in temperature that may exist between alternative sides and orientation of the plant's canopy. Given the direction of solar movement, measurements should be taken facing north and facing south to account for variances in canopy measurements for inter canopy shading. Users of IR guns should keep in mind that the lens of the IR gun should never come in contact with the sun, which can cause readings and measurement to be inaccurate.

7.2.3 THERMAL INFRARED INSTRUMENTS: ADVANCES AND WORLDWIDE USE

The advancement in technology in thermal infrared instruments has led from handheld to wired, wireless and thermal imagery instruments. The wired and wireless thermal infrared sensors allow direct and constant recording of temperatures of a specific body. Therefore, there can be continuous monitoring of crop canopy temperature. These infrared sensors have proven to be successful in center pivot irrigation systems. Wired infrared sensors were mostly used in the agriculture sector. It indicated crop water stress in commercial peach orchard in Spain and in Nebraska due to the decline in crop irrigation water supplies, which force farmer to use water more efficiently. It was used to schedule irrigation by establishing upper and lower crop water stress baseline for two major crops, corn and soybeans. The infrared thermometer was of model IRTS-P precision infrared thermocouple sensor (Apogee instruments, Inc. Logan, UT).

7.2.4 IR RAIN SENSOR

The Hydreon Optical Rain Sensor (Model RG-11) operates with the presence of raindrops hitting the outer surface with the use of beams of infrared light [28]. The sensor is extremely sensitive and virtually immune to false trips. It is unaffected by jostling or motion and is completely sealed off. The RG-11 sensor has been used for irrigation control measuring both rain accumulation and rain intensity. It prohibits the water cycle during a downpour even before the downpour accumulates. RG-11 has a built in microprocessor, the set points of system disable and enable are precise and optimal. The RG-11 infrared rain sensor consists of five modes, which must be manually set to perform specific duties automatically. Mode 4 is irrigation control, which inhibits watering upon the accumulation of 0.2 inches of water, and re-enables the system after the water has evaporated.

7.2.5 DATA LOGGER

Data logger samples environmental parameters relating to temperature, relative humidity, wind speed and direction, light intensity, water levels and water quantity over time. It is a compact device that is battery powered and equipped with microprocessor input channels and data storage. It is also programmed by software such as RTDAQ and LoggerNet [3]. Data loggers were used during the 1980s when personal computers were introduced. These devices were early adopted in Canada for fire weather index monitoring and snow avalanche monitoring. At present, it has been used for meteorological networks of hundreds of stations monitoring temperature, relative humidity, barometric pressures, solar radiation, precipitation and wind speed and directions.

Data loggers input the outputs form sensors, which are electrical signals in milliamps via the data logger channel and the microprocessor that performs arithmetic and logic operations. Adding to that, they are capable of communicating and transmitting data to a computer via the RS-232 serial port [27] and are calibrated on the field. For instance, Detroit water and sewage department used data loggers to monitor the flow of water to collect maximum day water use. Each data logger was calibrated with a twelve-point calibration curve that measures twelve data logger signals count response values between four to twenty milliamps. This was done also to reduce measurement errors of the data logger, which did so by 5% approximately less than 1% of actual flow.

The CR1000 Measurement and controller data logger [2, 29] has been used in South Dakota irrigation system by regulating automatic check gates for water delivery efficiency in crops. In semiarid region just north of Black Hill, west of South Dakota faces recurring drought to supply available water for irrigation. The Belle Fourche Irrigation District (BFID), contracted with RESPEC and in 2006 RESPEC, was automated the BFID canal check structures using sensors and the Campbell scientific data loggers CR1000 [2] and other models. Using the LoggerNet [3] program to enter the desired lateral flow into the data logger the lateral head gates will adjust automatically to deliver that flow of water.

7.2.6 PLC AND RELAY BOARD CONTROLLERS

PLC controllers have been incorporated in irrigation systems to control devices with moderate and high voltage: Moderate voltages between 7 and 48 volts, and high voltages between 110 and 220 volts. PLCs drastic development and affordable price has made it possible to use them as stand-alone irrigation controllers for daily irrigation. PLC controllers are programmed to satisfy peak water needs in crops. The programs are usually provided by manufactures. In addition to reducing wasted waters based on the drip irrigation using a PLC based adaptive irrigation system, it was found that farmers with drip and sprinklers irrigation systems used fixed schedule irrigation to water their crops and often waste water on cool and cloudy days at the beginning of growing seasons when crop water needs are minimal. The purpose of PLC controllers was developed for autonomous irrigation systems that use a single climate criterion to adapt daily irrigation depths to plant needs.

Criteria such a temperature, total radiation and total wind can be measured directly by PLCs which then adapts the irrigation scheduling to the observed conditions, leading to a reasonable saving in the amount of irrigation water.

Sophisticated irrigation controllers calculate daily evapotranspiration to establish the exact amount of water to be applied to crops. Evapotranspiration is the water lost by evaporation from soil and plant and by transpiration from the plant surface.

Relay boards were used in auto irrigation systems as well. These boards were used to interface with data loggers and solenoid valves. Relay boards were energized from control ports and distribute 24 volts of power to activate solenoids valves to irrigate crops. Also it can withstand a maximum switch over contact at 5 amps current at 220 volts. Most relay boards can operate eight relay contacts. It is capable of switching high current and offers isolation between the controller and controller device at the same time. In addition, it consists of dipswitches to allow manual activation of relay in cases of conducting maintenance work.

7.2.7   CROP EVAPOTRANSPIRATION AND REMOTE SENSING

Evapotranspiration (ETc) constitutes a major component of the hydrological cycle [4, 7]. The ETc is a fundamental and essential parameter for climate studies, weather forecasts and weather modeling, hydrological surveys and ecological monitoring and water resource management. The evaluating ETc and combing conventional meteorological ground measurements with remotely-sensed data has been widely studied and several methods have been implemented for this purpose. Crop evapotranspiration rate has become highly important to identify crop water stress, water deficiency, onset of irrigation and for determining the exact potential needs of crops for best yield. Remote sensed reflectance values can be used in combination with other detailed information for evaluating ETc of different crops. It has been found that saving irrigated water through remote sensing techniques could reduce farm irrigation cost and increase margin of net profit [7].

7.3 METHODOLOGY

This section discusses methods to accumulate information on the topic of infrared sensors to automate the current irrigation system used by local farmers not only in the developing country of Trinidad but the Caribbean as well. This was accomplished through literature review, interviews, researching formulas and their parameters, designing a UML interaction design of the functionality of the system, and building a prototype simulation.

7.3.1 LITERATURE REVIEW

To gain knowledge of the topic, online research white papers and journals on infrared technology with respect to irrigation systems was reviewed. Google Search Engine and Google Scholar were also used to acquire valid knowledge and initiatives that have been undertaken and published by private or government agriculture sectors.

7.3.1.1 Extracting the Information

Google search engine was used with phrases such as "infrared sensing and crop irrigation," and "using infrared to automate crop irrigation." This eventually led to "Remote Sensing in Crop irrigation." Google search engine had recommend articles from Google Scholar, which was used to search desired information. See list of references for further details.

7.3.1.2 Justification

Organizations or government bodies provide interesting information on the benefits and uses of IR technology. These white papers are a form of casual investigation that have been undertaken to identify whether it can improve food production in developing countries and can be highly recognized by government bodies. By comparing and contrasting designs and methodologies internationally and if any locally, it was investigated to improve the traditional way farmers harvest the leafy vegetables in developing countries in the Caribbean.

7.3.2 INTERVIEWS

Interview with engineers and pioneers were conducted both internationally and locally on the use of IR technology in the field. Emails or smart mobile telephones with applications such as Skype or magic jack were used to communicate with international engineers or pioneers. Local interviews were conducted via emails, telephone and face to face with farmers, engineers or technicians.

The purpose of the interview was: To compare the approaches being undertaken both internationally and locally in the agriculture sector; and to identify the technology used to design a complete system in the automation of crop irrigation to improve its conventional methods. What component either hardware or software or both are to be used to interface and interact between devices and users? It was verified whether these devices and application can aid in assessing the projects objectives or if further research is needed. This provided useful information. Engineers and pioneers have experience in the field, which can provide a useful guide towards the project objectives.

7.3.3 SOFTWARE AND HARDWARE

Hardware components or devices along with software applications were investigated through research papers and interviews, to identify the hardware and software needed to collect and store readings from on-site sensors to perform calculations for evapotranspiration, CWSI and controls to other components.

Software and hardware are fundamentals in the aid of designing an automated irrigation system. Not only reduce but eliminate the labor work of

watering crops and wasting water. The software and hardware aids in estimating when crops require water under crop water stress index and maintain
a moisture level in soil to keep crops cool at hot temperatures of the day.
The software can help to calculate crop evapotranspiration and CWSI; and
control devices based on readings from on-site sensors and water valves.
The hardware can be used for storage data and signaling parameters.

7.3.4 FORMULAS AND EQUATIONS: FIELD PARAMETERS

From the literature, formulas were identified to calculate evapotranspiration, crop water stress index and soil moisture depletion as well as for
suitability in the aid of crop water stress and soil moisture stress. The
technology and software used in this research aids in obtaining climate
readings such as temperature, solar radiation and the presence of rain.
The formulas or models can help to calculate evapotranspiration and crop
water stress index (CWSI) for lettuce.

7.3.5 UML ACTIVITY DIAGRAM

An UML activity diagram and layout diagram were drawn using Microsoft
Visio. The UML activity diagram was drawn to describe the behavior of
controls, and data flows. The Layout diagram was drawn to show how the
system can be implemented, the setup and connected to devices to collect
readings from sensors via data logger, where calculations are made and
signaling of the water valves to open and close via a relay board.

This will help in designing the simulation prototype as a start up and
can be used in the process of redesign and to aid in any field trails. These
are used as blue prints for any technical design that needs to be looked
over or redesign to satisfy a need.

7.3.6 SIMULATION PROTOTYPE

A simulation model was built using Java Graphic User Interface and using
suitable classes: javax.swing, java.awt and threads for the design and functionality. The functionality of infrared technology and other devices were

demonstrated in its aide to replenish crops. In order to implement and test the system, research was conducted done to understand the entire process to obtain appropriate results. In addition, it required purchase of hardware and software components, which incurred a significant cost. Therefore, application program with java programming was developed to simulate various activities to replace the need to purchase and setup equipments by designing a GUI interface of the system.

7.3.6.1 Design Structure for Simulation

Java graphics was used to display a rectangular bed with six rows of crops, along with IR sensors for temperature readings and rain detection. A pyranometer was added to measure solar radiation from the sun to calculate potential evaporation and it was connected to the data logger, the control DC device and computer. Subsurface drip irrigation lines were run to show the watering of crops at a subsurface level. The solenoid valves were triggered to allow the flow of water to the crops. Finally the PVC connection and activation of the water pump were allowed.

Experimental layout: Bed with six rows of lettuce crop; IR sensors, IR Rain Sensor and Pyranometer; Data logger, Computer station and DC controller board; Drip Lines, electronic valve, water pump and cross junctions; Readings and calculations: Screen shot.

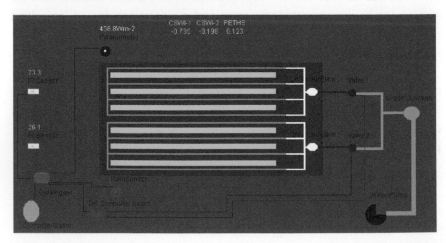

Thread was used to show the blinking action of measurements being recorded from measuring devices. Along with the control station to perform arithmetic and logic calculations, controls command to show when crop is under water stress using the lettuce CWSI and potential evapotranspiration (PET) using the Hargreaves-Samani methods (PETHS). In addition to the movement of the sun from sun rise to sunset, the temperatures changes were recorded.

The weblink <http://www.weather.com/weather/hourbyhour/l/TDXX0116:1:TD> provided data on temperature to establish values for the sun in this simulation [22]. It was also noticed that ambient temperature was either four or five degrees lower that what a person would actually feel in the open environment. These readings were used for the IR temperature sensors to measure temperature of the lettuce crop in the simulation. Furthermore, solar radiation was calculated using <http://www.engr.scu.edu/~emaurer/tools/calc_solar_cgi.pl> to calculate solar radiation (watts per meter square and millimeters per day). The screen shot to calculate extraterrestrial radiation is given below:

Calculation of extraterrestrial solar radiation (to horizontal surface at top of atmosphere)

Output can be watts per square meter (W/m2) or equivalent evaporated water depth (mm/d)

Latitude (decimal degrees, southern hemisphere is negative): 30

Month: 4 (January=1, February=2....)

Day (of month): 16

Select units for output: W/m2

Submit

Extraterrestrial solar radiation = 436.57 W/m2

Intermediate calculations:

Sunset hour angle (degrees) = 95.58
Maximum possible daylight hours = 12.7

Equation references: Duffie, J.A. and W.A. Beckman, Solar Engineering of Thermal Processes, Wiley, New York, as summarized in Maidment, Handbook of Hydrology, 1993.

Barataria and Arouca is the area where most farmers plant leafy vegetables as well as lettuce, using the tradition crop irrigation system. Following values were used in the simulation:

- **Infrared sensors:** Three degrees higher, e.g., sun temperature = 23°, then IR sensor measurement = 26°.
- **Pyranometer**: One value higher, e.g.: Assumed starting value of 455.96 Wm², then measuring value was 456.96 Wm².

It was assumed that well-watered crop displays a drop in canopy temperature by three degrees to be used in the simulation. This was done because no published research was found on the state of canopy temperature of a well-watered crop. In addition during rainfall, solar radiation measurement was decreased by two degrees.

The simulation was run to give an output resulting CWSIs within 28 degrees since lettuce has an upper limit of 28 degrees. The CWSI within this range was averaged to establish an integrated CWSI to assist the simulation to trigger irrigation. Also, condition was set so that irrigation will occur when calculated CWSI is greater than iCSWI and IR sensor readings is greater than or equal to 25 degrees since some lettuce crops grows best at 25 degrees and under. Sample output is shown below:

```
If((deg >= -1} && (deg <=33)) {
SunDeg = deg + IRpoints. nextDouble ( ) ;
addTo = deg + IRpoints. nextDouble ( ) + 3 ;//- canopylimits
        - rainTemp;
addTo1 = deg + IRpoints. nextDouble ( ) + 3 ;//-
         canopylimits - rainTemp;
```

CWSI1	CWSI2	MaxAir Temp	MinAir Temp
-0.129	-0.080	28.519	27.481
-0.129	-0.080	28.519	27.481
-0.129	-0.080	28.519	27.481
-0.129	-0.080	28.519	27.481
-0.110	-0.164	28.587	27.481
-0.110	-0.164	28.587	27.481
-0.110	-0.164	28.587	27.481
-0.110	-0.164	28.587	27.481
-0.263	-0.136	28.819	27.481
-0.263	-0.136	28.819	27.481
-0.263	-0.136	28.819	27.481
-0.263	-0.136	28.819	27.481
-0.049	-0.221	28.819	28.390
-0.049	-0.221	28.819	28.390
-0.049	-0.221	28.819	28.390
-0.049	-0.221	28.819	28.390
-0.142	-0.235	28.819	28.583

CWSI1	CWSI2	MaxAir Temp	MinAir Temp
-0.142	-0.235	28.819	28.583
-0.142	-0.235	28.819	28.583
-0.142	-0.235	28.819	28.583
-0.052	-0.150	28.819	28.215
-0.052	-0.150	28.819	28.215
-0.052	-0.150	28.819	28.215
-0.052	-0.150	28.819	28.215
-0.155	-0.204	28.819	28.617
-0.155	-0.204	28.819	28.617
-0.155	-0.204	28.819	28.617
-0.155	-0.204	28.819	28.617
-0.142	-0.125	28.819	28.366
-0.142	-0.125	28.819	28.366
-0.142	-0.125	28.819	28.366
-0.142	-0.125	28.819	28.366
-0.105	-0.191	28.819	28.448
-0.105	-0.191	28.819	28.448
-0.105	-0.191	28.819	28.448
-0.105	-0.191	28.819	28.448

In addition, solar radiation was converted from watts per meter square to equivalent evaporation in mm per day, so as to facilitate PETHS in mm/day. Sample outcome is shown below:

```
RA = addTo2 / 2454000;
RA = RA * 60 * 60 * 24;
PETHS = 0.0023 * RA * (SunDeg - 17.8) * (MaxA - MinA) ;
```

Max Solar	Min Solar	PETHS
458.540	458.056	0.092
458.540	458.056	0.092
458.540	458.056	0.092
458.540	458.056	0.092
458.540	458.056	0.092
458.540	458.179	0.096

Max Solar	Min Solar	PETHS
458.540	458.179	0.096
458.540	458.179	0.096
458.540	458.007	0.163
458.540	458.007	0.163
458.540	458.007	0.163
458.540	458.007	0.163
458.803	458.007	0.123
458.803	458.007	0.123
458.803	458.007	0.123
458.803	458.007	0.123
458.803	458.007	0.123
458.803	458.666	0.154
458.803	458.666	0.154
458.803	458.666	0.154
458.803	458.666	0.154
458.844	458.666	0.160
458.844	458.666	0.160
458.844	458.666	0.160
458.844	458.666	0.160
458.844	458.788	0.115
458.844	458.788	0.115
458.844	458.788	0.115
458.844	458.788	0.115
458.844	458.788	0.115
458.844	458.044	0.147
458.844	458.044	0.147
458.844	458.044	0.147
458.844	458.044	0.147
458.873	458.044	0.161
458.873	458.044	0.161
458.873	458.044	0.161
458.873	458.044	0.161

These values and conditions were inputted to bring the simulation as close as possible to the actual implementation of the system. Integrated CWSI was calculated to cater needs of different countries in the Caribbean as well as to trigger the irrigation system, since no published work was found for iCSWI for lettuce in the Caribbean. As a result, different countries will establish their own iCWSI for lettuce because the weather pattern will not be the same throughout all islands of the Caribbean. Also over water of crops id not recommended, as it will result in reduction of in yield.

7.4   RESULTS

Most published documents on the uses of Infrared technology for crop irrigation were based on the data in the United States. IR technology has been widely used in crop irrigation scheduling and site-specific irrigation scheduling not just to preserve water but also to calculate a pre-established crop water stress index to detect water stress. In other words, the crop requires water and also helps identify the require amount of water to irrigate the crop such as water use efficiency (WUE) and irrigated water use efficiency (IWUE). These IR radiometers or sensors detect canopy and ambient temperatures in the field. These readings are needed to calculate crop water stress index (CWSI) and evapotranspiration [10]. Not only do IR sensors detect crop water stress but also when crops are affect by some disease or fungus. Some research studies showed that the investigators used infrared sensors from Apogee and Campbell scientific SI-111 Infrared Radiometer [25] with standard field of view, in addition to data loggers and other devices to measure air temperature, radiation and wind speed. Research has indicated that lettuce crop has a lower limit range between 12 to 23 degrees or 20 to 25 degrees and an upper limit range of 28 degrees.

These technologies can help to optimize the existing irrigation systems used by farmers in the Caribbean. These can improve water scheduling so that crops will not suffer for water or can be watered twice a day morning and evening. These technologies gather the necessary meteorological patterns required to establish crop water stress and estimate the required water to replenish crops. This can improve crop production and can help farmers who may not be able to visit the fields throughout the day.

7.4.1   INTERVIEWS

7.4.1.1   Thermal Diagnostic Limited: Mr. Sonny James

Infrared technology is widely used in the medical and industrial sectors of Trinidad. It is used in medical centers to detect breast cancer and industrial sectors; for refractory in industrial boilers and heaters to detect wasted heat, cracks and void which can cause failures or danger to employees working on site. The device is a handheld Infrared thermography what captures an imagery of the object that shows any sign of materials under high pressure. The data is analyzed using quantitative methods examining the pixels on the thermography image. Software used to examine the targeted areas is provided by manufactures along with the device.

Mr. Sonny James was every willing to accept the interview over the phone even though he was very busy with work. His crop irrigation system uses a graphical imagery to identify the crop areas in need of water. Mr. James told that this technology can also benefit farmers, where data is analyzed using quantity methods by examining pixels. Farmers make a huge income from crops and can invest in technology to improve the irrigation systems. The imagery can be useful to examine what area of corps requires water more than other areas. On the other hand, this will require some level of training for farmers.

7.4.1.2   United States Department of Agriculture

The United States Department of Agriculture (USDA) has used infrared thermometers in crop experiments both wired/remote thermometers and wireless thermometers. Based on the interview sent via email, infrared sensors can be used in stationary irrigation systems such as drip line and sprinklers systems. Wired/remote infrared thermometers were used, before wireless infrared thermometers were manufactured. Wireless remote sensors can replace the data logger because it is connected to a computer network. A graphical user interface can be built for the wireless sensors using visual studio. Whereas wired remote sensor requires both computer and data logger. Manufactures provide their own software/GUI such as PC200W [1] and LoggerNet [3]. Targeted temperature of a specific body is collected via IR thermometers. The readings from IR thermometers may vary from manufacturer specifications. In agriculture, they are mostly used to measure canopy temperature, which aids in calculating an integrated crop water stress index (iCSWI). A pre-established integrated crop water stress index threshold is used for different crops grown in the United States. The crop water stress index can be used formula for calculating a thermal stress index, a theoretical method by Jackson based on the energy balance approach [10].

The pre-established iCSWI is used in center pivot irrigation systems. This irrigation system moves across a field calculating a mean iCSWI and compares it with the pre-established threshold and signal irrigation to plot area once the mean iCSWI is greater than the threshold.

On the other hand where soil moisture is concern, canopy temperature is directly related to crop water stress and indirectly related to soil water content. Therefore, as the stomata closes, latent heat of evaporation increases thus canopy temperature increases. The USDA has found that canopy temperatures detect crop water stress earlier than soil water content. Also if a plant cannot uptake water, the crop is unable to transpire thus resulting to an increase in canopy temperature. Therefore, it is much easier to monitor crop canopy temperature at a larger scale than soil water content due to the fact that thermal measurements have shown to detect crop water stress earlier than soil water content. Soil water sensing can be used as a redundant approach to evaluate whether the crop needs water or diseased stressed and indicate the amount of water to apply. Furthermore the components, that control the irrigation system to turn on or off, require both hardware and software. Programmable Logic Controllers are used in center pivot irrigation systems that send signals to particular solenoid on the pivot lateral. Whereas with data loggers, the control is via a relay board that opens and closes the valves. These controller devices also comes with software to perform the task by manufactures and allow users to write their own program for the task they which to see executed.

Many of the published articles in the United Stated Department of Agriculture do not provide the information on hardware and software devices in order to execute an automated irrigation system. This interview with USDA was able to provide in-depth knowledge of hardware and software required to implement an auto irrigated system. For instance, data loggers require relay boards to control solenoids and manufacturers provide the necessary software to program these devices. This interview provided better in sight of the mechanism that controls the solenoid valves and also when to water at the right time. The crop water stress index (CWSI) has been widely used in many of the published articles along with evapotranspiration. These two formulas seem to be the most dominant formulas in crop irrigation research.

7.4.1.3　Campbell Scientific Australia

Campbell Scientific is an organization that provides quality technology for industrial, commercial and government sectors. These technologies include sensors, gauges, data loggers, multiplexers for connecting multiple sensor and other control devices. In addition to enclosures, it supplies data loggers for measurements, battery packs and control boards. Details of such devices can be found at their website [24 to 27 29]: http://www.campbellsci.com.au/.

Based on the interview with Kahill Mitchell, that Data Logger CR1000 [2] can be connected directly to a computer using a USB to RS232 cable. This cable plugs into the RS232 port [http://campbellsci.com.au/17394-cable] on the data logger and a USB port on the computer. This allows the transfer of data to and from the data logger to the computer via a software program such as LoggerNet [3].

In addition to open source software, PC200W [1] can be used to program a data logger with an embedded application called "ShortCut." This can program the data logger by selecting sensors, the unit measurements are: average readings or samples, max and min readings as well as evapotranspiration. It also includes a long list of sensors and gauges but not all as well as data logger's types.

Finally the SDM-CD8S (Eight-Channel Solid-State DC Controller) is a mini device that expands a data logger capability to control more devices [8]. It is a synchronous device for measurement. This can control electronic solenoids and valves. This device also requires a program from the data logger program to send instructions of controls. CRBasic is an embedded software that comes with some data logger programs such as LoggerNet [3], which provide the appropriate code to control the

SDM-CD8S [8]. Campbell Scientific offers to do CRBasic programming for users if they are unable to do the programming themselves or finds someone experience in CRBasic.

Application Engineer Kahill Mitchell was willing to assist in answering questions for the research project. Campbell Scientific has the technology and software necessary for farmers to enhance their existing crop irrigation systems. Many organizations and government sectors in different countries have used their services and technology to carry out the automation of different tasks to suit their needs. This technology will benefit farmers during summer, spring and fairly autumn seasons in the Caribbean. As a result farmers will be able to conserve and preserve more water for their crops during the summer seasons.

7.4.1.4 Lettuce Farmers

Most lettuce farmers water the crops twice a day, during the early morning period before the ambient temperature increases and late in the evening when the sun is setting. Some farmers prefer not to work in the hot blazing sun and therefore leave the site and return in the evening. As a result they don't water crops during hot to peak temperatures of the day. As one farmer said he doesn't water crops during hot to very hot times of the day because the water and heat on plants will scorch the crop. This can affect crop growth. He prefers that the crop get its daily sunlight after the first morning irrigation to keep crops crispy, in this case lettuce [See Appendix A].

Most farmers interviewed regurgitated that they prefer not to work during the day because of the heat when were the question "why they don't water crops at peak temperatures of the day?" Most farmers used a drip line irrigation system called "*Porous Soaker Hose System*" made from recycle automobile tires (Figures 7.1 and 7.2). This is a plastic long tube fill with tiny holes. Based on the observation of farmer's irrigation system in operation, it was observed that a mist is emitted from those tiny holes on the tube. This allows water to be easily absorbed in the soil. In addition, the crop got sprinkled by the mist, which gradually builds up to large water molecules. As result of this, large water molecule remains on the leaf of lettuce crop when the irrigation system is turned off. The water remains will gradually adjust to ambient temperatures from the sun. Therefore, the

FIGURE 7.1 Kevita Alexander' farm.

FIGURE 7.2 Michael Diet's Farm.

remaining water molecules on the lettuce can scorch the lettuce crop as liquid molecules are easily energized when heated.

7.4.2 HARDWARE AND SOFTWARE

7.4.2.1 PC200W versus LoggerNet

Advance irrigation systems require software applications to store parameters, data from valves, sensors, gauges and thermocouples. Data loggers used in this chapter are provided with software applications such as PC200W [1] or LoggerNet [3]. These applications allow the user to select the data logger model for most of the data loggers. This makes it easier for the user to select which data logger needs to be purchased. Then user selects all the sensors, gauges or other measuring devices that should be connected to the data logger from a list provided in the GUI.

PC200W [1] is a free package design for first time users. This application allows user to set clock, monitor measurements and retrieve data. The software application has a default setting to record measurements from all the measuring devices connected to the data logger at an interval of one second. The extent of PC200W, which comes with a program Short-Cut that can be used to program the data logger at the user's choice. For example, the user can select the measurements he or she wishes to record such as: average, total, minimum, maximum or sample temperatures or radiation to be displayed in a similar excel spreadsheet. There is even the option to create own formulas as well as displaying the readings. These recording can be set in intervals either in seconds, minutes or hours and works well on a windows platform. Furthermore PC200W comes with a wiring diagram for every measuring device or other devices so that the user knows how to connect these devices to the data logger, along with full description and diagrams of every measuring device and its unit of measurements [Figure 7.3].

On the other hand LoggerNet is proprietary software that comes with all and more features than that of PC200W [1]. LoggerNet contains all the measuring devices and control peripherals, and a software ShortCut (CR Basic and Edlog programming software). LoggerNet also contains the same EZSetup Wizard as PC200W [1] with additional options. This allows

Port label and wiring panel for CR1000 by Campbell Scientific

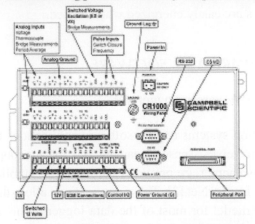

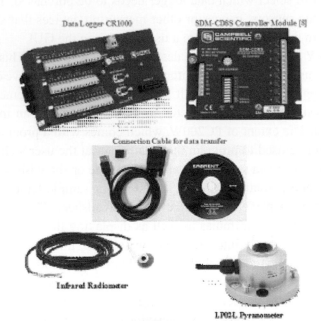

Data Logger CR1000

SDM-CD8S Controller Module [8]

Connection Cable for data transfer

Infrared Radiometer

LP02L Pyranometer

Hydreon Optical Rain Sensor, Model RG-11

FIGURE 7.3   Hardware for CR1000 by Campbell Scientific.

users to set up, configure, and retrieve data from a data logger via various communication ports. The LoggerNet server runs in the background handling all the communications, storing data and provides information to manage data logger network. The client application connects to the server to access information collected from the data loggers. LoggerNet store data in a file on the computer for further analysis. It also contain tools to resolve problems if they arrive. What make LoggerNet ideal are the embedded programs. CRBasic and Edlog works for a range of data loggers (Table 7.1).

TABLE 7.1   Features Between PC200W [1] and LoggerNet [3]

PC200W Features	LoggerNet Features
Provides basic tools (clock set, program download, monitor measurements, retrieve data, terminal emulation, etc.)	Creates custom datalogger programs using Edlog or CRBasic.
	Retrieves data using any of our telecommunication options.
Compatible with most Campbell Scientific dataloggers	Displays or graphs real-time or historic data.
	Processes data files using Split.
Communicates with mixed-array, table, and PakBus® datalogger operating systems	Builds custom display screens to view data or control flags/ports.
	Saves data in formats (including CSV and XML) that can be imported into third-party analysis packages.
PC200W allows to:	LoggerNet allows to:
• Send programs to the data loggers • Collect data • Set the datalogger clock • Access a terminal emulation mode	• Configure the server to communicate with multiple data loggers via a variety of communications hardware • Create custom datalogger programs using Short Cut, Edlog, or the CRBasic Editor • Connect to a datalogger to check or set the clock, send programs, or perform other administrative functions • Display or graph data • Build a custom display screen to view data or control flags/ports • Collect data on demand or schedule • Process data files using Split • Save data in formats (including CSV and XML) that can be imported into third-party analysis packages.

TABLE 7.1 Continued

PC200W Features	LoggerNet Features
Data Processing and Display	**Status Monitor**
PC200W collects and stores data in comma separated files on your PC. It includes a View program that displays data files in tabular or graphical form. For mixed-array data loggers, a simple Split program is included that separates arrays in a data file into separate files.	The Status monitor is used to view the communication and data collection status of the overall datalogger network.
	RTMC Development, RTMC Run-Time
	RTMC is used to create custom displays of real-time data, flags, and ports. It provides digital, tabular, graphical, and Boolean data display objects, as well as alarms.
Datalogger Programming	**Network Planner**
PC200W includes Short Cut^2 for creating simple datalogger programs, setting variable units to be measured and time intervals to record data.	LoggerNet 4 includes the Network Planner, a new tool for designing your PakBus datalogger network. First, PakBus devices are selected from a list and placed on the network design palette.
	The Network Planner calculates the optimum settings for each device in the network and then allows you to send these settings to the device, or save them for later download via the Network Planner or the Device Configuration Utility.
	Troubleshooter
	Troubleshooter helps you discover the cause of communication problems. Troubleshooter can be customized to display only the warnings of interest.
	Transformer
	The Transformer tool converts Edlog programs to CRBasic programs. Specifically, it can convert a CR510 or CR10X program to a CR1000, CR800, or CR850 program, or a CR23X program to a CR3000 program.
Contains most measuring and control devices not all.	Contains all measuring and control devices
Allow users to create their own formulas or equations.	Allow users to create their own formulas or equations.
Requirements	**LOGGERNET Specifications**
Windows XP, Vista, 7, or 8.	PC Operating System: Windows 8, 7 (32 and 64 bit), Vista, or XP

CRBasic works for the advance data logger while Edlog works for retired data loggers. The CRBasic program can be used to give instructions for sensors measurement, peripheral controls, data storage and peer-to-peer data transfer. The CRBasic program contains an editor that checks for program validity and offers many user-configurable options to make editing long programs easier. Edlog on the other hand perform similar functions but doesn't have an editor. Instead it has a build-in precompiler that provides error checking and warns of potential problems in the program (Figures 7.4–7.7). Appendices B to D indicate the Java Programming that was used in this chapter.

PC200W was downloaded for free. The software is very simple and straight-forward to install. It has all the sensors and gauges a user will need. It provides pictures for all sensors, gauges and data loggers so that users make the right selection. The PC200W application consisted of peripheral control boards to control other external devices: valves and solenoids to trigger the irrigation, as well as CRBasic to program the control board.

PC200W Open Source Software

Display Result of being programmed by ShortCut

FIGURE 7.4 Continued

Default display results of measurements based on one second intervals

View 4.1 - [C:\Campbellsci\PC200W\SB15271239_Sec5.dat (No Graph Associated) 120 Records]
File Edit View Window Help

TIMESTAMP	RECORD	PTemp	AirTemp C	RH Percent	SolarRad Wm2	Pressure hPa	BattVolt	WindDirect Avg	WindSpeed ms Avg	RainTot mm	Rain mm
2010-07-19 15:37:25	0	23.18	23.36	49.07	0.503	1015	10.84	213.3	0	0	0
2010-07-19 15:37:30	1	23.18	23.33	49.03	0.503	1015	10.81	213.3	0	0	0
2010-07-19 15:37:35	2	23.18	23.39	49.17	0.503	1015	10.84	213.3	0	0	0
2010-07-19 15:37:40	3	23.18	23.36	49.14	0.503	1015	10.83	213.3	0	0	0
2010-07-19 15:37:45	4	23.18	23.29	49.14	0.503	1015	10.83	213.3	0	0	0
2010-07-19 15:37:50	5	23.18	23.33	49.1	0.503	1015	10.83	213.3	0	0	0
2010-07-19 15:37:55	6	23.18	23.33	49.17	0.503	1015	10.81	213.3	0	0	0
2010-07-19 15:38:00	7	23.18	23.33	49.17	0.503	1015	10.83	213.3	0	0	0
2010-07-19 15:38:05	8	23.18	23.33	49.2	0.503	1015	10.83	213.3	0	0	0
2010-07-19 15:38:10	9	23.18	23.36	49.17	0.503	1015	10.83	213.3	0	0	0
2010-07-19 15:38:15	10	23.18	23.33	49.17	0.503	1015	10.83	213.3	0	0	0
2010-07-19 15:38:20	11	23.18	23.33	49.17	0.503	1015	10.83	213.3	0	0	0
2010-07-19 15:38:25	12	23.18	23.39	49.2	0.503	1015	10.85	213.3	0	0	0
2010-07-19 15:38:30	13	23.18	23.33	49.2	0.503	1015	10.85	213.3	0	0	0
2010-07-19 15:38:35	14	23.18	23.33	49.24	0.503	1015	10.85	213.3	0	0	0
2010-07-19 15:38:40	15	23.18	23.36	49.17	0.503	1015	10.85	213.3	0	0	0
2010-07-19 15:38:45	16	23.18	23.33	49.2	0.503	1015	10.83	213.3	0	0	0
2010-07-19 15:38:50	17	23.18	23.36	49.27	0.503	1015	10.85	213.3	0	0	0
2010-07-19 15:38:55	18	23.18	23.33	49.27	0.503	1015	10.85	213.4	0	0	0
2010-07-19 15:39:00	19	23.18	23.33	49.31	0.503	1015	10.85	213.3	0	0	0
2010-07-19 15:39:05	20	23.18	23.29	49.34	0.503	1015	10.84	213.3	0	0	0
2010-07-19 15:39:10	21	23.18	23.26	49.34	0.503	1015	10.83	213.3	0	0	0
2010-07-19 15:39:15	22	23.18	23.33	49.41	0.503	1015	10.85	213.3	0	0	0
2010-07-19 15:39:20	23	23.18	23.33	49.34	0.503	1015	10.84	213.3	0	0	0
2010-07-19 15:39:25	24	23.18	23.39	49.44	0.503	1015	10.85	213.3	0	0	0
2010-07-19 15:39:30	25	23.18	23.29	49.44	0.503	1015	10.85	213.3	0	0	0
2010-07-19 15:39:35	26	23.18	23.36	49.44	0.503	1015	10.83	213.3	0	0	0
2010-07-19 15:39:40	27	23.18	23.29	49.54	0.503	1015	10.85	213.3	0	0	0
2010-07-19 15:39:45	28	23.18	23.29	49.54	0.503	1015	10.84	213.3	0	0	0
2010-07-19 15:39:50	29	23.18	23.33	49.51	0.503	1015	10.81	213.3	0	0	0
2010-07-19 15:39:55	30	23.18	23.29	49.54	0.503	1015	10.81	213.3	0	0	0
2010-07-19 15:40:00	31	23.18	23.26	49.44	0.503	1015	10.84	213.3	0	0	0
2010-07-19 15:40:05	32	23.18	23.29	49.37	0.503	1015	10.85	213.3	0	0	0
2010-07-19 15:40:10	33	23.18	23.29	49.48	0.503	1015	10.85	213.3	0	0	0
2010-07-19 15:40:15	34	23.18	23.33	49.44	0.503	1015	10.84	213.3	0	0	0
2010-07-19 15:40:20	35	23.18	23.26	49.44	0.503	1015	10.85	213.3	0	0	0
2010-07-19 15:40:25	36	23.18	23.26	49.44	0.503	1015	10.84	213.3	0	0	0
2010-07-19 15:40:30	37	23.18	23.26	49.41	0.503	1015	10.85	213.3	0	0	0
2010-07-19 15:40:35	38	23.18	23.26	49.31	0.503	1015	10.85	213.3	0	0	0
2010-07-19 15:40:40	39	23.18	23.26	49.37	0.503	1015	10.84	213.3	0	0	0
2010-07-19 15:40:45	40	23.18	23.26	49.34	0.503	1015	10.85	213.3	0	0	0

Graphed view of results

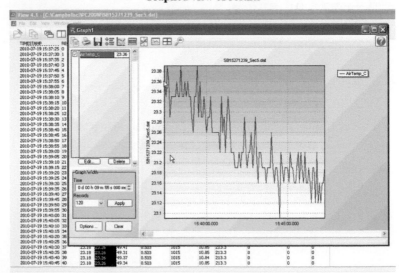

FIGURE 7.4 Software applications. (a) PC200W Open Source Software; (b) Display Result of being programmed by ShortCut; (c) Default display results of measurements based on one second intervals; (d) Graphed view of results.

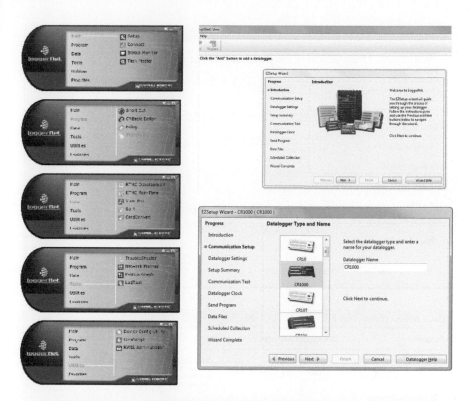

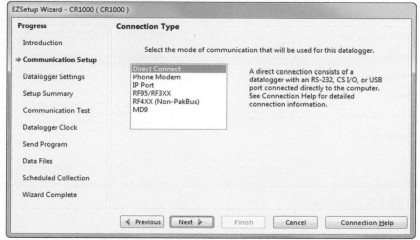

FIGURE 7.5 Continued

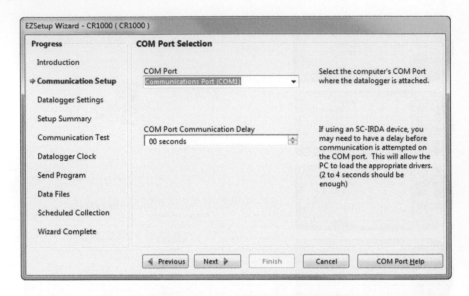

FIGURE 7.5   Continued

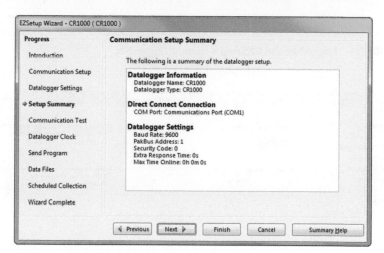

FIGURE 7.5 Continued

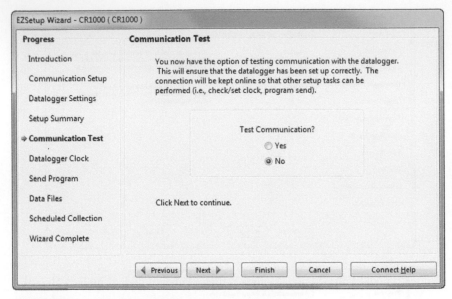

FIGURE 7.5   LoggerNet and EZSetup wizard of LoggerNet.

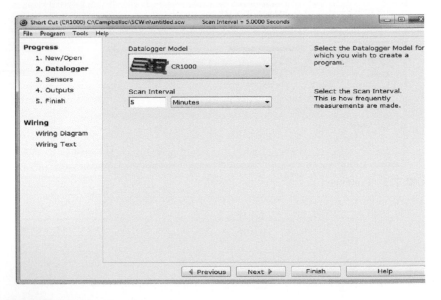

FIGURE 7.6   Continued

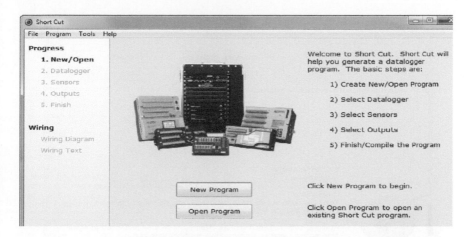

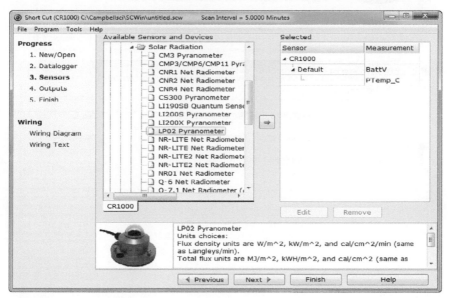

FIGURE 7.6   Continued

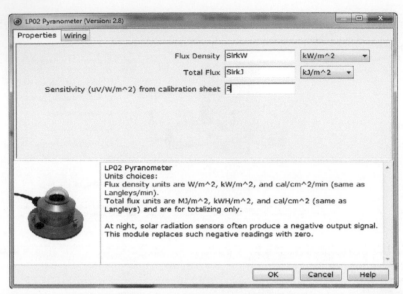

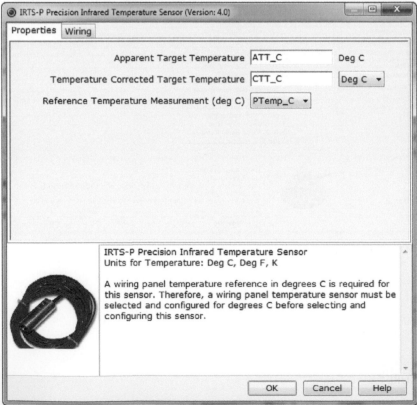

FIGURE 7.6   Continued

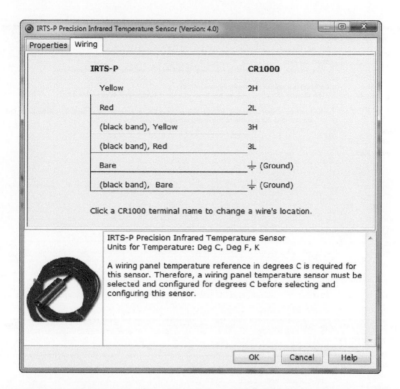

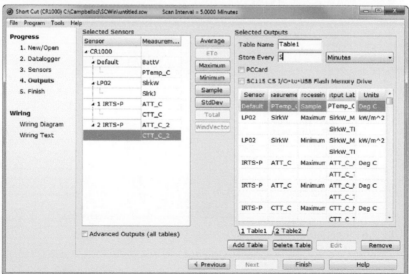

FIGURE 7.6   Continued

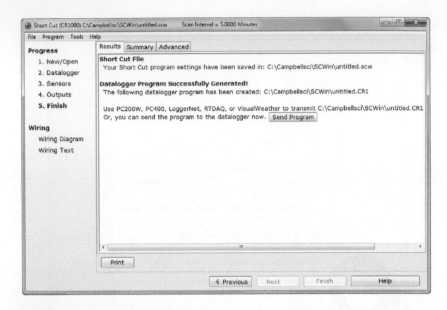

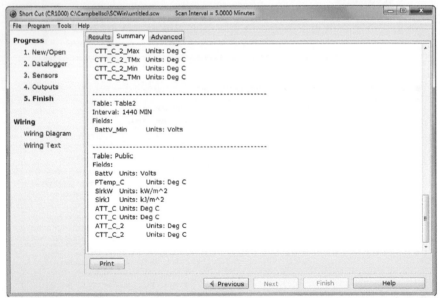

FIGURE 7.6   Continued

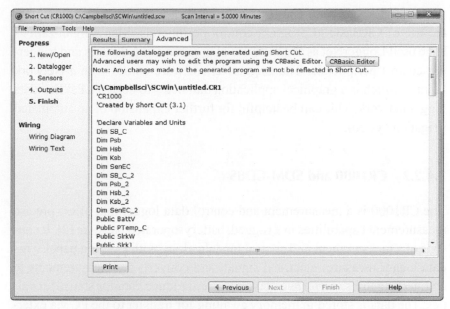

FIGURE 7.6 Programming Data logger.

FIGURE 7.7 CRBasic Programming.

LoggerNet software application is recommended for automation with the use of CRBasic to program control peripherals. Many organizations have used LoggerNet as the software of programming devices and triggering control peripherals. The LoggerNet software comes with a Network Planner which is a graphical application that helps to design a PakBus data logger network. This can be helpful for further expansion of an automated irrigation system.

7.4.2.2   CR1000 and SDM-CD8S

The CR1000 is a measurement and control data logger that offers precise measurement capabilities in a rugged, battery-operated package [2]. It consists of a measurement and control module along with a wiring panel. This data logger measures electrical signals and converts the measurements to engineering units, perform calculations, and reduce data for statistical values. The data is stored in memory awaiting for transfer to the PC via external storage devices or data cables via communication ports. Data is copied from the data logger to the computer. It does not move the data from data logger to PC. This is done so that multiple users can have access to the same data logger without compromising the data or coordinating data collection activities. The RS-232 and CS I/O ports are integrated with the CR1000 writing panel to facilitate data collection. A USB-to-RS-232 [27] cable is used to connect the CR1000 to a PC to accommodate data transfer. The CR1000 is normally powered by a 12-volt power supply and charging regulator battery. In addition, it supplies 5 and 12 volts convenient power distribution to sensors and peripherals. The CR1000 [2] can control four electronic peripherals and supply them with 5 to 12 volts each.

The SDM-CD8S [8] is an eight channel solid-state DC control module that expands the CR1000 data logger [2] to control more control peripherals. It has eight DC voltage outputs and returns that can be switched on or off manually or under the data logger control. The SDM-CD8S can be powered by the data logger or by an external power supply. In most cases, it is powered by an external power supply. The SDM-CD8S powered control peripherals at 6 to 26 volts. Using the data logger software with the embedded CRBasic application to program the SDM-CD8S [8] control module is useful to automate and control irrigation when necessary. The SDM-CD85 has an override switch and an individual rocker switch for each of its outputs. The override switch is used for any service or maintenance that needs to be done among controlled devices.

The CR1000 data logger is capable of performing lots of task and processes [2]. It has been used widely in various countries and is well equipped for the automation of tasks. On the other hand, it is limited by the voltage distribution of control peripherals. Most control peripherals require more voltage than a 12 volts supply. For example, electronic valves and solenoids require voltage >12 V, such as, voltage of 18 volts to 26 volts. Thus the SDM-CD8S will expand the CR1000 to control peripherals of higher voltage. The SDM-CD8S can supply the control peripherals with its own power source and allow the CR1000 [2] data logger to perform its controls and commands over the control peripherals.

7.4.3   EFB-CP SERIES VALVE AND XFS SUBSURFACE DRIP LINE

The EFB-CP is a brass irrigation valve that can withstand harsh conditions in non-potable water situations. This is due to the diaphragm which is made of chlorine and chemical resistance material. The EFB-CP series valve is an electronic remote control valve which requires 24 volts supply to operate and can even operate at 19.6 volts. It also contains a self-flushed stainless steel screen in the valve inlet to filter out grid and prevent clogging of the hydraulic control ports and assure reliable operations. In addition to has a Reverse flow and Fluid resistor features, which prevents flooding, water waste, landscape damage and damage to the valve itself. The reverse flow feature ensures that the valve will fail in the closed position if tear or rip in the diaphragm occurs. The fluid resistor slows down the flow of water through the valve, reducing closing speed and preventing water hammer and system damage. Lastly but not least it also comes with a manual internal and external bleed to turn off the flow of water manually (Figure 7.8).

The XFS subsurface drip line is used to supply water to the roots (Figure 7.8). The emitters are installed in-line at 30 cm spacing to produce even flow of water. The emitters are designed to reduce in-line pressure loss and prevent root intrusion with its patent-pending The copper shield technology causes ions to release to bond to the root causing the tip of the root to become blunt, thus preventing further growth and keeping the

Brass Valve EFB –CP Series

Inside the EFC-CP Brass Valve

XFS Subsurface Drip line

XFS subsurface drip line with copper shield technology

XFS subsurface drip line allows debris to pass

FIGURE 7.8   Valves and subsurface drip lines.

emitter hole free from roots. The emitters are also designed to compensate pressure variation between 8.5 to 60 psi. This allows consistent flow of water over an entire lateral length ensuring higher uniformity for increase in reliability (Figure 7.9).

The solenoid valves are reliable to aid in irrigation and preventative measures. It is even more reliable to save water for farmers who store water in barrels, drums and tanks to water the crops. These technologies

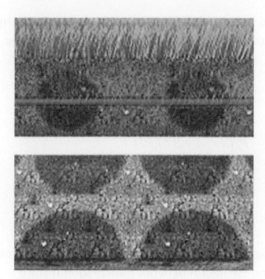

FIGURE 7.9   XFS subsurface drip line versus lateral flow of water.

can improve the way farmers replenish the crops in Small Islands of the Caribbean. Based on the many videos that were watched on YouTube, it was observed that many users install XFS subsurface drip line at least six inches underground. This was the key advantage for farmers in the Caribbean to prevent the crop from being scorched due to heat. It also prevents the irrigation with hot water to crops. Due to the high temperatures in the Caribbean region, drip lines on the ground surface are in direct contact with sunlight that can also heat up water left inside the drip lines causing the crop to be irrigated with hot water. As a result farmer reaps the benefit from its lateral flow of water underground, supplying water directly at the root zone of the crop (Figure 7.9).

7.4.4   FORMULAS/EQUATIONS FOR THIS STUDY

7.4.4.1   CWSI Formulas

Crop water stress index is widely for many different types of crops. In addition, the United States Department of Agriculture (USDA) uses this formula for crop irrigation to detect stress in crops. USDA has established a mean integrated crop water stress index (iCWSI) so that if the mean is greater than the threshold, a signal is sent to trigger the irrigation. The CWSI was defined by Hongyan Gao et al. [5] is as follows:

$$CWSI = \frac{(T_c - T_a) - (T_c - T_a)ll}{(T_c - T_a)ul - (T_c - T_a)ll} \qquad (1)$$

where,

$$(T_c - T_a)ll = A + B * VPD$$

$$VPD = 0.611 * e^{17.27*Ta/Ta*237.3} * (1 - \frac{RH}{100})$$

$$(T_c - T_a)ul = A + B * VPG \qquad (2)$$

On the other hand, CWSI formula for lettuce was narrowed down by a model based on BP neural network information fusion:

$$CWSI = \frac{(T_c - T_a) - 3.7544}{5.0785} \qquad (3)$$

This was accomplished via a model of difference in temperature between canopy and air by data analysis:

$$(T_c - T_a) = 4.8287 - 1.4647 * VDP \qquad (4)$$

Based on an experiment carried out using Multi Spectral Imaging Technology, when there was full water supply, VDP were 5.86, the difference in temperature between canopy and air was minimum. Thus the lower limits of CWSI. (Tc − Ta)ll was −3.7544. On the other hand when lettuce was severely stressed, canopy temperature reached maximum during the experiment. This was regarded as the upper limits of CWSI, so that (Tc − Ta)ul was 1.3241. This gives us a "Lettuce CWSI model":

$$CWSI_L = \frac{(T_c - T_a) - 3.7544}{5.0785} \qquad (5)$$

7.4.4.2 Evapotranspiration Equations

Another widely used formula for evapotranspiration includes parameters, such as: P for precipitation, I for irrigation water applied, F for flux across the lower boundary of the control volume and ΔS for change in soil-water stored in the profile. Evapotranspiration is another means used to measure crop water requirements, when to irrigate and how much water to irrigate:

$$ET = P + I + F \Delta S - R \tag{6}$$

A simpler evapotranspiration equation is the estimated potential evapotranspiration using the Hargreaves-Samani method (PETHS). This equation has been used to calculate PET for many crops in Trinidad; and includes: Ra for solar radiation, T for the average daily temperature, Tmax for maximum temperature and Tmin for minimum temperature.

$$\text{PETHS} = 0.0023 \times Ra \times (T + 17.8) \times (Tmax - Tmin) \tag{7}$$

In Eq. (7), it was observed that the equations are presented without stating whether or not a measuring device is used to obtain a specific value. It only presents the parameter its units. For example, CWSI equation requires that the lower and upper limits are calculated separately and then entered into the CWSI equation. The same goes for evapotranspiration. Some of the parameters have its own equations to calculate the value separately, before in putting it into the ET equation. The equations used in this chapter require measurements with temperature and solar radiation sensors.

7.4.5 EXPERIMENT AND SIMULATION

By creating separate *Java Classes*, one can reduce the amount of unnecessary coding and improve understanding of methods and variables. Thus classes such as "WorkStations" and "IRSensors" were created to call graphic methods and Color variables to display and show what measuring devices are active. Having separate *Java Classes* allows code from different classes to be reused or to reassign a value. This makes coding simple and is not too complex for other programmers to understand.

7.4.5.1 Calculations of iCWSI

An estimated integrated crop water stress index was calculated for both irrigated rows of crops. These values were taken from first run of the simulation without irrigation within a limit of 28°, which is an upper limit for lettuce crop.

iCWSI-1 = [(0.129*4) + (0.110*4) + (0.263*4) + (0.049*4) + (0.142*4) + (0.052*4) + (0.155*4) + (0.142*4) + (0.105*4)]/36 = **0.1274**

iCWSI-2 = [(0.080*4) + (0.164*4) + (0.136*4) + (0.221*4) + (0.235*4) + (0.150*4) + (0.204*4) + (0.125*4) + (0.191*4))/36 = **0.167**

a. Scenario 1 – Sunny Day

Due to the long list of results, a sample Table 7.2 was developed to show changes in measurements and activation of solenoid valves.

Irrigation cycles between morning and evening periods were based on the activation of solenoid valves (Figure 7.10)

Based on the iCWSI = 0.1274, it showed significant activation of valve to irrigate crops in the first three rows of crops. The iCWSI = 0.1274 uses a greater amount of water to irrigate crops compared to iCWSI = 0.1673. The iCWSI = 0.1673 had no irrigation for temperature at 31° and a little irrigation at 33° based on the readings taken from the second infrared sensor for the second three rows of crops. Most irrigation took place at lower temperature than that at higher temperature.

The result of no irrigation for temperatures of 31° and little irrigation at 33° was unexpected. This shows that crops either are well watered and transpiring or crops are under water stress. Since this is a simulation and not the actual implementation, a visual observation will best determine the outcome. Most irrigation at temperatures 25° to 26° was also unexpected. Due to fact that lettuce crops grows best at temperature 25°. This may result in over watering of crops.

b. Scenario 2 – Rainy Day

Again due to large data on results, a sample Table 7.3 was developed to show changes in measurements and deactivation of solenoid valves during the rainfall.

No irrigation took place during rainfall. This is due to the control of the rain sensor to turn off the irrigation cycles. In addition, the CWSI values in Table 7.3 show extreme low CWSI calculations. Thus CWSI drops significantly during rainy days. Temperature in the simulation drop significantly to 19.931 and 19.815. Since this is a simulation, again temperature in Trinidad during rainy season does not drop below 20°. Thus this is the

TABLE 7.2 Sample of Temperature Changes

Cswi1	Cswi2	Max Air	Min Air	Max R1	Min R1	Max R2	Min R2	Max Solor	Min Solor	PETHS	Valve1 On	Valve2 On
-0.067	-0.171	23.929	23.491	26.904	23.389	26.770	26.375	17.591	17.096	0.098	0	0
-0.067	-0.171	23.929	23.491	26.904	23.389	26.770	26.375	17.591	17.096	0.098	1	0
-0.759	-0.110	23.929	23.474	26.904	23.372	26.770	26.670	17.591	17.219	0.102	1	1
-0.759	-0.110	23.929	23.474	26.904	23.372	26.770	26.670	17.591	17.219	0.102	1	1
-0.088	-0.736	23.929	23.103	26.904	26.410	26.770	23.117	17.591	17.047	0.172	1	1
-0.088	-0.736	23.929	23.103	26.904	26.410	26.770	23.117	17.591	17.047	0.172	1	1
-0.093	-0.099	24.606	24.533	27.816	24.894	27.783	26.457	18.855	17.693	0.021	1	0
-0.093	-0.099	24.606	24.533	27.816	24.894	27.783	26.457	18.855	17.693	0.021	1	0
-0.892	-0.316	24.916	24.533	27.816	24.142	27.783	27.065	18.895	17.693	0.119	1	0
-0.892	-0.316	24.916	24.533	27.816	24.142	27.783	27.065	18.895	17.693	0.119	1	0
-0.683	-0.156	24.916	24.331	27.862	24.617	27.804	27.294	18.895	18.050	0.159	1	0
-0.683	-0.156	24.916	24.331	27.862	24.617	27.804	27.294	18.895	18.050	0.159	0	0
-0.683	-0.156	24.916	24.331	27.862	24.617	27.804	27.294	18.895	18.050	0.159	0	0
-0.228	-0.163	25.475	24.331	28.069	24.617	28.402	27.294	19.637	18.050	0.397	0	0
-0.228	-0.163	25.475	24.331	28.069	24.617	28.402	27.294	19.637	18.050	0.397	0	0
-0.228	-0.163	25.475	24.331	23.069	24.617	28.402	27.294	19.637	18.050	0.397	0	0
-0.207	-0.762	25.905	25.414	28.873	28.610	28.844	25.791	19.963	19.898	0.182	0	0
-0.207	-0.762	25.905	25.414	28.873	28.610	28.844	25.791	19.963	19.898	0.182	0	0
-0.010	-0.070	26.161	25.414	29.866	28.610	28.844	25.791	20.920	19.898	0.300	0	0

TABLE 7.2 Continued

Cswi1	Cswi2	Max Air	Min Air	Max R1	Min R1	Max R2	Min R2	Max Solor	Min Solor	PETHS	Valve1 On	Valve2 On
-0.010	-0.070	26.161	25.414	29.866	28.610	28.844	25.791	20.920	19.898	0.300	0	0
-0.575	-0.091	26.996	26.141	29.877	26.974	29.914	29.433	20.990	20.481	0.336	1	1
-0.575	-0.091	26.996	26.141	29.877	26.974	29.914	29.433	20.990	20.481	0.336	1	1
-0.125	-0.813	27.748	26.141	30.866	26.974	29.914	27.373	21.846	20.481	0.804	1	1
-0.125	-0.813	27.748	26.141	30.866	26.974	29.914	27.373	21.846	20.481	0.804	1	1
-0.199	-0.322	27.887	27.481	30.907	30.633	30.942	30.004	21.924	21.605	0.204	0	0
-0.199	-0.322	27.887	27.481	30.907	30.633	30.942	30.004	21.924	21.605	0.204	0	0
-0.129	-0.080	28.519	27.481	31.618	30.633	31.867	30.004	22.981	21.605	0.588	0	0
-0.129	-0.080	28.519	27.481	31.618	30.633	31.867	30.004	22.981	21.605	0.588	0	0
-0.696	-0.191	28.819	28.448	31.897	28.668	31.885	31.233	22.981	22.746	0.207	1	0
-0.696	-0.191	28.819	28.448	31.897	28.668	31.885	31.233	22.981	22.746	0.207	0	0
-0.696	-0.191	28.819	28.448	31.897	28.668	31.855	31.233	22.981	22.746	0.207	0	0
-0.032	-0.088	29.213	28.448	32.803	28.668	32.523	31.233	23.893	22.746	0.480	0	0
-0.032	-0.088	29.213	28.448	32.803	28.668	32.523	31.233	23.893	22.746	0.480	0	0
-0.739	-0.206	29.888	29.843	32.981	29.844	32.550	32.261	23.952	23.249	0.029	0	0
-0.739	-0.206	29.888	29.843	32.981	29.844	32.550	32.261	23.952	23.249	0.029	1	0
-0.739	-0.206	29.888	29.843	32.981	29.844	32.550	32.261	23.952	23.249	0.029	1	0
-0.321	-0.165	30.886	29.843	33.009	29.844	33.802	32.261	24.082	23.249	0.756	0	0
-0.321	-0.165	30.886	29.843	33.009	29.844	33.802	32.261	28.082	23.249	0.756	0	0

TABLE 7.2 Continued

Cswi1	Cswi2	Max Air	Min Air	Max R1	Min R1	Max R2	Min R2	Max Solor	Min Solor	PETHS	Valve1 On	Valve2 On
-0.618	-0.207	30.886	30.371	33.961	30.985	33.802	33.074	24.798	24.059	0.359	1	0
-0.618	-0.207	30.886	30.371	33.961	30.985	33.802	33.074	24.798	24.059	0.359	1	0
-0.618	-0.207	30.886	30.371	33.961	30.985	33.802	33.074	24.798	24.059	0.359	0	0
-0.278	-0.314	30.906	30.371	33.961	33.247	33.802	33.065	24.798	24.504	0.395	0	0
-0.278	-0.314	30.906	30.371	33.961	33.247	33.802	33.065	24.798	24.504	0.395	0	0
-0.200	-0.114	30.906	30.271	33.961	33.011	33.892	33.445	24.798	24.126	0.440	0	0
-0.200	-0.114	30.906	30.271	33.961	33.011	33.892	33.445	24.798	24.126	0.440	0	1
-0.211	-0.771	31.993	30.271	34.674	33.011	33.892	31.831	25.454	24.126	1.431	0	1
-0.211	-0.771	31.993	30.271	34.674	33.011	33.892	31.831	25.454	24.126	1.431	0	1
-0.211	-0.771	31.993	30.271	34.674	33.011	33.892	31.831	25.454	24.126	1.431	0	0
-0.072	-0.210	31.993	31.324	34.715	33.011	34.011	31.831	25.629	24.126	0.533	0	0
0.008	0.006	31.993	31.137	34.933	31.741	34.934	34.921	25.991	25.488	0.669	0	0
0.008	0.006	31.993	31.137	34.933	31.741	34.934	34.921	25.991	25.488	0.669	1	0
-0.739	-0.082	32.078	31.137	34.933	32.082	35.415	34.921	26.762	25.488	0.827	1	0
-0.739	-0.082	32.078	31.137	34.933	32.082	35.415	34.921	26.762	25.488	0.827	1	0
-0.138	-0.815	32.525	31.137	35.576	32.082	35.415	32.140	26.881	25.488	1.263	1	1
-0.822	-0.238	32.673	31.137	35.576	32.254	35.415	35.221	26.881	26.617	1.398	1	1
-0.169	-0.117	32.978	32.193	35.846	35.087	35.839	35.353	26.881	26.721	0.694	0	0
-0.169	-0.117	32.978	32.193	35.846	35.087	35.839	35.353	26.881	26.721	0.694	0	1

TABLE 7.2 Continued

Cswi1	Cswi2	Max Air	Min Air	Max R1	Min R1	Max R2	Min R2	Max Solor	Min Solor	PETHS	Valve1 On	Valve2 On
-0.089	-0.676	33.434	32.193	36.736	35.087	35.839	33.758	27.233	26.721	1.215	0	1
-0.089	-0.676	33.434	32.193	36.736	35.087	35.839	33.758	27.233	26.721	1.215	0	1
-0.288	-0.291	33.930	32.747	36.766	36.041	36.835	36.021	27.966	27.866	0.187	0	0
-0.288	-0.291	33.930	33.747	36.766	36.041	36.835	36.021	27.966	27.866	0.187	0	0
-0.152	-0.043	33.930	32.416	36.766	35.396	36.835	35.951	27.966	26.502	1.349	0	0
-0.152	-0.043	33.930	32.416	36.766	35.396	36.835	35.951	27.966	26.502	1.349	0	0

TABLE 7.3 Sample of Temperature Changes

Cswi1	Cswi2	Max Air	Min Air	Max R1	Min R1	Max R2	Min R2	Max Solor	Min Solor	PETHS	Valve1 On	Valve2 On
0.097	0.145	33.930	21.393	36.766	24.653	36.835	24.410	27.966	15.257	1.581	0	0
0.097	0.145	33.930	21.393	36.766	24.653	36.835	24.410	27.966	15.257	1.581	0	0
0.494	0.461	33.930	21.126	36.766	22.371	36.835	22.538	27.966	13.172	1.290	0	0
0.494	0.461	33.930	21.126	36.766	22.371	36.835	22.538	27.966	13.172	1.290	0	0
0.557	0.661	33.930	20.912	36.766	21.839	36.835	21.309	27.966	12.815	1.194	0	0
0.557	0.661	33.930	20.912	36.766	21.839	36.835	21.309	27.966	12.815	1.194	0	0
0.555	0.587	33.930	20.514	36.766	21.451	36.835	21.288	27.966	12.707	1.064	0	0
0.468	0.532	33.930	20.278	36.766	21.657	36.835	21.332	27.966	12.226	0.951	0	0
0.556	0.530	33.930	20.458	36.766	21.391	36.835	21.519	27.966	12.357	1.018	0	0

TABLE 7.3 Continued

Cswi1	Cswi2	Max Air	Min Air	Max R1	Min R1	Max R2	Min R2	Max Solor	Min Solor	PETHS	Valve1 On	Valve2 On
0.556	0.530	33.930	20.458	36.766	21.391	36.835	21.519	27.966	12.357	1.018	0	0
0.561	0.522	33.930	19.285	36.766	20.191	36.835	20.389	27.966	11.122	0.556	0	0
0.561	0.522	33.930	19.285	36.766	20.191	36.835	20.389	27.966	11.122	0.556	0	0
0.561	0.522	33.930	19.285	36.766	20.191	36.835	20.389	27.966	11.122	0.556	0	0
0.453	0.386	33.930	19.103	36.766	20.555	36.835	20.897	27.966	11.587	0.515	0	0
0.629	0.508	33.930	19.752	36.766	20.312	36.835	20.927	27.966	11.456	0.729	0	0
0.629	0.508	33.930	19.752	36.766	20.312	36.835	20.927	27.966	11.456	0.729	0	0
0.507	0.589	33.930	19.377	36.766	20.559	36.835	20.142	27.966	11.796	0.623	0	0
0.507	0.589	33.930	19.377	36.766	20.559	36.835	20.142	27.966	11.796	0.623	0	0
0.533	0.560	33.930	19.578	36.766	20.623	36.835	20.487	27.966	11.542	0.677	0	0
0.533	0.560	33.930	19.578	36.766	20.623	36.835	20.487	27.966	11.542	0.677	0	0
0.524	0.396	33.930	18.247	36.766	19.342	36.835	19.992	27.966	10.392	0.167	0	0
0.524	0.396	33.930	18.247	36.766	19.342	36.835	19.992	27.966	10.392	0.167	0	0
0.454	0.495	33.930	18.086	36.766	19.536	36.835	19.325	27.966	10.158	0.106	0	0
0.454	0.495	33.930	18.086	36.766	19.536	36.835	19.325	27.966	10.158	0.106	0	0
0.538	0.561	33.930	18.911	35.766	19.931	36.835	19.815	27.966	10.530	0.404	0	0
0.538	0.561	33.930	18.911	36.766	19.931	36.835	19.815	27.966	10.530	0.404	0	0

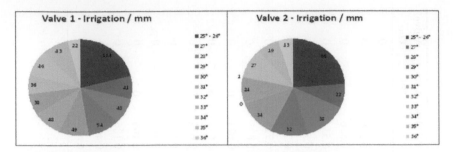

FIGURE 7.10   Solenoid vales 1 and 2 to activate the irrigation cycle.

best lower limit for lettuce crop that can be acquired in Trinidad. Other countries in the Caribbean may differ.

7.5   DISCUSSIONS

A review of published papers provided more than adequate information on the uses and benefits of infrared thermometer in irrigation. IR technology is a stepping-stone and key element in enhancing and improving the watering of many crops. Also, IR technology can establish and determine an integrated crop water stress index for many crops. The automation of irrigation improves the water use efficiency, providing irrigation when needed and the right amount of water without affecting crop yield, than manual irrigation. This allows water saving. A simplified formula for *Crop Water Stress Index* for lettuce crop was developed and PET values were calculated.

Most published papers referred to the Infrared thermometer as the main measuring device. On the other hand when calculating CWSI and ET, additional measurements of solar radiation and wind speed are required. Also, the IR technology has been mainly used for center pivoted irrigation systems. Only a few published papers indicate its use in drip irrigation systems in developing countries.

7.5.1   INTERVIEWS

Through interviews, the author was able to:

- Gather adequate information on the advancements of crop irrigations in developing countries.

- Could know if information on these advancements can be applied to drip irrigation systems.
- Identify the appropriate technology for automation of irrigation.
- Identify whether developing countries in the Caribbean has the technology to design their own automated system of crop irrigation or requires its system to be outsourced.
- To get a clear idea on the technologies that can be used in project implementation in this chapter.
- To know the compatibility between devices and how these communicate and interact with each other.

Interview with Susan helped to establish whether such a system can be implemented and to acquire a basic idea on how it can be implemented. In addition to gathering data on canopy temperature and other meteorological patterns, author was familiar with the integrated crop water stress index. Susan also stated that canopy temperature can detect crop water stress index earlier than soil moisture content. This led to development of formula for crop water stress index for lettuce. Interview with Mitchell was very helpful in identifying the technology and tools needed to program a system to perform automated tasks and display readings and results. This helped to develop the simulation program for lettuce in Java Classes.

Base on the interview with farmers, only one farmer knew why he does not water his crops during peak temperatures of the day. While other farmers say they prefer not to attend to crops during the blazing sun when author asked the question "Why it is that you don't water crops during peak temperatures of the day or during the hours of 12 noon and 2 pm?" Author was expecting to get different answers instead getting repeating replies by most lettuce farmers. The interviews were conducted to gather additional information on the effects of lettuce crop being irrigated as ambient temperature increases during daylight hours. Additional information may have been beneficial to the project on the effects of lettuce being watered during hot temperatures.

7.5.2   SOFTWARE

LoggerNet is loaded with features for a user to program the data logger and control devices without the need for additional separate programs. LoggerNet consists of a full package of embedded software. What is interesting about the User can program the data logger without the need for writing code, based on choosing selections from a range of choices for each device they wish to acquire readings from and even write their own formulas using an embedded program called *ShortCut*. For instance if user selects a *SI-111 Precision Infrared Radiometer* [25], he or she can choose measurement readings from a range of units such as Kelvin, Degrees Celsius or Fahrenheit. In addition, it comes with a wiring diagram so that user can easily connect the sensor to the data loggers without being confused. Symbols are provided both on the GUI and the data logger itself so that the user won't be lost. Also the user can choose a min, max and average readings to be viewed or analyzed. LoggerNet can control, display reading and network data loggers and the measuring devices connected to the data logger all at once. In addition to *Network Planner*, embedded software can assist user in designing own network of irrigation systems that can be controlled by many data loggers without any system mishaps.

Even though LoggerNet is ideal to automate irrigation system with the main embedded programs such as *CRBasic* and *Edlog* to control the *SDM-CD8S* an eight channel solid state DC control module or other control modules so that crop can be replenished. Still, farmers will require help from the expert in these programming languages. These program languages require written code in order to program the *SDM-CD8S* control module so that crops are replenished at the required parameters. Agricultural organization will have to provide assistance to farmers or outsource the programming expertise from Campbell Scientific so that farmers can have their irrigation system automated and working according to their satisfaction.

7.5.3 HARDWARE

The *CR1000* datalogger [2] is well equipped to accommodate many sensors, control modules, multiplexers for additional sensors and a lot more other peripherals that can interact with it. The *CR1000* datalogger can also control valves, relays, solenoids and other electronic devices without a control module once its power supply is 12 volts and lower. It collects all the information on control of devices. In addition to its many means of communication, peripherals (such as direct, internet, Ethernet, Wi-Fi, GSM and 3G for hand held mobiles and smart devices, fiber optics and field) display to give real time readings. This will be beneficial for the users who wish to expand their operation.

The *SDM-CD8S* control module was chosen because of its capabilities of expanding the *CR1000* data logger [2] to control more peripherals at power supply with higher DC voltage. One can make servicing and maintenance without resorting to cease operation of the data logger. User can use the rocker switch to switch to manual mode and select individually the electronic device they wish to service without compromising other electronic devices that are in operation.

The *EFB-CP* Series brass irrigation valve was chosen because of: Its protective and preventative features; ability to prevent flooding, water waste, landscape damage and the buildup of debris; its chemical and chlorine resistant materials, in order to withstand harsh conditions of non-potable water. Thus farmers do not have to worry about losing water or wasted water if there is a mechanical failure or other failures that may occur in the solenoid valve. Furthermore the *EFB-CP* Series brass irrigation valve has a manual internal external bleed which is an additional preventative method for conditions where the electronic solenoid is malfunctioning so that farmers can manually turn off the flow of water.

The *XFS subsurface drip in-line irrigation hose* is best suited for irrigation compared to the traditional porous soaker hose system. The *XFS Subsurface Drip Line* provides water near the root zone over an entire lateral length covering more area underground than on the surface and also prevents crops from becoming scorched. The porous soaker hose system is laid on the ground surface to irrigates crops so that water has to be absorbed into the soil. During the hot blazing sun, water would evaporate at a steady rate. Whereas the XFS subsurface drip line prevents the evaporation of water from the soil. Therefore, underground soil is kept moist for an extended period of time to keep crops replenished and provide readily available water.

The major problem was supplying farmers with the resources mentioned here. Farmers may have to take it into their own account and invest in such a system but will still require experts to assist them. Agricultural suppliers only provide the basic tools and resources they need to cultivate land. The *Ministry of Food Production* in Trinidad may reimburse farmers a percentage of how much was spent by a farmer based on farmer's incentives. Organization such as CARDI and Ministry of Food Production focuses on crops that produce a larger production base and marketing segment. Research and testing conducted in these organizations are mainly for pest and pesticides, soil testing, disease attacks, surveys, administration, water management and operations and maintenance. Automation of crop irrigation has not been taken into serious consideration.

7.5.4 FORMULAS AND EQUATIONS

The crop water stress index equation for lettuce is much easier to input values compared to the first crop water stress index equation (1). Author was able to identify static values for the difference between upper and lower limits. Thus reducing the need to perform separate complex calculations, it may require additional measuring devices to read different meteorological parameters. Therefore, the *CWSI* for lettuce crops only requires the difference between canopy and ambient temperatures. Similarly for PET with the Hargreaves-Samani method requires only temperature and solar radiation data that can be taken from the sensors and entered into the equation without the need for additional separate calculations.

PET using the Hargreaves-Samani method requires solar radiation in mm. Therefore, reading from the pyranometer was needed to be converted from Watts per meter square to equivalent evaporation in mm/day. Finding an appropriate conversion on the Internet is sometimes time consuming. Nevertheless one can follow an appropriate example. As a result solar radiation and PET must be in same units. It is noted that the pyranometer from Campbell Scientific only provides measurements in watts per meter square, kilowatts per meter square or calorie per centimeter square per minute.

7.5.5 SIMULATION/PROTOTYPE

The simulation prototype shows the work and function of an automated irrigation system and how one can benefit from it. In addition to displaying readings measured from infrared sensors and Pyranometer every one second, it also displays calculations for crop water stress index and a potential evapotranspiration using the Hargreaves-Samani method (PETHS) every one second. The simulation also gives an output of table of results for users to analyze just like PC200W or LoggerNet software. It also shows the activation of solenoids to allow the flow of water to irrigate crops and the DC Controller board to control the solenoid valves. This simulation gives an idea of a simple setup of what a real automated irrigation system will look like. Data for sunny days was taken from a weather channel website [22]. In addition data for solar radiation (mm/day and watts/m^2) was taken from http://www.engr.scu.edu/~emaurer/tools/calc_solar_cgi.pl.

Due to lack of information on research papers and available time, the simulation had some limitations. It did not model various weather patterns: cloudy, partly cloudy, overcast and windy. It could only model sunny and rainy days. Temperature was assumed to drop by two degrees. Adding to that, no published papers were found to give a fair idea on the drastic changes in temperature on the account of a rainy day. In addition acquiring sample data of the temperature and solar radiation changes for each of these weather patterns to input in the simulation was quite challenging.

7.5.6   RECOMMENDATIONS

- A farmer's workshop the improvement of crop irrigation systems is needed by local agriculture centers on. The technology and tools required to design such a system as well as the benefits farmers will gain from these improvements.
- Research development bodies such as Caribbean Agricultural Research and Development Institute (CARDI) should:

  - Assist agricultural centers in providing training and other services for farmers who wish to implement their own auto irrigation system.
  - Provide technical training, installation of hardware components and measuring devices.
  - Provide software and programming services, such as data logger programming and control devices programming using CRBasic.
  - Supply farmers the technology and resources needed to upgrade their current irrigation system using data loggers, measuring devices, control devices and new and improved drip lines or sprinkler systems.

- Outsource the technologies required from companies such as Campbell Scientific or Rain Bird. These companies not only provide but improve their services by creating new innovations.
- Start introducing the *XFS subsurface drip line and other irrigation components* to farmers and explain the benefits of using it. For instance with

an *XFS subsurface drip line*, it can be installed on the surface or subsurface. Subsurface installation can provide water directly at the root zone of crops.

7.6 CONCLUSIONS

There is a great need for a more diverse irrigation system in Trinidad and Tobago. This research study investigated the advances in crop irrigation systems for local farmers. Many farmers use the traditional drip and sprinklers systems, which require farmers to be on site and manually turn on or off irrigation. In addition, farmers prefer to water crops during morning and evening period. This shows that crop may or will be starving for water during between these periods. This research included: Literature review of experiments and implementations; interviews with specialists, engineers, local technicians and local farmers; researching the tools and technologies needed; researching appropriate equations and formulas/simulation of an automated irrigation system. This research project showed that there is a way to automate irrigation systems and improve the watering needs of crops based on a Crop Water Stress Index and adequate technologies.

Additional meteorological data is required to simulate different weather patterns: cloudy, overcast and partly cloudy. Adding to that over the years some farmers have experimented to improve production and watering but still require manual labor using a hydro phonics or aqua phonics systems. By using a bucket of water and other chemicals for growth and nutrients mixed together. An electric pump is placed in the bucket to pump water through the tubes and the tubes recycle the water in the bucket at an open end. The *Java program* has a lot of room to simulation various weather pattern and also show automation can be applied to hydro phonics and aqua phonics systems.

7.7 SUMMARY

The objective of this academic research project was to investigate how a more effective method could be advanced to grow crops, particularly the cultivation and harvesting of lettuce within the local agricultural community of Trinidad. Emphasis was placed on that variety of crops that are highly dependent on a continuous irrigation supply. Of these, there are many leafy type vegetables that require frequent watering. The hypothesis was that lettuce like others in this category will exhibit similar analyzes and interpretations to guide the research. To this end, a focus on applied techniques for translating meteorological patterns into infra-red signaling became integral part of this study. It was anticipated that the impact would be highly received as the adoption of more contemporary approaches would serve to improve the traditional culture of watering crops as practiced by farmers.

Results of the literature survey illustrated that there was a widespread acceptance for more economical methods of crop irrigation such as the drip-line and sprinkler irrigation systems; highly advanced central pivot irrigation system popular within the United States of America and other countries. However, investigations conducted with farmers revealed a lack of knowledge pertinent to this area as the mainstay of sources of information. Additionally, there was a strong sense of hesitation on their part as is familiar with many ICT-based initiatives. Here, the line of defense that pervaded discussions were framed in the chant that an adequate supply of water was mandatory to satisfy the demands of modern-day technologies proposed by the solution under review.

In pursuing further and in every effort to support the research with a working version of a system, a prototype was developed using the Java programming language. This allowed the opportunity for a more meaningful retrospection of the role automation could play and the potential for enhancing the dispensation of water to fields and beds of lettuce production without the intervention of a farmer. As expected and reported in the testing, the simulation does have a combination of constraints given the various parameters applied in this version: *ambient temperature, canopy temperature and solar radiation.* Based on these three combinations, it was proven that the CSWI is a determinant of the triggering of the dispensation of water to the crop.

Finally, the report sheds light that the contributions made in this effort can only be further enhanced when adequate training and technical assistance is provided to the local farmer. Agencies such as the Agricultural Society of Trinidad and Tobago (ASTT), the Caribbean Agricultural Research and Development Institute (CARDI), among others, need to place greater emphasis on superior alternatives and apply innovative techniques for the development and sustenance of the sectors they serve particularly from a technological perspective.

ACKNOWLEDGMENTS

Author expresses special thanks to his advisor, *Anjan Lakhan,* for guiding me through the initial and continuing stages of the project, *"Investigating the approach for translating meteorological patterns into infra-red signaling for variable dispensation in crop irrigation systems."* He is also grateful to *Susan O' Shaughnessy*, PhD Research Agricultural Engineer at the Conservation and Production Research Laboratory (CPRL), *Kahill Mitchell*, Application Engineer at Campbell Scientific Australia, *Sonny James*, Certified Infrared Thermographer at Thermal Diagnostics Limited, and *Megh R. Goyal*, PhD, PE for their superior knowledge and expertise. His sincere thanks to *lettuce farmers*.

KEYWORDS

- **Agricultural Society of Trinidad and Tobago**
- **Automated irrigation**
- **Belle Fourche irrigation district**
- **Boom sprayer**
- **Campbell Scientific**
- **Canopy temperature**
- **Caribbean**
- **Caribbean Agricultural Research and Development Institute**
- **Computer based simulation**
- **Conservation & Production Research Laboratory**
- **Crop irrigation**
- **Crop temperature**
- **Crop water requirement**

- **Plant color**
- **Ponds**
- **Potential evapotranspiration (PET) using the Hargreaves-Samani methods**
- **Programmable logic controller**
- **Rainfall**
- **Rivers**
- **Soil moisture**
- **Soil water potential**
- **Solar energy**
- **Streams**
- *Trinidad*
- **UML state activity diagram**
- **Vapor pressure deficit**
- **Water consumption**
- **Water pump**
- **Water stress**
- **Water supply**
- **Water use efficiency**
- **Workstations**

REFERENCES

A. Technical Information

1. Campbell Scientific Inc. *PC200W Version 4.* Retrieved March 17th, 2015. http://s.campbellsci.com/documents/au/product-brochures/b_pc200w.pdf
2. Campbell Scientific, Inc. (2000–2013). *CR1000 Measurement and Control System (Revision: 5/13) Instruction Manual.* Retrieved March 21th, 2015. http://s.campbellsci.com/documents/au/manuals/cr1000.pdf
3. Campbell Scientific, Inc. (1999–2015). *LoggerNet verion 4.3 Instruction Manual (Revision 3/15).* Retrieved March 20th, 2015. http://s.campbellsci.com/documents/au/manuals/loggernet.pdf

4. DeGannes, A., Kamau Ra Heru, Aziz Mohammed, Compton Paul, Jervis (2014). *Tropical Greenhouse Growers Manual for the Caribbean*. Retrieved December 12th. http://www.cardi.org/cfc-pa/files/downloads/2014/01/TROPICAL-GREENHOUSE-GROWERS-MANUAL.pdf.

5. Gao, Hongyan, Hanping Mao, & Xiaodong Zhang (2011). Inspection of lettuce water stress based on multi-sensor information fusion technology. In: *D. Li, Y. Liu, and Y. Chen (Eds.):* CCTA 2010, Part II, IFIP AICT, 345, 53–60. Retrieved April 4th: http://download-v2.springer.com/static/pdf/281/chp%253 A10.1007%252F978-3-642-18336-2_7.pdf?token2=exp=1428076929~acl=%2Fst atic%2Fpdf%2F281%2Fchp%25253A10.1007%25252F978-3-642-18336 2_7.pdf *~hmac=6c6154f1047d73b5885e5fff86c4286e2b1af41d0336dbb90ad47ef292d1.

6. Geo Scientific, L. (2005). *Data logger fundamentals for environmental monitoring applications*. Retrieved February 28th, 2015. http://www.geoscientific.com/technical/tech_references_pdf_files/Data_Logger_Fundamentals.pdf.

7. Hadjimitsis, Diofantos G., & Giorgos Papadavid (2013). *Remote sensing for determining evapotranspiration (Chapter 2)*. Retrieved January 1st, 2015. http://cdn.intechopen.com/pdfs-wm/45169.pdf

8. James, S. (2006). *The fundamental of refractory inspection with infrared thermography*. Retrieved December 12th, 2014. http://www.tdlir.com/article_2_2006_james.pdf

9. Li, Daoliang, Yande Liu, & Yingyi Chen (Eds.). (2010). *Computer and computing technologies in agriculture IV*. Retrieved April 4th, 2015. https://books.google.tt/boo ks?id=i_928nCnCQMC&pg=PA58&lpg=PA58&dq=what+is+the+cwsi+of+lettuce &source=bl&ots=hklp7S7FTk&sig=GdtNNlr3-CJJN4Dq_IJhtJyCRVU&hl=en&sa =X&ei=3BkeVe6EFMqYgwTlzIGYDg&ved=0CCcQ6AEwAw#v=onepage&q=w hat%20is%20the%20cwsi%20of%20lettuce&f

10. Moran, M. S., T. R. Clarke, Y. Inoue and A. Vidal. (1994). Estimating crop water deficit using the relation between surface- air temperature and spectral vegetation index. *Remote Sens. Environ.,* 49, 246–263. Retrieved December 25th, 2014. http://afrsweb.usda.gov/SP2UserFiles/Place/53442010/RemoteSensing/WCLPUB-1735-Moran.pdf

11. Nogueira, L. C., Dukes, M. D., Haman, D. Z., Scholberg, J. M., & Cornejo, C. (2002). Data acquisition system and irrigation controller based on CRX10 datalogger and TDR sensors. *Soil and Crop Science Society of Florida, Proceedings, Volume 62*. Retrieved January 2nd, 2015: http://www.abe.ufl.edu/mdukes/pdf/drip/Nogueira-et-al-SCSSFL-62(2003)38–46.pdf

12. O'Shaughnessy, S. A., Evett, S. R., & Colaizzi, P. D. (2014). *Infrared thermometry as a tool for site-specific irrigation scheduling*. Retrieved December 1st: http://www.cprl.ars.usda.gov/wmru/pdfs/OShaughnessy%20et%20al%20%20 %282014%29%20IR%20thermometry%20as%20a%20tool%20for%20SS%20 irrig%20sched.pdf

13. O'Shaughnessy, Susan A., Steven R. Evett, Paul D. Colaizzi, & Terry A. Howell (2008). Thermal indices for center pivot irrigation scheduling, *SWMRU Publications*. Retrieved May 7th, 2015: http://www.cprl.ars.usda.gov/wmru/pdfs/Soil%20

Water%20measurement%20and%20Thermal%20Indices%20for%20Center%20 Pivot%20Irrigation%20Scheduling.pdf

14. Pinter, Paul, J. Jr., Jerry L. Hatfield, James S. Schepers, Edward M. Barnes, & Susan Moran, M. (2003). *Remote Sensing for Crop Management,* 69, 647–664, Retrieved January 1st, 2015: http://asprs.org/a/publications/pers/2003journal/june/2003_ jun_647–664.pdf

15. Rain Bird. (2012). *GB and EFB-CP Series Valves (Tech Spec).* Retrieved March 31st, 2015, from: http://www.rainbird.com/: http://www.rainbird.com/documents/turf/ts_ GB_EFB-CP.pdf

16. Rogalski, A. (2012). History of infrared detectors. *Opto Electronics Review,* 20(3), 279–308.

17. Shahidian, S., Serralheiro, R. P., Teixeira, J. L., Santos, F. L., & Oliveira, M. R. G. (2009). *Drip irrigation using a PLC based adaptive irrigation system.* Retrieved February 16th, 2015: http://repositorium.sdum.uminho.pt/bit-stream/1822/18683/1/28–911.pdf

18. *Task G: Temporary Flow, Comprehensive Water Master Plan DWSD Contract No. CS-1278: Final Report.* (2001). Retrieved February 2nd, 2015: http://www.dwsd.org/downloads_n/about_dwsd/masterplan_freshwater/ Task_G_Flow_Monitoring.pdf

19. Taylor & Francis (2010). History highlights and future trends of infrared sensors. *Journal of Modern Optics,* 57(18), 1663–1686. Retrieved December 12, 2014: http://www.consorziocreo.it/downloads/pubblications/papers/2010_Corsi_JMo-dOpt_v57pp1663–1686.pdf

B. Images

20. http://www.rainbird.com/landscape/products/valves/EFB-CP.htm, accessed: 4/12/15. EFB-CP Series Brass Valves

21. http://www.rainbird.com/landscape/products/dripline/XFS.htm, accessed: 4/12/15. XFS Subsurface Drip line

22. http://www.weather.com/weather/hourbyhour/l/TDXX0116:1:TD, accessed: 4/16/2015 Weather Temperature Readings – Screen Shot

23. Farmers and their crops were taken from a Smartphone.

24. https://www.campbellsci.com/lp02-l, accessed: 4/12/2015. LP02-L Pyranometer

25. https://www.campbellsci.com/si-111, accessed: 4/12/2015. SI-111 Infrared Radiometer with Standard Field of View

26. https://www.campbellsci.com/sdm-cd8s, accessed: 4/12/2015. SDM-CD8S 8-Channel Solid-State DC Controller

27. https://www.campbellsci.com/17394-cable, accessed: 4/12/15. 17394 Converters, USB to DB9 Male RS-232.

28. http://www.rainsensors.com, accessed: 4/21/2015 Hydreon Optical Rain Sensor – Model RG-11

29. https://www.campbellsci.com/cr1000-datalogger, accessed: 4/12/2015. CR1000 Measurement and Control System.

APPENDIX A – LETTUCE FARMERS

Kevita Alexander – Lettuce Farmer, Arouca, Trinidad

Interview planning checklist items	Answers
How often do you water your lettuce crop?	Twice a day.
What times of the day do you water your lettuce crops?	Early in the morning and late in the evening, before sunrise and sets.
Why it is that you don't water crops during peak temperatures of the day or during the hours of 12 noon and 2 pm?	No, I don't do so because at these times. I prefer to not to work in the hot sun. That is why I work at cool periods during the day, morning and evening, attending and watering crops.
What irrigation system do you used to water your lettuce crops?	Drip Line.

Michael Diet – Lettuce Farmer, Arouca, Trinidad

Interview Planning Checklist Items	Answers
How often do you water your lettuce crop?	Twice a day, morning and evening.
What times of the day do you water your lettuce crops?	Between 6 am and 7 am in the morning and between 5 pm and 7 pm in the evening.
Why it is that you don't water crops during peak temperatures of the day or during the hours of 12 noon and 2 pm?	Weather too hot to attend to crops and water crops.
What irrigation system do you used to water your lettuce crops?	Drip Line.

Farmers in Barataria: Lettuce Farmer – (other farmers that give the same replies) Barataria, Trinidad

Interview Planning Checklist Items	Answers
How often do you water your lettuce crop?	Twice a day early in the morning and late in the evening when the sun is setting.
What times of the day do you water your lettuce crops?	Between 6 am and 7 am in the morning and between 5 pm and 7 pm in the evening.
Why it is that you don't water crops during peak temperatures of the day or during the hours of 12 noon and 2 pm?	Sun too hot, that's why I water crops early in the morning and late in the evening. Don't be on site around those times. Weather does be too hot to attend to crops.

Interview Planning Checklist Items	Answers
What irrigation system do you used to water your lettuce crops?	Drip Line.

Johnny Baldeo – Lettuce Farmer, Barataria, Trinidad

Interview Planning Checklist Items	Answers
How often do you water your lettuce crop?	Twice a day early in the morning and late in the evening when the sun is setting.
What times of the day do you water your lettuce crops?	Between 6 am and 7 am in the morning and between 5 pm and 7 pm in the evening.
Why it is that you don't water crops during peak temperatures of the day or during the hours of 12 noon and 2 pm?	The reason why I don't water my crop during peak temperature of the day is because of the water and heat of the sun can scorched the crop. So I allow it to get its day of sunlight for it to remain crispy. Water and heat can burn the leaf of the lettuce.
What irrigation system do you used to water your lettuce crops?	Drip Line.

APPENDIX B – UML ACTIVITY DIAGRAM

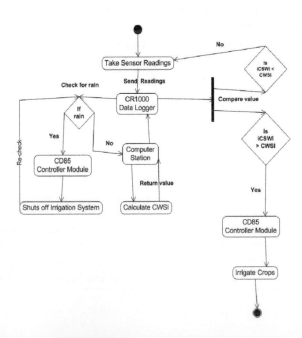

APPENDIX C – LAYOUT DIAGRAM OF IMPLEMENTATION

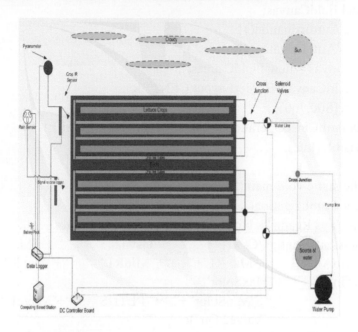

APPENDIX D – IRRIGATION SIMULATION CODE

```java
import java.awt.BorderLayout;
import javax.swing.JFrame;
import javax.swing.JPanel;
/**
*@author Kevin Rodriguez
* SRN: 101035635
* Date Created: 10/1/2015
* Date Modified: 5/12/2015
* Description: The Running Class, Displays a Graphical User Interface of
  an automated irrigation system,using Infrared Sensors and Pyranometer.
* This shows a simple implementation and function of how an automated
  irrigation system works, in sunny and rainy conditions.
* In addition performing calculation and operation using CWSI to irrigate
  crops at the right parameters.
*/
```

```
public class AutoIrrigation extends JFrame{
    JPanel fieldCanvas;
    public AutoIrrigation(){
    super("AutoIrrigation Systems");
    fieldCanvas = new FieldCanvas();
    add(fieldCanvas, BorderLayout.CENTER);
    setSize(800,700);
    setDefaultCloseOperation(JFrame.EXIT_ON_CLOSE);
    setVisible(true);
}

    public static void main(String args []){
        new AutoIrrigation();
    // Displays Readings being recorded by the data logger.
    System.out.println("Cswi1"+"  "  +"Cswi2"  +"  "  +"MaxAir"+"
    "+"MinAir"+"    "+"MaxR1"+"    "+"MinR1"+"    "+"MaxR2"+    "
    "+"MinR2"+" "+"MaxSolor"+
                    "  "+"MinSolor"+" "+"PETHS"+" "+"Valve1  On"+"
                    "+"Valve2 On");
    }
    }
import java.applet.Applet;
import java.awt.Color;
import java.awt.Graphics;
import java.awt.List;
import java.awt.event.ActionEvent;
import java.awt.event.ActionListener;
import java.util.ArrayList;
import java.util.Random;
import javax.swing.JPanel;
import javax.swing.Timer;
/**
*@author Kevin Rodriguez
* SRN: 101035635
* Date Created: 10/1/2015
* Date Modified: 5/12/2015
```

```
* Description: The FieldCanvas class, this is the main class that does
  all the heavy lifting. This class make calls to WorkStations class and
  IRSensors class.
* It draws all the Sensors, plain field, crop bed, lettuce crops, drip line
  irrigation system, computer station for gathering data, processing and
  sending
* commands to the Control Module (SDM-CD8S). Which in return con-
  trols the electronic solenoid valves to irrigate the crops. In addition to
  using Threads to
* show the changes in weather patterns and the changes in temperature
  as the sun rise and sets. As well as display rainfall and the irrigation of
  crops using an
* Integrated Crop Water Stress Index values which is compared with on
  calculated Crop Water Stress Index values every one second to trigger
  the irrigation system.
* In addition to calculating the Potential Evapotranspiration using the
  using the Hargreaves-Samani methods.
*/
public class FieldCanvas extends JPanel{
    // Global variables for sensors, workstation and controls
    Sensors sensor1,sensor2,sensor3,senrLens1,senrLens2,senrLens3,plyo
    Meter,rainSensor;
  WorkStations dataLogger,computerBaseStation,controlBoard,
      waterPump,crossJunction,crossj_1,crossj_2,valve1,valve2,
      xfsDripLine1,xfsDripLine1_2,xfsDripLine1_3,xfsDripLine1_4,xfs
      DripLine1_5,
      xfsDripLine1_6,xfsDripLine1_7,xfsDripLine1_8,xfsDripLine1_9,x
      fsDripLine1_10,
pipeLine_1,pipeLine_2,pipeLine_3,pipeLine_4,pipeLine_5,pipeLine_6,
pipeLine_7,pipeLine_8,
pipeLine_9,pipeLine_10,pipeLine_11,pipeLine_12,pipeLine_13,pipeLin
e_14,pipeLine_15,pipeLine_16,pipeLine_17,pipeLine_18,
      pipeLine_19;
  FieldCanvas bedOfCrops,PipeLines1,PipeLines2;
  Color field,readings, PlySensor,Rain,pl,rd;
  static Random IRpoints;
```

```java
public String IRreadings,IRreadings1,IRreading2,degrees;
private int bedPositionx, bedPositiony,startSun,bedSizex, bedSizey;
static  int  move,temp,reach,re,re2,re3,re4,re5,rainTemp,valveOn1,valve
On2,RainSen,deg;
static double addTo,addTo1,addTo2,Cswi1,Cswi2,finalCswi1,finalCswi
2,canopylimits,canopylimits1,mean,mean1,
PETHS,MaxA,MinA,MaxR1,MaxR2,MinR1,MinR2,MaxSolar,MinSolar
,SunDeg,tempMA,tempmA,tempMR1,tempMR2,tempmR1,tempmR2,
    tempMS,tempmS,extrs,RA;
public FieldCanvas(){
super();
field = new Color(0,51,0);
this.setBackground(field);
// Create Infrared Sensors
sensor1 = new Sensors(40,398,20,10,0,0);
sensor2 = new Sensors(40,497,20,10,0,0);
sensor3 = new Sensors(65,447,20,10,0,0);
senrLens1 = new Sensors(46,401,5,3,0,0);
senrLens2 = new Sensors(46,500,5,3,0,0);
senrLens3 = new Sensors(71,450,5,3,0,0);
// Create Dripline Installation row 1
xfsDripLine1 = new WorkStations(183,360,325,3,0,0);
xfsDripLine1_2= new WorkStations(183,388,325,3,0,0);
xfsDripLine1_3= new WorkStations(183,416,325,3,0,0);
xfsDripLine1_4= new WorkStations(183,444,325,3,0,0);
xfsDripLine1_5= new WorkStations(183,360,3,84,0,0);
// Create Dripline Installation row 2
xfsDripLine1_6= new WorkStations(183,458,325,3,0,0);
xfsDripLine1_7= new WorkStations(183,486,325,3,0,0);
xfsDripLine1_8= new WorkStations(183,514,325,3,0,0);
xfsDripLine1_9= new WorkStations(183,542,325,3,0,0);
xfsDripLine1_10= new WorkStations(183,458,3,84,0,0);
// Pyranometer and Rain Sensor
plyoMeter = new Sensors(170,320,20,20,0,360);
rainSensor = new Sensors(190,575,20,20,0,360);
//Measurements, Controls and Calculations
```

```
dataLogger = new WorkStations(50,550,30,20,17,360);
computerBaseStation = new WorkStations(30,600,30,40,30,360);
controlBoard = new WorkStations(150,620,30,10,10,360);
// Water, Connections and Pipelines
waterPump = new WorkStations(650,605,40,40,90,270);
crossJunction = new WorkStations(720,430,30,25,0,360);
crossj_1 = new WorkStations(545,396, 20, 15, 0, 360);
crossj_2 = new WorkStations(545, 495, 20, 15, 0, 360);
pipeLine_1 = new WorkStations(690,620,45,8,0,0);
pipeLine_2 = new WorkStations(735,445,8,183,0,0);
pipeLine_3 = new WorkStations(680,440,55, 5,0,0);
pipeLine_4 = new WorkStations(680, 409, 5, 100,0,0);
pipeLine_5 = new WorkStations(627,405,58, 5,0,0);
pipeLine_6 = new WorkStations(627, 504, 53, 5,0,0);
pipeLine_7 = new WorkStations(540,403,10,3,0,0);
pipeLine_8 = new WorkStations(540,360,3,84,0,0);
pipeLine_9 = new WorkStations(508, 360, 35, 3,0,0);
pipeLine_10 = new WorkStations(508, 360, 35, 3,0,0);
pipeLine_11 = new WorkStations(508,388,35, 3,0,0);
pipeLine_12 = new WorkStations(508,416,35, 3,0,0);
pipeLine_13 = new WorkStations(508,444,35, 3,0,0);
pipeLine_14 = new WorkStations(540,502,10,3,0,0);
pipeLine_15 = new WorkStations(540,458,3,84,0,0);
pipeLine_16 = new WorkStations(508,458,35, 3,0,0);
pipeLine_17 = new WorkStations(508,486,35, 3,0,0);
pipeLine_18 = new WorkStations(508,514,35, 3,0,0);
pipeLine_19 = new WorkStations(508,542,35, 3,0,0);
// Create Selenoid Valves
valve1 = new WorkStations(620,398,15,15,0,360);
valve2 = new WorkStations(620,497,15,15,0,360);
// Create Bed for Crops
bedOfCrops = new FieldCanvas(170,350,400,200);
// Create Sun and Sun Movement
Thread sunMove = new Thread(new weatherPatterns());
sunMove.start();
Thread input = new Thread(new DeviceRreadings());
```

```
input.start();
Thread IrSenorRead = new Thread(new RetreiveIRreadings());
IrSenorRead.start();
Thread waterCrops = new Thread(new IrrigateCrops());
waterCrops.start();
}
public FieldCanvas(int x,int y,int sizex,int sizey){
    this.bedPositionx=x;
    this.bedPositiony=y;
    this.bedSizex=sizex;
    this.bedSizey= sizey;
}
// @Override
@Override
public void paintComponent(Graphics g){
super.paintComponent(g);
//Display Sun and Sun Temperature
SunDegrees(g);
DrawSun(g);
Rainny(g);
// Black Outline of Crops Bed along with Brown as the soil color.
bedOfCrops.DrawBed(g);
// Readings and Calculations
sensorReadings(g);
DisplayMeasurementReadings(g);
// Draw Infrared Sensors for recording temperature readings.
sensor1.DrawIRSensor(g);
sensor2.DrawIRSensor(g);
senrLens1.IRLens(g);
senrLens2.IRLens(g);
// Draw XFS Dripline Irrigation
xfsDripLine1.IrrigationLines1(g);
xfsDripLine1_2.IrrigationLines1(g);
xfsDripLine1_3.IrrigationLines1(g);
xfsDripLine1_4.IrrigationLines1(g);
xfsDripLine1_5.IrrigationLines1(g);
```

```
xfsDripLine1_6.IrrigationLines2(g);
xfsDripLine1_7.IrrigationLines2(g);
xfsDripLine1_8.IrrigationLines2(g);
xfsDripLine1_9.IrrigationLines2(g);
xfsDripLine1_10.IrrigationLines2(g);
// Draw Plyometer for radiation readings and Rain Sensor
plyoMeter.DrawPyranometer(g);
rainSensor.DrawRainSensor(g);
// Draw Controls and Recording Devices
dataLogger.DrawDatalogger(g);
computerBaseStation.DrawComputerBaseStation(g);
controlBoard.DrawControlBoard(g);
// Draw Connections and Pipelines
waterPump.DrawWaterPump(g);
crossJunction.DrawCrossJunction(g);
crossj_1.DrawCrossJunction(g);
crossj_2.DrawCrossJunction(g);
pipeLine_1.PipeLines3(g);
pipeLine_2.PipeLines3(g);
pipeLine_3.PipeLines3(g);
pipeLine_4.PipeLines3(g);
pipeLine_5.PipeLines3(g);
pipeLine_6.PipeLines3(g);
pipeLine_7.PipeLines1(g);
pipeLinc_8.PipeLines1(g);
pipeLine_9.PipeLines1(g);
pipeLine_10.PipeLines1(g);
pipeLine_11.PipeLines1(g);
pipeLine_12.PipeLines1(g);
pipeLine_13.PipeLines1(g);
pipeLine_14.PipeLines2(g);
pipeLine_15.PipeLines2(g);
pipeLine_16.PipeLines2(g);
pipeLine_17.PipeLines2(g);
pipeLine_18.PipeLines2(g);
pipeLine_19.PipeLines2(g);
```

```
// Draw Selenoid Valves
valve1.DrawValve1(g);
valve2.DrawValve2(g);
}
// This method display the Bed with Lettuce Crops Only
public void DrawBed(Graphics g1){
// Bed Outline in Black
g1.setColor(Color.BLACK);
g1.drawRect(getBedPositionx(), getBedPositiony(), getBedSizex()+3,
getBedSizey()+3);
// Crops Bed
Color Bed = new Color(51,25,0); // To acquired brown color for bed
g1.setColor(Bed);
g1.fillRect(getBedPositionx()+2, getBedPositiony()+2, getBedSizex(),
getBedSizey());
// Lettuce Crops
Color lettuce = new Color(102,255,102);
g1.setColor(lettuce);
g1.fillRect(190,369, 300,10);
g1.fillRect(190,398, 300,10);
g1.fillRect(190,427, 300,10);
g1.fillRect(190,468, 300,10);
g1.fillRect(190,497, 300,10);
g1.fillRect(190,527, 300,10);
}
// Draw Sun
public void DrawSun(Graphics g3){
degrees = String.valueOf(deg);
Color sun = new Color(255,255,102);
g3.setColor(sun);
g3.fillArc(750-move,20, 40, 40, 0, 360);
g3.drawString( degrees, 770-move, 20);
}
// Draw Rain Fall
public void Rainny (Graphics g11){
g11.setColor(rd);
```

```
if(move >=790 && move < 890){
g11.fillArc(100, 100, 10, 10, 0, 360);
g11.fillArc(150, 100, 10, 10, 0, 360);
g11.fillArc(200, 100, 10, 10, 0, 360);
g11.fillArc(250, 100, 10, 10, 0, 360);
g11.fillArc(300, 100, 10, 10, 0, 360);
g11.fillArc(350, 100, 10, 10, 0, 360);
g11.fillArc(400, 100, 10, 10, 0, 360);
g11.fillArc(450, 100, 10, 10, 0, 360);
g11.fillArc(500, 100, 10, 10, 0, 360);
g11.fillArc(550, 100, 10, 10, 0, 360);
g11.fillArc(600, 100, 10, 10, 0, 360);
g11.fillArc(650, 100, 10, 10, 0, 360);
g11.fillArc(100, 150, 10, 10, 0, 360);
g11.fillArc(150, 150, 10, 10, 0, 360);
g11.fillArc(200, 150, 10, 10, 0, 360);
g11.fillArc(250, 150, 10, 10, 0, 360);
g11.fillArc(300, 150, 10, 10, 0, 360);
g11.fillArc(350, 150, 10, 10, 0, 360);
g11.fillArc(400, 150, 10, 10, 0, 360);
g11.fillArc(450, 150, 10, 10, 0, 360);
g11.fillArc(500, 150, 10, 10, 0, 360);
g11.fillArc(550, 150, 10, 10, 0, 360);
g11.fillArc(600, 150, 10, 10, 0, 360);
g11.fillArc(650, 150, 10, 10, 0, 360);
g11.fillArc(100, 200, 10, 10, 0, 360);
g11.fillArc(150, 200, 10, 10, 0, 360);
g11.fillArc(200, 200, 10, 10, 0, 360);
g11.fillArc(250, 200, 10, 10, 0, 360);
g11.fillArc(300, 200, 10, 10, 0, 360);
g11.fillArc(350, 200, 10, 10, 0, 360);
g11.fillArc(400, 200, 10, 10, 0, 360);
g11.fillArc(450, 200, 10, 10, 0, 360);
g11.fillArc(500, 200, 10, 10, 0, 360);
g11.fillArc(550, 200, 10, 10, 0, 360);
g11.fillArc(600, 200, 10, 10, 0, 360);
```

```
g11.fillArc(650, 200, 10, 10, 0, 360);
    }
}
// showing the rise and fall in sun temperature
public void SunDegrees(Graphics g9){
        re = 33;
        re3 = 396;
        if (move == 1){
            deg = 23;
            extrs = 455.96;
            re2 = re2 + re;
        }if(move == re2){
            deg = deg + 1;
            extrs = extrs + 0.868;
            re2 = re2 + re;
        }
        if(move == re3){
            deg = deg -2;
            extrs = extrs - 0.868;
            re4= re3 + re;
        }
        if(move ==re4){
            deg = deg-2;
            extrs = extrs - 0.868;
            re4=re4+re;
        }
    }
/*
    String mo = String.valueOf(reach);
    g9.setColor(Color.BLACK);
    g9.drawString(mo,230,30);
 */
    }
// Draw Readings for Sensors
    public void sensorReadings (Graphics g3){
        g3.setColor(Color.YELLOW);
        g3.drawString(IRreadings,sensor1.getPositionx(),sensor1.
        getPositiony()-25);
```

```
        g3.drawString(IRreadings1,sensor2.getPositionx(),sensor2.
        getPositiony()-25);
        g3.drawString(IRreading2 +"Wm-2", plyoMeter.getPositionx(),
        plyoMeter.getPositiony()-25);
        }
// Displaying, Perform Calcuations and Printing Readings
public void DisplayMeasurementReadings(Graphics g10){
String Cswi1Read = String.valueOf(String.format("%.3f",Cswi1));
String Cswi2Read = String.valueOf(String.format("%.3f",Cswi2));
String PETHSRead = String.valueOf(String.format("%.3f",PETHS));
String SMaxA = String.format("%.3f",MaxA);
String SMaxR1 = String.format("%.3f",MaxR1);
String SMaxR2 = String.format("%.3f",MaxR2);
String SMaxSolar = String.format("%.3f",MaxSolar);
String SMinA = String.format("%.3f",MinA);
String SMinR1 = String.format("%.3f",MinR1);
String SMinR2 = String.format("%.3f",MinR2);
String SMinSolar = String.format("%.3f",MinSolar);
tempMA = SunDeg;
if(tempMA > MaxA){
MaxA = tempMA;
}
tempMR1 = addTo;
if(tempMR1 > MaxR1){
MaxR1 = tempMR1;
}

tempMR2 = addTo1;
if(tempMR2 > MaxR2){
MaxR2 = tempMR2;
}
tempMS = addTo2;
if(tempMS > MaxSolar){
MaxSolar = tempMS;
}
tempmA =SunDeg;
if((tempmA != 0.000) && (tempmA < MaxA)){
```

```
MinA = tempmA;
}
tempmR1 = addTo;
if((tempmR1 != 0.000) && (tempmR1 < MaxR1)){
MinR1 = tempmR1;
}
tempmR2 = addTo1;
if((tempmR2 != 0.000) && (tempmR2 < MaxR2)){
MinR2 = tempmR2;
}
tempmS = addTo2;
if((tempmS != 0.000) && (tempmS < MaxSolar)){
MinSolar = tempmS;
}
if(valve1.dv1.equals(new Color(255,128,0))) {
    valveOn1 = 1;
}else {
valveOn1 = 0;
}
if(valve2.dv2.equals(new Color(255,128,0))) {
    valveOn2 = 1;
}else {
valveOn2 = 0;
}
  Cswi1 = ((addTo - SunDeg) - 3.7544) / 5.0785;
  Cswi2 = ((addTo1 - SunDeg) - 3.7544) / 5.0785;
  RA = addTo2 / 2454000;
  RA = RA * 60 * 60 * 24;
  PETHS = 0.0023 * RA *(SunDeg - 17.8) * (MaxA - MinA);
  System.out.println(Cswi1Read+   "   "+Cswi2Read+"   "+SMaxA+"
  "+SMinA+" "+SMaxR1+" "+SMinR1+" "+SMaxR2
              +"   "+SMinR2+"   "+SMaxSolar+"   "+SMinSolar+"
              "+PETHSRead+" "+valveOn1+" "+valveOn2);
  g10.setColor(Color.ORANGE);
  g10.drawString("CSWI-1",295, 285);
  g10.drawString("CSWI-2",345, 285);
```

```java
    g10.drawString("PETHS", 395, 285);
    g10.drawString(Cswi1Read,300,300);
    g10.drawString(Cswi2Read,350,300);
    g10.drawString(PETHSRead,400,300);
}
public int getBedPositionx() {
    return bedPositionx;
}
/**
 * @param bedPositionx the bedPositionx to set
 */
public void setBedPositionx(int bedPositionx) {
    this.bedPositionx = bedPositionx;
}
/**
 * @return the bedPositiony
 */
public int getBedPositiony() {
        return bedPositiony;
}
/**
 * @param bedPositiony the bedPositiony to set
 */
public void setBedPositiony(int bedPositiony) {
    this.bedPositiony = bedPositiony;
}
/**
 * @return the bedSizex
 */
public int getBedSizex() {
    return bedSizex;
}
/**
 * @param bedSizex the bedSizex to set
 */
public void setBedSizex(int bedSizex) {
```

```
    this.bedSizex = bedSizex;
}
/**
* @return the betSizey
*/
public int getBedSizey() {
  return bedSizey;
}
/**
* @param betSizey the betSizey to set
*/
public void setBedSizey(int bedSizey) {
  this.bedSizey = bedSizey;
}
// Movement of the sun and rain fall display
class weatherPatterns implements Runnable{
  @Override
public void run(){

  int n = 0;
while(move < 890) {
  try{
    move = move +1;
    reach = reach +1;
    repaint();
    Thread.sleep(300);
  }
  catch (InterruptedException ex){
  }
  try{
    rainTemp =0;
    rainSensor.signalRain = Color.DARK_GRAY;
    if(move>=790 && move < 890){
    rainSensor.signalRain = Color.CYAN;
    rd = Color.gray;
    rainTemp = rainTemp +2 ;
```

```
        repaint();
        Thread.sleep(1000);
        rainSensor.signalRain = Color.DARK_GRAY;
        rd = Color.cyan;
        }
    }
    catch (InterruptedException ex){
    }
 }
}
}// End Of Thread For Weather Patterns
// Show blinking colors of readings being gathered
  class DeviceRreadings implements Runnable{
    @Override
    public void run(){
    while(move < 890) {
    try{
        sensor1.ir = Color.YELLOW;
        sensor2.ir = Color.YELLOW;
        sensor3.ir = Color.YELLOW;
        plyoMeter.solarInputs = new Color(102,0,51);
        repaint();
        Thread.sleep(1000);
        sensor1.ir = Color.RED;
        sensor2.ir = Color.RED;
        sensor3.ir = Color.RED;
        plyoMeter.solarInputs = new Color(204,0,102);
    }
    catch (InterruptedException ex){
    }
    try{
    mean1 = 0;
    mean = 0;
    dataLogger.inputs = new Color(102,51,0);
    computerBaseStation.dcbs = new Color(255,204,153);
    repaint();
```

```java
    Thread.sleep(1000);
    dataLogger.inputs = new Color(153,76,0);
    computerBaseStation.dcbs = new Color(255,153,153);
    }
    catch(InterruptedException ex1){
    }
  }
}
}// End Of Thread For Device Readings
// Display On Site Readings From Sensors
  class RetreiveIRreadings implements Runnable{
    @Override
    public void run(){
      double decPoint [] = {0.1,0.2,0.3,0.4,0.5,0.6,0.7,0.8,0.9};
    for (int count = 0; count < 1; count++){
      for(int i = 0 ; i <= decPoint.length; i++) {
      IRpoints = new Random(decPoint.length);
        }
      }
    while(move < 890) {
  try{
    if((deg >= -1) && (deg <= 33)){
  SunDeg = deg + IRpoints.nextDouble();
  addTo = deg + IRpoints.nextDouble() + 3 - canopylimits - rainTemp;
  addTo1 = deg + IRpoints.nextDouble() + 3 - canopylimits1 - rainTemp;
  addTo2 = extrs + IRpoints.nextDouble() + 2 - rainTemp;
  IRreadings = String.valueOf(String.format("%.1f",addTo));
  IRreadings1 = String.valueOf(String.format("%.1f",addTo1));
  IRreading2 = String.valueOf(String.format("%.1f",addTo2));
    }
      Thread.sleep(1000);
    }
    catch (InterruptedException ex){
    }
  }
 }
}
```

```
}// End Of Thread For Retrevie Infrared Readings
// Peforming Calculations In Order To Irrigate The Lettuce Crops At The
Righ Parameters.
  class IrrigateCrops implements Runnable{
    @Override
    public void run(){
    while(move < 890) {
      try{
          canopylimits =0;
          canopylimits1 =0;
          if((Cswi1>=-0.1673) && (addTo >= 25)){
          waterPump.setArcy(360);
          controlBoard.DCC = Color.lightGray;
          valve1.dv1 = new Color(255,128,0);
          crossJunction.cros_j = new Color(51,153,255);
          crossj_1.cros_j = new Color(51,153,255);
          pipeLine_7.pipclinc1 = new Color (51,153,255);
          xfsDripLine1.pl1= new Color (204,255,255);
          canopylimits = canopylimits + 3;
          }
          else if((Cswi2>=-0.1274) && (addTo1 >= 25)){
          waterPump.setArcy(360);
          controlBoard.DCC = Color.lightGray;
          valve2.dv2 = new Color(255,128,0);
          crossJunction.cros_j = new Color(51,153,255);
          crossj_2.cros_j = new Color(51,153,255);
          pipeLine_14.pipeline2 = new Color (51,153,255);
          xfsDripLine1_6.pl2= new Color (204,255,255);
          canopylimits1 = canopylimits1 + 3;
          }
          else{
          waterPump.setArcy(270);
          controlBoard.DCC = Color.DARK_GRAY;
          valve1.dv1 = new Color(51,0,0);
          valve2.dv2 = new Color(51,0,0);
          crossJunction.cros_j = new Color(51,153,255);
```

```java
                    crossj_1.cros_j = Color.WHITE;
                    crossj_2.cros_j = Color.WHITE;
                    pipeLine_14.pipeline2 = Color.WHITE;
                    xfsDripLine1_6.pl2 = new Color (255,128,0);
                    pipeLine_7.pipeline1 = Color.WHITE;
                    xfsDripLine1.pl1 = new Color (255,128,0);
                    }
                    Thread.sleep(1000);
                }
            catch (InterruptedException ex){
            }
        }
    }
}// End Of Thread For Irrigate Crops
    }
import java.awt.Color;
import java.awt.Graphics;
/**
*@author Kevin Rodriguez
* SRN: 101035635
* Date Created: 10/1/2015
* Date Modified: 5/12/2015
* Description: The WorkStation Class is a methods class to call computing
    and controls devices as well as hardware components for the irrigation
    system.
* In addition to calling variables for Color in order to show that the device
    is functioning, recording measurements and performing controls
* by blinking colors. This also helps minimize excessive coding.
*/
    public class WorkStations {
    private int positionx;
    private int positiony;
    private int sizex;
    private int sizey;
    private int arcx;
    private int arcy;
```

```java
public Color inputs,pl1,pl2,dwp,dcbs,pipeline1,pipeline2,pipeline3,cros_j,
    dv1,dv2,DCC;
public WorkStations(int x, int y, int sizex, int sizey, int arcx, int arcy){
    this.positionx=x;
    this.positiony=y;
    this.sizex=sizex;
    this.sizey=sizey;
    this.arcx=arcx;
    this.arcy=arcy;
}
// Method to Draw Pyranometer to measure solar radiation
public void DrawDatalogger(Graphics g3){
    g3.setColor(inputs);
    g3.fillRoundRect(getPositionx(), getPositiony(), getSizex(), getSizey(),
    getArcx(),getArcy());
    g3.setColor(Color.BLACK);
    g3.drawString("Datalogger",getPositionx()+10, getPositiony()+40);
}
    // Method to Draw Draw Computer Base Station
public void DrawComputerBaseStation(Graphics g4){
g4.setColor(dcbs);
g4.fillRoundRect(getPositionx(), getPositiony(), getSizex(), getSizey(),
getArcx(),getArcy());
// Label Computer
g4.setColor(Color.BLACK);
g4.drawString("ComputerStation",getPositionx(), getPositiony()+50);
// Connection from computer to water pump
g4.setColor(Color.BLACK);
g4.drawLine(52,570, 52, 600);
}
// Method to Draw DC Controller Board (SDM-CD8S) to irrigat crops
public void DrawControlBoard(Graphics g5){
g5.setColor(DCC);
g5.fillRoundRect(getPositionx(), getPositiony(), getSizex(), getSizey(),
getArcx(),getArcy());
g5.setColor(Color.BLACK);
```

```
g5.drawString("DC Controller Board",getPositionx(), getPositiony()-10);
// Connection From DC Controller Board To Data Logger
g5.drawLine(140, 620, 150, 620);
g5.drawLine(140, 571, 140, 620);
g5.drawLine(77, 571, 140, 571);
}
// Method to Draw Water Pump to show the activation of the pump
public void DrawWaterPump(Graphics g6){
//Water Pump
dwp = new Color(0,0,102);
g6.setColor(dwp);
g6.fillArc(getPositionx(), getPositiony(), getSizex(), getSizey(),getArcx
(),getArcy());
g6.setColor(Color.BLACK);
g6.drawString("Water Pump", getPositionx()+10, getPositiony()-10);
}
// Method to Draw Cross Junction to show the division of water flow
public void DrawCrossJunction(Graphics g7){
   //cros_j = Color.WHITE;
   g7.setColor(cros_j);
   g7.fillArc(getPositionx(), getPositiony(), getSizex(), getSizey(),getArcx(),
   getArcy());
   //Labe Cross Junction
   g7.setColor(Color.BLACK);
   g7.drawString("Cross Junction", getPositionx()-30, getPositiony()-10);
}
// Method to Draw PipeLines Connections
public void PipeLines1(Graphics g12){
//pipeline = Color.WHITE;
g12.setColor(pipeline1);
g12.fillRect(getPositionx(), getPositiony(), getSizex(), getSizey());
}
// Method to Draw PipeLines Connections
public void PipeLines2(Graphics g13){
//pipeline2 = Color.WHITE;
g13.setColor(pipeline2);
```

```
g13.fillRect(getPositionx(), getPositiony(), getSizex(), getSizey());
}
// Method to Draw PipeLines Connections
public void PipeLines3(Graphics g14){
pipeline3 = new Color (51,153,255);
g14.setColor(pipeline3);
g14.fillRect(getPositionx(), getPositiony(), getSizex(), getSizey());
}
// Method to Draw Electronic Solenoid Valve1 to show the activation of
valve1
public void DrawValve1(Graphics g8){
  //dv1 = new Color(51,0,0);
  g8.setColor(dv1);
  g8.fillArc(getPositionx(), getPositiony(), getSizex(), getSizey(),getAr
  cx(),getArcy());
  g8.setColor(Color.BLACK);
  g8.drawString("Valve 1", getPositionx(), getPositiony()-10);
  //Connection from valve one to cross junction
  g8.setColor(dv1);
  g8.fillRect(565,403,60,3);
  // Connection From DC Controller to selenoid valve 1
  g8.setColor(Color.BLACK);
  g8.drawLine(180, 630, 410, 630);
  g8.drawLine(410, 600, 410, 630);
  g8.drawLine(410,600,650,600);
  g8.drawLine(650, 410, 650, 600);
  g8.drawLine(632, 410, 650, 410);
}
// Method to Draw Electronic Solenoid Valve2 to show the activation of
valve2
public void DrawValve2(Graphics g9){
  // dv2 = new Color(51,0,0);
  g9.setColor(dv2);
  g9.fillArc(getPositionx(), getPositiony(), getSizex(), getSizey(),getAr
  cx(),getArcy());
  g9.setColor(Color.BLACK);
```

```
    g9.drawString("Valve 2", getPositionx(), getPositiony()-10);
    //Connection form valve two to cross junction
    g9.setColor(dv2);
    g9.fillRect(565,502,60,3);
    //Connection from DC Controller Board to Selenoid Valve 2
    g9.setColor(Color.BLACK);
    g9.drawLine(180, 620, 400, 620);
    g9.drawLine(400, 590, 400, 620);
    g9.drawLine(400,590,625,590);
    g9.drawLine(625, 512, 625, 590);
    }
// Method to Draw XFS subsurface drip line hose for the first three lanes
of lettuce
public void IrrigationLines1(Graphics g10){
    //Drip Lines Irrigation Hose
    g10.setColor(pl1);
    g10.fillRect(getPositionx(), getPositiony(), getSizex(), getSizey());
    }
// Method to Draw XFS subsurface drip line hose for the second three
lanes of lettuce
public void IrrigationLines2(Graphics g11){
    //Drip Lines Irrigation Hose
    // pl2 = new Color (255,128,0);
    g11.setColor(pl2);
    g11.fillRect(getPositionx(), getPositiony(), getSizex(), getSizey());
    }
/**
    * @return the positionx
    */
public int getPositionx() {
    return positionx;
    }
/**
    * @param positionx the positionx to set
    */
public void setPositionx(int positionx) {
```

```
    this.positionx = positionx;
}
/**
 * @return the positiony
 */
public int getPositiony() {
    return positiony;
}
/**
* @param positiony the positiony to set
*/
public void setPositiony(int positiony) {
    this.positiony = positiony;
}
/**
* @return the sizex
*/
public int getSizex() {
    return sizex;
}
/**
* @param sizex the sizex to set
*/
public void setSizex(int sizex) {
    this.sizex = sizex;
}
/**
* @return the sizey
*/
public int getSizey() {
    return sizey;
}
/**
* @param sizey the sizey to set
*/
public void setSizey(int sizey) {
```

```
    this.sizey = sizey;
}
/**
* @return the arcx
*/
public int getArcx() {
    return arcx;
}
/**
* @param arcx the arcx to set
*/
public void setArcx(int arcx) {
    this.arcx = arcx;
}
/**
* @return the arcy
*/
public int getArcy() {
    return arcy;
}
/**
* @param arcy the arcy to set
*/
public void setArcy(int arcy) {
    this.arcy = arcy;
}
}
import java.awt.Color;
import java.awt.Graphics;
import java.util.Random;
/**
*@author Kevin Rodriguez
* SRN: 101035635
* Date Created: 10/1/2015
* Date Modified: 5/12/2015
* Description: The Sensors class is a methods class as well to call the sen-
sors that takes measurements and detects rain falls.
```

```
* In addition to calling variables for Color in order to show that the device
is functioning, reading measurements from the sensors
* by blinking colors. This also helps minimize excessive coding.
*/
public class Sensors {
  private int positionx;
  private int positiony;
  private int sizex;
  private int sizey;
  private int arcx;
  private int arcy;
  public Double addTo;
  public String IRreadings;
  public Color ir,solarInputs,signalRain;
  public Random IRpoints;
  public Sensors(){
  }
  public Sensors(int x, int y, int sizex, int sizey, int arcx, int arcy){
    this.positionx=x;
    this.positiony=y;
    this.sizex=sizex;
    this.sizey=sizey;
    this.arcx = arcx;
    this.arcy = arcy;
  }
  public void DrawIRSensor(Graphics g){
  g.setColor(ir);
  g.fillRect(getPositionx(), getPositiony(), getSizex(), getSizey());
  g.setColor(Color.BLACK);
  g.drawString("IR Sensor",getPositionx(), getPositiony()-10);
    //Sensor 1
  g.drawLine(20, 408, 40, 408);
  g.drawLine(20, 408, 20, 560);
  g.drawLine(20, 560, 50, 560);
  //Sensor 3
  g.drawLine(55, 507, 55, 550);
  }
```

```java
public void IRLens(Graphics g2){
Color lens = new Color(192,192,192);
g2.setColor(lens);
g2.fillRect(getPositionx(), getPositiony(), getSizex(), getSizey());
}
// Method to Draw Pyranometer to measure solar radiation
public void DrawPyranometer (Graphics g1){
// Pyranometer
g1.setColor(solarInputs);
g1.fillArc(getPositionx(), getPositiony(), getSizex(), getSizey(),getArcx(),
getArcy());
g1.setColor(Color.BLACK);
g1.drawString("Pyranometer",getPositionx(), getPositiony()-10);
g1.setColor(Color.yellow);
g1.fillArc(getPositionx()+7,        getPositiony()+7,        getSizex()-15,
getSizey()-15,getArcx(),getArcy());
//Pyranometer Connection
g1.setColor(Color.BLACK);
g1.drawLine(170, 325, 125,325);
g1.drawLine(125, 325, 125, 550);
g1.drawLine(77,550,125,550);
}
// Method to Draw Rain Ssensor to detect rain fall
public void DrawRainSensor(Graphics g2){
  g2.setColor(signalRain);
  g2.fillArc(getPositionx(), getPositiony(), getSizex(), getSizey(),getArcx(),
  getArcy());
  g2.setColor(Color.BLACK);
  g2.drawString("RainSensor",getPositionx(), getPositiony()-10);
  g2.setColor(Color.BLUE);
  g2.fillArc(getPositionx()+7,        getPositiony()+7,        getSizex()-15,
  getSizey()-15,getArcx(),getArcy());
  //Rain Sensor Connections
  g2.setColor(Color.BLACK);
  g2.drawLine(155, 575, 195, 575);
  g2.drawLine(155, 568, 155, 575);
```

```java
    g2.drawLine(77, 568, 155, 568);
}
/**
* @return the positionx
*/
public int getPositionx() {
   return positionx;
}
/**
* @param positionx the positionx to set
*/
public void setPositionx(int positionx) {
   this.positionx = positionx;
}
/**
* @return the positiony
*/
public int getPositiony() {
   return positiony;
}
/**
* @param positiony the positiony to set
*/
public void setPositiony(int positiony) {
   this.positiony = positiony;
}
/**
* @return the sizex
*/
public int getSizex() {
   return sizex;
}
/**
* @param sizex the sizex to set
*/
public void setSizex(int sizex) {
```

```java
    this.sizex = sizex;
}
/**
 * @return the sizey
 */
public int getSizey() {
    return sizey;
}
/**
 * @param sizey the sizey to set
 */
public void setSizey(int sizey) {
    this.sizey = sizey;
}
/**
 * @return the arcx
 */
public int getArcx() {
    return arcx;
}
/**
 * @param arcx the arcx to set
 */
public void setArcx(int arcx) {
    this.arcx = arcx;
}
/**
 * @return the arcy
 */
public int getArcy() {
    return arcy;
}
/**
 * @param arcy the arcy to set
 */
public void setArcy(int arcy) {
```

```
    this.arcy = arcy;
}
/**
* @return the ir
*/
}
```

CHAPTER 8

SOLAR PHOTOVOLTAIC POWERED MICRO IRRIGATION SYSTEM IN AEROBIC RICE CULTIVATION

MANOJ K. GHOSAL

Professor, Department of Farm Machinery and Power, College of Agricultural Engineering and Technology, Orissa University of Agriculture and Technology, Bhubaneswar-751003, Odisha, India, E-mail: mkghosal1@rdiffmail.com

CONTENTS

In this chapter: 1 US$ = 60.00 Rs.

8.1   INTRODUCTION

Sustainability of rice production in the present context of fast growing population is a big challenge before all of us with respect to achieving food security and controlling over the increased concerns of water scarcity, energy crisis and global warming due to the emission of greenhouse gases through anthropogenic activities particularly in the agricultural sector. In order to attain self sufficiency in food grain, the production and productivity of rice crop alone play a crucial role for India, Asia and the world as well, as rice is the principal food of more than 60% of the world's population and around 90% of the rice area worldwide is in Asia and low land rice fields produce about 75% of the world's rice supply. Agriculture too in Odisha to a considerable extent means growing rice. It is a staple food for almost entire population of Odisha and therefore, the state economy is directly linked with the improvements in production and productivity of rice. Rice in the state covers about 70% of the cultivated area and is grown in an area of 4.0 million hectares covering 65% rainfed and 35% irrigated areas. Rice eco-system in our state basically comprises of rainfed upland (19.1%), irrigated kharif (27.4%), medium land (12.4 %), shallow low land (22.5 %), semi deep (7.9%), deep (3.4 %) and irrigated rabi (7.3%). The state has three systems of rice cultivation; dry, semi-dry and wet.

The dry system accounts for about 18% of the total rice area and the rest is shared by semi-dry and wet systems. Kharif rice is grown in about 93% of the total area under rice cultivation in which up land, medium land and low land constitute 20%, 40% and 40%, respectively. Rainfed areas mostly in kharif season are vulnerable to the vagaries of monsoon due to erratic, inadequate and sometimes continuous heavy rainfall along with the frequent occurrence of natural calamities like flood, cyclone, dry spells, moisture stress, and drought resulting into low and fluctuating yield of rice in our state. Rice production practices mostly followed in the state are through wet and low land cultivation, covering about 80% of the total rice area and the methods of cultivation arc usually transplanting of seedlings and broadcasting of sprouted seeds in the puddle soil. A newly coming up method of rice cultivation is the system of rice intensification (SRI), which is though now in immature stage, but slowly gaining momentum among the farming community of our state due to the strong initiatives being taken by the government through the implementation of centrally sponsored schemes like *National Food Security Mission* (NFSM) and bringing green revolution in eastern India (BGREI) to meet the problems of water stress and the poor as well as fluctuating yield from the rice crop. Though SRI method of cultivation consumes about 25–30% of less water compared to traditional flooded condition, yet the channel around the crop bed needs to be flooded with water regularly in order to keep the soil around the plant wet and saturated all the time.

Labor requirement is almost same as the usual traditional practice due to preparation of good seedbed, raising mat type nursery, transplanting of uprooted seedlings and intensive intercultural operation etc. Labor availability during peak period of transplanting is also a major concern. Hence, in the above three methods of cultivation, the field remains either in fully or partially flooded conditions most of the time of the growing season, creating favorable environment for emission of methane, which is one of the principal and potent greenhouse gases relative to global warming potential of 25 times higher than that of CO_2 and accounts for one-third of the current global warming phenomenon. Rice cultivation is a major source of atmospheric CH_4 and contributes about 10–20% of total methane emission to the atmosphere. The environmental experts have now recognized and expressed in *Intergovernmental Panel on Climate Change* (IPCC, 2010) that the submerged rice fields are the most significant contributors of atmospheric methane. According to the current estimate, the production of rice needs to be enhanced by around 70% to meet the demands for ever-increasing population by 2030 and thus making rice cultivation, a potential major cause of growing atmospheric methane.

The above three methods of rice cultivation also require a considerable amount of water during growing season. It is estimated that on an average 4000 L of water is needed to produce 1 kg of rice under conventional rice production system. Although it varies from 3000 to 5000 L per kg of yield including transpiration, evaporation, percolation and seepage. Transpiration alone requires about 500–1000 L of water per kg of rice. Evapotranspiration requires an average of 1500 L/kg (ranging from 1000–2000 L/kg), which is about same as for other cereals such as wheat, maize etc. Water crisis is the greatest threat to crop cultivation, due attention needs to be focused on its conservation and judicious use in the agricultural sector. Inadequate rainfall, depletion of groundwater and misuse of water in agriculture have brought down the per capita availability of water by 40–60% in many Asian countries including India. The share of water allocated to irrigation is also likely to decrease by 15–20% in the near future due to shrinking water resources and competition from other sectors like domestic consumption, industries, municipality, tourism, recreation, environment etc. Hence, effective management and utilization of water is of today's paramount importance and is high time now to consider more crops per drop of water.

Poor yield from kharif paddy due to some unavoidable and uncontrollable reasons as mentioned earlier and less coverage of area during rabi season (about 7% of total rice area in Odisha) due to lack of assured irrigation facilities have aggravated the concerns of achieving targeted production in the next decade along with environmental protection and have forced the practitioners, planners as well as policy makers to go for a novel technology relating mostly to a water saving approach for rice production. The losses occurring regularly from kharif rice cannot be compensated unless more area is covered during rabi season and which may not be possible in the near future due to uncertainties and requirement of huge expenditures for achieving required level of irrigation potential in the state of Odisha. Hence, water saving technology may be a viable proposition for rice production system in order to cover more area under rice cultivation particularly during rabi season with the available water resources. Increasing cost of labor and restricted supplies of irrigation water would definitely force the practitioner to search for an alternative in shifting from flooded to non-flooded system of rice cultivation resulting into transition from anaerobic to aerobic soil conditions.

Aerobic rice production system may be a such promising approach, which needs to be introduced and promoted in water-deficient regions and would be a viable adaptation strategies for farmers who want to continue growing rice under water-short environment, especially in rabi season. By the way, it would be possible to sustain self-sufficiency by stabilizing production and productivity with the uncertainty in required yield in kharif rice and increasing more area during rabi season by following precise water management practices with the available assured water supply.

Aerobic rice cultivation, where fields remain unsaturated or nearer to saturation throughout the season like an upland crop, offers an opportunity to produce rice with less water. Aerobic and upland rice are both grown under aerobic conditions, however the former is under controlled water management system but latter is not. Aerobic rice is a new cropping system in which specially developed varieties are directly seeded in well-drained, un-puddled and unsaturated soil conditions for most of the crop-growing period. The rice is grown like an upland crop with adequate inputs and supplementary irrigation when rainfall is insufficient. Aerobic rice varieties are bred by combining the draft resistance of upland varieties with high yielding characteristics of low land cultivars. Some of the suitable varieties like CRDhan-200 (pyari), Naveen and Annada are also available in our state Odisha. Aerobic rice is therefore grown with the use of external inputs such as supplementary irrigation and fertilizers with an aim to maintain the productivity at par with the traditional production system without compromise with yield decline. Also due to unsaturated soil condition, there is the possibility that aerobic rice emits lesser methane gas into the atmosphere, thus keeping environment safe besides saving water. Studies show that water saving would be in the range of 35–40% by curtailing conventional irrigation water of 1700–1800 ha-mm to 1000–1200 ha-mm in aerobic rice. Water use efficiency and water productivity would further be improved in aerobic rice cultivation if micro irrigation system be incorporated with the conventional irrigation practices.

The micro irrigation in general and drip irrigation in particular has received considerable attention from researchers, policy makers, economists etc. for its perceived ability to contribute significantly to ground water resources development, agricultural productivity, economic growth and environmental sustainability. Also to sustain the aerobic rice cultivation, micro irrigation may be adopted which contributes to the judicious application of water resulting into its optimum savings in terms of water use efficiency and water productivity. Efficient and precise fertilizer utilization through fertigation and saving of energy (electric/fossil fuels) by the decrease in annual working hours of pump sets with reduced water consumption in drip irrigation are some of the other advantages in addition to precise water management of a precious input commodity in crop cultivation. However, ground water is the only source of water being used for drip method of irrigation in India and is mostly pumped through electrically and diesel/petrol/kerosene fuels operated pump sets. The fossil fuels operated water lifting devices (pump sets) are not only polluting the environment but also becoming very costly day by day and beyond the capacity of the majority of marginal and small farmers of the state for use in irrigation purposes. It is worthwhile to mention here that the installation of shallow tube well (56%) followed by deep tube well (31%) and dug well (13%) by the individual farmer during 2011–12 under subsidy scheme of Government of Odisha favors the growth of more diesel/gasoline operated pump sets in the state and may increase tremendously in the near future due to strong initiatives being taken by the Government for assured irrigation. Similarly, in Odisha only 1.7 % of the total electricity consumption was used for irrigation and agricultural purposes in 2011–12 compared to national average of 24.97%. Moreover, electricity consumption in agricultural sector shows a declining trend both in terms of absolute units as well as its share in total electricity consumption between 1992–93 (5.6%)

and 2011–12 (1.7%). The decline in electricity consumption is in-spite of the fact that about 81 % of villages are already electrified during 2011–12. This clearly indicates the inefficiency in transmission and distribution of electricity for agricultural purposes. The dependence of energy both through fossil fuels and grid connected electricity, may therefore, not be a viable solution for increasing mechanized water lifting devices to sustain assured irrigation in rural areas.

Currently, our state has around 1.38×10^5 (35%) grid based (electric) and 2.47×10^5 (65%) diesel irrigation pump sets. However, erratic grid supply of electricity and high cost of diesel pumping continue to remain as a big problem for the farmers. The growing mismatch between demand and supply of energy and electricity in particular is posing challenges especially to the farmers in remote areas. Developing a grid system is often too expensive, because rural villages are generally located too far away from the existing grid lines. In addition, the electric tariff is increasing in every year and thus increasing the cost of water pumping operation. Further, the increasing unreliability of monsoon rains is forcing farmers to look at alternate fuels such as diesel for operating irrigation pump sets. Even if, petroleum based fuels are available in the nearby towns, transporting them to remote areas is also very difficult. There are no roads and supporting infrastructure in many remote villages of the state. The rising cost of using diesel for powering irrigation pump sets is often beyond the means of small and marginal farmers. Consequently, the lack of required amount of water often leads to poor growth of plants, thereby, reducing yields and income. Hence, use of conventional diesel/gasoline powered pumping systems poses an economic risk to the farmers. Scientific studies reveal that timely and required amount of water availability for the crop favors the increase in the yield by 10–15%. The burning of petroleum fuels also creates threats by polluting environment and causing global warming by releasing a considerable amount of CO_2 into the atmosphere. The continuous exhaustion of limited stocks of conventional energy sources and their environmental impacts have, therefore, forced researchers, planners, and policy makers to search for the reliable, environment-friendly and cost effective energy resources to power water pumping system in a sustainable manner. Hence to keep pace with the growing demands of energy and acute shortage of water, solar photovoltaic water pumping device may be integrated with drip irrigation system for rice cultivation in addressing the issues of food security under climate change scenario.

The use of solar photovoltaic systems may provide adequate solution not only for all energy related problems of the present agrarian society but also perform excellently in terms of productivity, reliability, sustainability and environmental protection ability. With the cost of solar PV system falling steadily and price of diesel/gasoline soaring, solar PV pumping may be an appropriate and economically feasible option. In order to make the system compact, the required number of PV modules for the set up, can be made foldable when not in use and be stretched when in use. The device may be mounted over a lightweight trolley rickshaw to make it portable and thus eliminating the fear of theft as against of permanent installation of the unit outside. Thus an integrated approach of water saving technology and reliable source of energy along with the reduced greenhouse gasses emissions from the conventional method of rice cultivation has been thought up in the proposed study to achieve food security of the nation.

The objectives of the study were therefore as follows;

  a. Developing a portable and affordable solar photovoltaic powered water-pumping device integrated with drip irrigation system for aerobic rice cultivation especially during rabi season.
  b. Feasibility and performance study of solar photovoltaic powered drip irrigation system in aerobic rice cultivation.
  c. Comparative study of methane emission in anaerobic and aerobic rice field conditions.
  d. Techno-economic study of the system developed.

8.2 DIFFERENCES BETWEEN AEROBIC AND UPLAND RICE

Upland rice is grown in rainfed and naturally well-drained soils that are usually on sloping land with erosion problems, drought-prone, and poor in physical and chemical properties. Upland rice varieties are low-yielding but drought and low fertility- tolerant, thus giving low but stable yields under the adverse environmental conditions of uplands. However, high levels of fertilizer application and supplemental irrigation to upland rice lead to lodging and thus reduce yield.

Aerobic rice is targeted at more favorable environments where land is flat or terraced, and soil can be frequently brought to water field capacity by rainfall or supplemental irrigation, or where land is sloping but frequent rainfall can keep soils moist throughout the growing season. Aerobic rice can be a replacement of lowland rice wherever available water is insufficient for lowland rice but sufficient for aerobic rice. Both aerobic and upland rice are adapted to aerobic soil conditions, but aerobic rice varieties are more input-responsive and higher yielding than traditional upland ones.

8.3 REVIEW OF RESEARCH ON SOLAR PHOTOVOLTAIC WATER PUMPING SYSTEM IN RICE CULTIVATION

To combat against the present scenario of water stress and energy shortage particularly in the agricultural sector and to reduce methane emission from the conventional method of rice cultivation in the submerged fields, a suitable technology is the need of the hour in achieving food security for the burgeoning human population in the world. Literatures pertaining to the improved water-use efficiency measures, secured supply of energy and mitigation attempts for methane emission in rice cultivation have been referred to evolve a viable technology of addressing the above issues in one approach for the rice farming community under the climate change scenario. Studies on solar powered drip irrigation system in aerobic rice cultivation are very scarce. However, micro-irrigation, particularly drip irrigation has been successful in Rajasthan under RKVY, NHM and NMMI schemes in which subsidy facilities have been provided by MNRE for installation of about 15,000 solar water pumps in the state. But those solar water pumps have not been used for rice cultivation. Hence, use of solar pump for drip irrigation mostly in aerobic method of rice cultivation may be a feasible proposition in agricultural sector. The following literatures provide an insight to move forward in assessing the feasibility and viability of a solar photovoltaic power micro irrigation system in aerobic rice cultivation in order to face the present day's challenges of water scarcity, energy deficiency and food security of our country.

Epule et al. [4] discussed various mechanisms through which methane emissions can be reduced in rice fields and at the same time rice productivity can be maintained to achieve win-win sustainability. The main methodological concept for win-win sustainability describes a scenario in which development and environmental protection go hand in hand. The approaches for their study were that the current production of rice should not be done at the expense of wide scale methane emissions. On the other hand, the desire to reduce methane emission should not be at the expense of rice production. The activities of methane producing bacteria go a long way to set the stage for methanogenesis. Several mechanisms such as water management, cultivar selection, soil additives and soil amendments were suggested to reduce methane emission without compromising the yield decline. Most of these methods are 90% efficient but have a 10% uncertainty. The study therefore recommends that a synergetic approach in which a combination of these techniques is used for obtaining stabilized yield in rice cultivation.

Alberto et al. [1] investigated the seasonal and annual variability of sensible heat flux (H), latent heat flux (LE), evapotranspiration (ET), crop coefficient (Kc) and crop water productivity (WPET) under two different rice environments, flooded and aerobic soil conditions, using the eddy covariance (EC) technique during 2008–2009 cropping periods. Average for four cropping seasons, flooded rice fields had 19% more LE than aerobic fields whereas aerobic rice fields had 45% more H than flooded fields. This resulted in a lower Bowen ratio in flooded fields (0.14 ± 0.03) than in aerobic fields (0.24 ± 0.01). The results of this investigation showed significant differences in energy balance and Evapotranspiration between flooded and aerobic rice ecosystems. Aerobic rice is one of the promising water-saving technologies being developed to lower the water requirements of the rice crop to address the issues of water scarcity. This information should be taken into consideration in evaluating alternative water-saving technologies for environmentally sustainable rice production systems.

Kato et al. [9] studied and compared the potential productivity of aerobic rice and flooded rice using high-yielding varieties at two locations in Japan in two successive years. In aerobic fields, the total amount of water supplied (irrigation plus rainfall) was 800–1300 mm. The soil water potential at 20-cm depth averaged between 15 and 30 kPa each growing season, but frequently reached 60 kPa. The average yield under aerobic conditions was similar to or even higher than that achieved with flooded conditions (7.9 t/ha in 2007 and 9.4 t/ha in 2008 for aerobic versus 8.2 t/ha for flooded). The average water productivity under aerobic conditions was 0.8–1.0 kg grain/m^3 water, slightly higher than common values in the literature. The super-high-yielding cultivar Takanari achieved yields greater than 10 t/ha with no yield penalty under aerobic conditions in 3 out of 4 experiments. They have concluded that the high-productivity rice cultivation in aerobic soil is a promising technology for water conservation. With continued breeding, future aerobic rice varieties will possess large numbers of spikelets and sufficient adaptation to aerobic conditions such that they will consistently achieve yields comparable to the potential yield of flooded rice.

Katsura and Nakaide [10] investigated the factors that determine grain weight in aerobic culture. They grew four rice varieties in non-puddled, unsaturated (aerobic) soils with a soil water potential at 20-cm depth kept above 60 kPa and in continuously flooded culture in two years. They found a significant variety × water interaction in grain weight in 2009; weights under aerobic culture were 6% and 13% larger than under flooded culture in Sasanishiki and IRAT109, respectively, versus 4% and 10% smaller in Habataki and Takanari. There was no significant variety × water interaction in grain weight in 2010. However, an excessive increase in spikelet number per unit area in Takanari under aerobic culture in 2009 reduced the source capacity per spikelet and single husk size, decreasing grain weight. In 2010, frequent soil drying under aerobic culture during the late reproductive period (around 20 days preceding heading) reduced single husk size, thereby decreasing grain weight. They found that sink activity and source capacity per plant could be both higher under aerobic culture during the ripening period, producing larger grain weight at a soil water potential above 40 kPa at a 20-cm depth relative to those under flooded culture. In contrast, greater drying under aerobic culture during the late reproductive period reduced single husk size, thereby reducing grain weight.

Meah et al. [12] studied on solar photovoltaic water pumping for remote locations in rural western US. They realized that *solar photovoltaic water pumping system* (SPVWPS) is a cost effective and environmental friendly way to pump water in remote locations where 24 hours electrical service is not necessary and maintenance is an issue. From their survey, it was indicated that a total of 88 number of solar photovoltaic water-pumping systems are being installed in all 23 countries of the United States, of which 75 systems are in operation till 2005. They have observed that drought affected areas like Wyoming, Montana, Idaho, Washington, Oregon and part of Texas could use solar photovoltaic water pumping systems to improve the water supply to livestock in remote locations. They have convinced that successful demonstration of these systems is encouraging other ranchers to try this relatively new technology as another viable water supply option. They concluded that SPVWPS had excellent performance in terms of productivity, reliability and cost effectiveness and the system could reduce the CO_2 emission considerably over its 25-year life span.

Kelley et al. [11] studied on the feasibility of solar powered irrigation in the United States. They developed a method for determining the technical and economic feasibility of photovoltaic power irrigation systems applicable to any geographic location and crop type in USA and applied it to several example cases. According to the opinion of authors, the results of technical feasibility analysis agreed with the results obtained from past studies and also showed that there is no technological barrier to implementation of PVP irrigation if land is available for installation of solar panels. The results of economic feasibility study suggested that the price of diesel has increased sufficiently within the last ten years to make PVP irrigation economically feasible, despite the high capital costs of photovoltaic systems. The authors concluded that as the price of the solar panels decreased, the capital costs would decrease making PVP systems even more economically attractive.

Hamidat et al. [7] studied on Small-scale irrigation with photovoltaic water pumping system in Sahara regions. The authors have developed a mathematical program to test the performance of photovoltaic arrays under Saharan climatic condition. Their work showed that it is possible to use a photovoltaic water pumping system for low heads for small scale-irrigation of crops in Algerian Sahara regions. Thus, the photovoltaic (PV) water pumping system could easily cover the daily water need rates for small- scale irrigation with an area smaller than 2 ha. They also concluded that the photovoltaic pumping system (PVPS) could improve the living condition of the farmer with the development of local farming and thus the rural depopulation would be brought to an end.

Singh [17] reported that rice fields are the most significant contributors of atmospheric methane accounting for 13–15% of the world's total anthropogenic methane emission. According to the International Rice Research Institute, Philippines, world's rice harvested areas have been increased by approximately 33% from 115.5 M ha in 1961 to 153.3 M ha in 2004. According to a current estimate, rice production will need to expand by around 70% over the next 25 years to meet the demands of the world's growing human population, making rice cultivation, a potential major cause of increasing atmospheric methane.

Pathak and Aggarwal [16] conducted field experiments in nine farmers' fields during kharif season (June to October) of 2009 and 2010 in three villages in Jalandhar district of Punjab, India for estimating GHG emission from direct seeded rice (DSR) and transplanted rice (TPR) crops. Both crops were grown following recommended package of practices and data on crop management were collected. There was very small methane emission ranging from 0 to 0.1 kg/ha/day in the DSR fields in different sites. In transplanted rice, methane emission ranged from 0.2 to 0.8 kg/ha/day. Cumulative methane emission in one crop-growing season was 0.6 to 1.5 kg/ha in the DSR and 42.4 to 57.8 kg/ha in TPR during 2009. During 2010, methane emission in the DSR and TPR ranged from 4.2 to 4.9 kg/ha and 56.0 to 56.5 kg/ha, respectively. Field experiments at Jalandhar, Punjab showed that the DSR reduces GHG emission and saves irrigation water, human labor and machine labor without any yield penalty compared to conventional puddled transplanted rice. Thus, the DSR seems to be a feasible alternative to conventional puddled transplanted rice for mitigating and adapting to climate change, saving water and labor and increasing farmers' income.

Datta et al. [3] studied the effects of inorganic fertilizers (N, P, K) on methane emission from the tropical rice field in the Central Rice Research Institute, Odisha, India. An experiment was conducted during the dry season (January–April) and wet season (July-November) of rice cultivation to study the effects of nitrogen (N), phosphorus (P) and potassium (K) fertilizer application on grain yield and methane (CH_4) emission. The experiment was carried out with five treatments (No fertilizer (control), N-fertilizer, P-fertilizer, K-fertilizer and N + P + K fertilizer) with three replicates of each under a completely randomized block design.

Significantly higher CH_4 emission was recorded from all plots during wet season. Among fertilizer applied plots, significantly higher CH_4 emission was recorded from N-fertilizer applied plots (dry season: 80.27 kg/ha; wet season: 451.27 kg/ha), while significantly lower CH_4 emission was recorded from N + P +K applied plots (dry season: 34.60 kg/ha; wet season: 233.66 kg/ha). Negative CH_4 emission was recorded during the fallow period, which may be attributed to higher methanotrophic bacterial population. Study suggests that the effects of P and K-fertilizer on CH_4 emission from rice field along with the CH_4 emission during the fallow period need to be considered to reduce the uncertainty in upscaling process.

Parthasarathi et al. [14] evaluated the yield potential of aerobic rice varieties with soil moisture content at or below field capacity and with the moderate application of fertilizers. Aerobic rice varieties produced yield as much as irrigated puddled rice varieties, which were traditionally grown in rice paddies. Yields were at par with the irrigated puddled rice with an average of 5.5–6 t/ha with 60% less water use.

Anand et al. [2] developed a system dynamics model for simulation of methane emissions from rice fields. The model was used to obtain trends and estimates for a time span of 20 years under different scenarios. Possible strategies to mitigate the methane emissions from this sector were suggested and analyzed. Methane emissions from rice fields are dependent on the quantity of rice produced, varieties of rice, flooded area under rice cultivation and fertilizer amendment. Baseline scenario for methane emissions and its accumulation in the atmospheric pool were generated. In the baseline scenario, the rate of methane emissions was projected to reach 2.57 Tg/year over the next 20 years, adding 29.30 Tg to the atmosphere. Reductions in the surplus production of rice, the use of alternate low methane emitting varieties of rice, effective water management will reduce the methane emissions.

Jain et al. [8] reported that methane with its current concentration of 1.72 ppm V in the atmosphere accounts for 15% of the enhanced greenhouse effect. The atmospheric concentration of CH_4 is increasing at 0.3 per cent/year. Lowland rice soil is considered to be one of the major contributors of atmospheric methane. Various soil, climate and management factors control methanogenesis, the geochemical process that occurs in all anaerobic environments in which organic matter undergoes decomposition, resulting in the formation of CH_4. Methane formed in soil escapes to the atmosphere through vascular transport and diffusion. Emission of methane from rice fields can be reduced by: (i) midseason drainage instead of continuous flooding (ii) use of cultivars with low emission potential (iii) use of low C:N organic manure and (iv) direct establishment of rice crop like dry direct seeded rice.

Patel et al. [15] studied the physiological attributes and yield performance of aerobic rice over conventional flood-irrigated rice in the valley land of North-East Hill (NEH) ecosystems of India. A field experiment was conducted at the experimental farm, ICAR Research Complex for NEH Region, Umiam (950 m msl), Meghalaya during rainy seasons of 2006 and 2007 under aerobic and flooded conditions with aerobic rice variety collected from IRRI, Philippines. Some important high-yielding varieties (HYVs) recommended for the region were also included in the study. The objectives of this study were: (i) to evaluate the influence of frequent mid-season drainage as a measure of water saving technique besides inducing the pre-conditioning effect on genotypes to withstand water stress during the subsequent growth period of crop ontogeny; (ii) to compare crop performance between aerobic and flooded rice management practices; and (iii) to identify attributes responsible for the yield gap between aerobic and flooded rice. The results revealed that the yield difference between aerobic (average yield, 1.67 t/ha) and flooded rice (average yield, 2.31 t/ha) ranged from 18.4 to 37.8% ($P < 0.05$) depending on varieties, highest difference being observed with rice hybrid DRRH 1. Cultivation of rice under aerobic condition resulted in 27.5% yield reduction over flooded rice. The study suggests that, variety Sahsarang 1 with its moderate values of photosynthesis rate, transpiration rate and water use efficiency (WUE) along with higher grain yield seems to be better choice for both stress (aerobic) as well as normal (flooded) condition. Aerobic rice varieties with minimum yield gap compared to flooded rice are the key for success of aerobic rice cultivation.

Singh et al. [18] discussed various ways of enhancing use efficiency and productivity of water in agricultural production system. These include: better utilization of stored soil moisture by adjusting time and method of sowing, reducing evaporation loss of soil moisture by mulching, supplemental and deficit irrigation provided to crops at critical growth stages, practice of mixed cropping systems involving deep rooted crops during kharif season and shallow rooted crops during rabi season, removal of nutrient constraints by supplying optimum fertilizer inputs, crop diversification in lowlands, reducing water use in rice cultivation, improved irrigation methods like drip and drip irrigation, improved planting patterns, integrated farming systems for flood-prone lowland areas, and multi-uses of water in agriculture by combining different farm enterprises like cropping, fishery and dairy. They have also discussed on the need of favorable public policies to create conducive socio-economic environments for enhancing water use efficiency in the agriculture sector.

Garg et al. [5] discussed regarding the uses the 2006 IPCC Guidelines and latest country specific emission factors to estimate Indian methane emissions at sectorial and district level for the years 1990, 1995, 2005 and 2008. The estimates showed that while methane emissions have increased steadily over past two decades, their share in India's aggregate GHG emissions has declined from 31% in 1985 to 27% in 2008 mainly due to relatively higher growth CO_2 emissions from the fossil fuels. The estimates for the year 2008 showed that: (i) agriculture sector, which employed two-thirds of India's population and contributed 17% of GDP, accounted for 23% of India's GHG emissions; (ii) 83% of country's methane emissions are contributed by enteric fermentation, manure use and rice production; and (iii) methane emissions from urban solid waste are steadily rising over the past two decades; their share in aggregate methane emissions has reached 8%. Resting on the disaggregated emissions estimates, they opined for using geographical and sectorial flexibilities to develop a roadmap for mitigation of methane emissions for India.

Narale et al. [13] presented the design and economic analysis of efficient solar PV water pumping system for irrigation of banana. The system was designed and installed in solar farm of Jain Irrigation System Limited (JISL), at Jalgaon (Maharashtra). The study area falls at 21° 05' N latitude, 75° 40'E longitude and at an altitude of 209 m above mean sea level. The PV system sizing was made in such a way that it was capable of irrigating 0.165 ha of banana plot with a daily water requirement of 9.72 m³/day and total head of 26 m. Also, the life cycle cost (LCC) analysis was conducted to assess the economic viability of the system. The results of the study encouraged the use of the PV systems for water pumping application to irrigate orchards. The installed system of solar PV water pumping system was capable of irrigating 0.165 ha area of banana crop within 6.02 hrs with a daily water requirement of 9.72 m³/day. The results of the study indicated that irrigating orchards in the remotest areas using PV systems was beneficial and suitable for long-term investments as compared to diesel powered engines as total cost (TC) of PV system considering life span of 20 years was found to be Rs. 1,38,958/- while TC of diesel engine was Rs. 7,60,029/-.

Gopal et al. [6] reviewed the research developments on renewable energy source water pumping systems referring 168 research papers across the globe. They concluded that renewable energy solar water pumping systems are identified as an alternative source for replacing conventional pumping methods. The integration of renewable energy sources with water pumping systems plays a major role in reducing the consumption of conventional energy sources and their environmental impacts, particularly for irrigation applications. The solar photovoltaic water pumping systems are the most widely used renewable energy solar water pumping systems for irrigation and domestic applications, followed by wind energy water pumping systems. The solar thermal and biomass water pumping systems are less popular due to their low thermal energy conversion efficiencies.

8.4 MATERIALS AND METHODS

The study was carried out in the following phases.

a. **In Phase-I**, survey work was conducted to collect information on the methods of rice cultivation, variety of seeds, crop duration, sources of water, availability of power and labor, use of different water pumping devices, use of manure and fertilizer, yield potential, cropping intensity etc. in and around the study area.

b. **In Phase-II**, soil samples from each field were collected and analyzed for organic carbon, p^H, electrical conductivity (EC) and texture. Meteorological data like temperature, relative humidity, rainfall, wind velocity, evaporation, solar radiation, etc., were collected for estimation of Evapotranspiration of rice crops to be grown in the experimental plots using standard procedure. The maximum plot size was limited to one acre (0.4 ha) only.

c. **Phase-III** comprised of layout of experimental plots for comparative studies of various rice growing methods and their statistical analysis. Procurement of solar photovoltaic water pumping device and installation of drip irrigation system for the experimental plots. Information on input solar energy and output electrical energy, depth of water level and pumping water level in the water source, the discharge rate and its characteristics, delivery head, average total hours of operation and maximum hours of operation in a day under clear sky condition were measured or collected. The cost of pumping of solar photovoltaic water pumping system was also assessed.

d. **Phase-IV** comprised of the comparative studies of four methods of rice cultivation like aerobic, SRI, conventional and alternate wetting and drying practices with respect to seed rate, water, labor and power requirement, quantity of manures and fertilizers, use of weedicide, pesticides, emissions of methane, carbon dioxide, nitrous oxide and yield of the crop. The proposed study would be undertaken by conducting experiments considering the following independent and dependent parameters.

e. **Phase-V** comprised of the transfer of the developed technology for collecting necessary feed backs for its popularization with respect to research, development, demonstration and commercialization which are the continuous processes in any developmental activity.

Design and development of solar photovoltaic (SPV) drip irrigation system was done for cultivating paddy in 1 acre (0.4 ha) of land to achieve secured irrigation and to improve water use efficiency (WUE) mostly in aerobic method of cultivation. The experiments were carried out during the year 2014 in the Central Farm of Orissa University of Agriculture and Technology (OUAT), Bhubaneswar – Odisha, which lies at the latitude of

20^0 15' N and longitude of 85^0 52' E and coming under warm and humid climatic condition. Paddy was cultivated in rabi and summer seasons.

8.4.1 SOIL

The soil type of the experimental site is sandy loam and slightly acidic in nature which is taxonomically grouped under the order alfisol. It is partly eroded due to high intensity of rainfall in the area. Geologically the soil is derived from laterite. The general slope of the land at the experimental site is 2%. The physical and chemical properties of soil of the experimental site are presented in Table 8.1.

8.4.2 CLIMATE

The climate of the study area is very pleasant and the weather is suitable for a wide range of crops available in Odisha. Generally the climate of the study area is humid and sub tropical in nature. The average annual rainfall at the site is about 1250 mm. Eighty percent of the total rainfall occurs during June to September in almost every year which is popularly known as monsoon rainfall. It experiences typical tropical weather conditions and succumbs to the heat and cold waves that sweep in from north India. The summer months from March to May are very hot and humid, and temperatures often rise above 40°C in May. The South West monsoon lashes Odisha in the second week of June, bringing relief to the parched

TABLE 8.1 Physical Properties of the Soil At the Experimental Site

Parameter	Value
Sand (%)	55.5
Silt (%)	26.2
Clay (%)	18.3
Bulk density (g cm^{-3})	1.55
Particle density (g cm^{-3})	2.64
Proctor moisture content (% by weight)	9.05

environment of the area. The study area receives maximum rainfall in July and August, whose average value is 220 mm per month.

8.4.3 DESIGN OF PHOTOVOLTAIC POWERED DRIP IRRIGATION SYSTEM

8.4.3.1 Water Requirement for Aerobic Rice Crop (1 acre or 0.4 hectare area)

$$W_r = (A \times PE \times P_c \times K_c \times w_a)/E_u \qquad (1)$$

where, W_r = peak water requirement (m³/day); A = crop area (m²); PE = Pan evaporation rate (mm/day) converted to m/day; P_c = Pan coefficient (0.7 to 0.9); K_c = crop coefficient (0.8 to 1); w_a = wetted area (%, 95% for drip irrigation); E_u = emission uniformity of drip irrigation (Approx. 0.9).

In this study, author used: A = 4,000 m²; PE = 5 mm/day; P_c = 0.9; K_c = 1; w_a = 0.95 and E_u = 0.9. Therefore, W_r = 19 m³/day (19,000 liters/day). Taking irrigation interval to be 2 days, net water requirement: W_{net} = 19,000/2 = 9,500 liters/day = 9.5 m³/day.

8.4.3.2 Sizing of PV Module for Water Requirements

$$E = (\rho\, g\, H\, V)/3.6 \times 10^6 \qquad (2)$$

where, E = hydraulic energy required (kWh/day); ρ = density of water (= 1000 kg/m³); g = gravitational acceleration (= 9.81 m/s²); H = total hydraulic head (m, = 30 m in this case); V = volume of water required (9.5 m³/day in this case under Section 8.4.1.1).

Putting all values, we get E = 0.77 kWh/day = 770 Wh/day. Assuming actual sun shine hours in a day = 4.5 hours, total wattage of PV module = 770/4.5 = 171.11 watt. Following assumptions were made:

a. Operating factor = 0.75–0.85 (PV panel mostly does not operate at peak rated power).
b. Pump efficiency = 70–80% (can be taken 75%).

    c.   Motor efficiency = 75–85% (can be taken 80%).

    d.   Mismatch factor = 0.75–0.85 (PV panel does not operate at maximum power point).

Considering system losses, wattage requirement = (Total PV panel wattage)/(pump efficiency × mismatch factor) = (171.11)/(0.75 × 0.8) = 285.18 watt. Considering operating factor for PV panel, wattage = (Total PV panel wattage after losses)/(operating factor × motor efficiency) = (285.18)/(0.8 x 0.8) = 445.60 watt. Number of 75 w_p solar PV panels required = 445.60/75 = 5.94 ≈ 6 modules. Power rating of motor = 445.60/746 = 0.59 hp. However, for drip irrigation system, 1 hp submersible pump with an array capacity of 1000 watt is required to pump 10,000 liters of water from total head of 30 m.

8.4.3.3 Tentative Cost Estimate For Solar Photovoltaic Powered Drip Irrigation System (Figure 8.1)

a.	Solar PV Module of 1000 watt @ Rs. 50 per w_p	= Rs.	50,000
b.	One hp DC motor with pump set	= Rs.	80,000
c.	Mounting structure	= Rs.	15,000
d.	Civil works/Balance of system	= Rs.	20,000
e.	Drip set up for 1 acre land	= Rs.	35,000
	Total	**= Rs.**	**2,00,000**

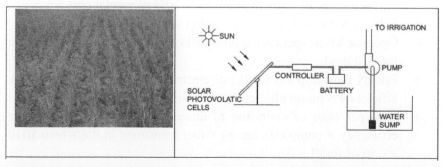

FIGURE 8.1 Experimental set-up for solar water pumping based drip irrigation in aerobic rice.

8.4.3.4   Hourly Cost of Operation of Various Water Pumping Devices

Assumptions for cost analysis

a. Cost of 1 hp electric pump set = Rs. 7,000
b. Cost of 1 hp diesel pump set = Rs. 10,000
c. Cost of 1 hp PV powered pump set = Rs. 1,00,000
d. Prevailing interest rate = 10%
e. Efficiencies of motor = 70–80% (70% taken)
f. Efficiencies of pump = 70–80% (70% taken)
g. Efficiencies of diesel engine = 30–40% (40% taken)
h. Useful life of PV panel = 20–25 years (can be taken 22 years)
i. Useful life of diesel engine pump set = 8 years
j. Useful life of electric pump set = 8 years
k. Maintenance cost of PV system with drip = 0.5% of total capital cost per year
l. Maintenance cost of diesel engine pump = 10% of total capital cost per year
m. Maintenance cost of electric pump = 10% of total capital cost per year
n. Annual working hours of diesel, electric pump sets and PV system = 500 hours
o. One hp engine consumes about 250 mL. diesel per hour (present cost of diesel Rs. 60/L)
p. One unit of electric energy (1 kWh) = Rs. 5.00
q. Salvage value of diesel pump set = 20% of capital cost
r. Salvage value of electric pump set = 20% of capital cost
s. Salvage value of PV powered pump set = 5% of capital cost
t. Operator's time spent in the proposed system be 1 h/day (labor charge Rs. 250/day)
u. Energy consumption (kWh) of electric pump set = (BHP)/(motor efficiency × pump efficiency) × 0.746 × 1 hour
v. Cost per hour of operation of diesel pump set = (BHP)/(motor efficiency × pump efficiency) × fuel consumed in liters/hour/BHP × cost of fuel/L

Hourly operating cost of PV powered water-pumping device with drip system

I. *Fixed Cost*

   a. Depreciation: $D = (C-S)/ (L \times H)$ where C= capital cost; S = Salvage Value; L = Useful life of device; H= Annual working hour. Putting the values of all necessary data, D = Rs. 17.27/hour
   b. Interest $(I) = (C + S)/(2) \times$ (Interest rate/100) $\times$ (1/H) = Rs. 21/hour
   c. Insurance and taxes and housing are not applicable
   d. Total fixed cost = 17.27 + 21 = Rs. 38.27/hour

II. *Variable Cost*

   a. Fuel cost = Nil
   b. Lubricants = Nil
   c. Repair and maintenance = (C) $\times$ (0.5/100) $\times$ (1/H) = Rs. 2/hour
   d. Operator's wages: Rs. 250/8 = Rs. 31.25/hour
   e. Total variable cost = 2 + 31.25 = Rs. 33.25/hour
      Total operation cost per hour = Total fixed cost/hour + Total variable cost/hour = **Rs. 71/hour** ---------------------------- A

Hourly operating cost of diesel pump set

I. *Fixed Cost*
   a. Depreciation: $D = (C-S)/ (L \times H)$ where C= capital cost; S = Salvage Value; L = Useful life of device; H= Annual working hour. Putting the values of all necessary data, D = Rs. 2.0/hour
   b. Interest $(I) = (C + S)/(2) \times$ (Interest rate/100) $\times$ (1/H) = Rs. 1.2/hour
   c. Insurance and taxes and housing are not applicable
      Total fixed cost = 2.0 + 1.2 = **Rs. 3.20/hour**

II. *Variable Cost*

   a. Fuel cost = (1)/(0.4 × 0.7) × 0.25 × 60 = Rs. 53.57/hour
   b. Lubricants = 20 % of cost of fuel = Rs. 10.71/hour
   c. Repair and maintenance = (C) $\times$ (10/100) $\times$ (1/H) = Rs. 2.0/hour
   d. Operator's wages Rs. 250/8 = Rs. 31.25/hour
   e. Total variable cost = 53.57 + 10.71 + 2.0 + 31.25 = **Rs. 97.53/hour**
      Total operation cost per hour = Total fixed cost/hour + Total variable cost/hour = **Rs. 100/hour** ---------------------------- B

Hourly operating cost of electric pump set

I. Fixed Cost

    a.   Depreciation: $D = (C-S)/(L \times H)$ where C = capital cost; S = Salvage Value; L = Useful life of device; H = Annual working hour. Putting the values of all necessary data, D = Rs. 1.4/hour

    b.   Interest $(I) = (C + S)/(2) \times$ (Interest rate/100) $\times (1/H)$ = Rs. 0.84/hour

    c.   Insurance and taxes and housing are not applicable

    d.   Total fixed cost = 1.4 + 0.84 = **Rs. 2.24/hour**

II. Variable Cost

    a.   Energy consumption (kWh) = $(1)/(0.7 \times 0.7) \times 0.746 = 1.52$ kWh

    b.   Electric energy cost = 1.52×5 = Rs. 7.6/hour

    c.   Lubricants = 20% of cost of fuel = Rs. 1.52/hour

    d.   Repair and maintenance = $(C) \times (10/100) \times (1/H)$ = Rs. 1.4/hour

    e.   Operator's wages Rs. 250/8 = Rs. 31.25/hour

    f.   Total variable cost = 7.6 + 1.52 + 1.4 + 31.25 = Rs. 41.77/hour
Total operation cost per hour = Total fixed cost/hour +
Total variable cost/hour = **Rs. 44/hour** ---------------------------- **C**

8.5   RESULTS AND DISCUSSION

Rice is one of the most important and major crops in Odisha and is grown in an area of 40×10^5 ha in kharif season and only 2.5×10^5 ha in rabi season, (Agricultural Statistics, 2012, Govt. of Odisha). Kharif rice is entirely monsoon-fed. The area for rice cultivation during rabi season has not been covered widely due to lack of assured irrigation causing most of the cultivable land to remain unutilized. The yield of rice from the areas during rabi season is not satisfactory due to erratic supply of grid electricity for operating irrigation pumps. Farmers are also not interested to use diesel pump sets due to frequent rise in the cost of diesel fuel. Hence captive power and water source along with water saving technology for rice cultivation are the only alternative for the resource poor farmers of the state. The harnessing of solar energy for electricity generation through PV system is a viable option in the state due to abundant availability of solar radiation in about 300 days in a year. The variety chosen for the study was CR Dhan-200 (Pyari). This variety of rice was cultivated for the present study in order to evaluate the effectiveness of the developed solar PV drip irrigation device with respect to production and productivity, without depending upon conventional source of energy and flooded system of watering practice. The cost of cultivating rice in an one acre of land was calculated in order to know the annual profits out of it and its expected pay-back period. Similarly, the mitigation of greenhouse gases with the use of the developed set-up was estimated and compared with traditional diesel and electric pump sets for its contribution in combating global warming and climate change and thus achieving sustainable agriculture.

8.5.1   COST-BENEFIT CALCULATION OF AEROBIC RICE CULTIVATION IN ONE ACRE (0.4 HA)

Cost of rice cultivation in one acre of land is shown in Table 8.2. **Benefits** were calculated as follows:

a. Yield of paddy in aerobic rice practice = 2.5 tons/acre
b. By-product yield = 1.5 tons/acre
c. Returns from paddy @ Rs. 13.50/kg = Rs. 33,750
d. Returns from by-product @ Rs. 2.50/ kg= Rs. 3,750
e. Total returns = 33,750 + 3,750= Rs. 37,500
f. Net gain = Rs. 37,500– Rs. 13,500 = Rs. 24,000 (kharif)
g. Net gain = Rs. 37,500– Rs. 13,500 = Rs. 24,000 (rabi)
h. Total gain from 1 acre of aerobic rice cultivation in a year = Rs. 48,000
i. Monthly income from aerobic rice cultivation with assured water supply = **Rs. 48,000/12 = Rs. 4000 per month**
j. **Simple payback period** = (Initial investment cost)/(Net annual gain) = **2, 00,000/48,000 = 4.1 years ≈ 4 years**

8.5.2   ESTIMATION FOR MITIGATION OF CO_2 EMISSION BY USE OF SOLAR PHOTOVOLTAIC POWERED WATER PUMPING SYSTEM

Diesel and electricity are the two mostly used fuels to operate diesel and electric pump sets, respectively for water pumping in irrigating cultivable land. Burning of diesel in the internal combustion engines and generation of electricity in power plants contribute to emission of greenhouse gases

TABLE 8.2 Cost of Rice Cultivation Per Acre

Name of operation	Implements used	No. of operations	Man-h/Ac	Operation cost (Rs.)	Input (kg)	Cost of input (Rs.)	Total cost (Rs.)
Tillage	Tractor drawn rotavator	1	2	1200	–	–	1200
Seed					20 kg		500
Direct line sowing (by rope method)		1	8	31.25/hour			250
Manures and Fertilizer	FYM	Once			1 tractor load	1500	1500
	Gromer	Twice			100 kg	1000	1000
	Potash	Twice			100 kg	1000	1000
Interculture (cono weeder)	Manual	Thrice	16	31.25/hour			1500
Plant protection	Knapsack sprayer	Thrice	2	31.25/hour	Pesticides	2000	2062
Irrigation	Solar PV powered drip system	45 (2 days interval)	1	Rs. 51/hour			2295
Harvesting	Hired Reaper	Once	1				750
Threshing	Pedal thresher	Once	50–60 kg/hr	Six man days			1500
						Subtotal, Rs. =	**13,557 ≈ 13,500**

to the atmosphere causing more to the present concerns of global warming and climate change. The use of pump sets has been accelerated due to the strong initiatives being taken by the Government to expand more areas under assured irrigation. The replacement of diesel and electric pump sets with a reliable solar photovoltaic powered water pumping system particularly in the irrigation sector would definitely reduce to a greater extent in the emissions of greenhouse gases to the atmosphere. The existing diesel and electric pump sets in Odisha is 2.47×10^5 and 1.38×10^5, respectively in the power rating range of 1–5 hp. Taking the average power rating of both diesel and electric pump sets as 3 hp, the amount of emissions of CO_2 are as follows:

1. One hp engine consumes about 250 mL of diesel per hour.
2. Burning of 1 L of diesel releases 3 kg of CO_2 to the atmosphere.
3. The average carbon dioxide emission for electricity generation from coal based thermal power plant is approximately 0.98 kg of CO_2 per kWh at the source. If the transmission and distribution losses under Indian condition are assumed as 40% and poor inefficient electric equipment losses of 20%, then the value of 0.98 will be 1.58 kg/kWh.
4. In case of thermal power plants, out of the total amount of thermal input, 30–40% output is in the form of generation of electricity and the rest 60–70% of the total input energy is dissipated into the environment in the form of heat energy, which also accounts for global climate change. In addition to this large amount of heat dissipation, a substantial amount of CO_2 along with particulate matter is released into the environment each year. In contrast, a PV power system does not dissipate such enormous amount of heat energy into the surrounding environment and saves huge amount of CO_2 emissions. Hence, PV system is an environment friendly option for power generation and should be preferred where there is no electricity or extension of grid power is a costlier option.
5. Annual working hours of diesel and electric pump sets can be taken 500 hours.
6. Annual CO_2 emissions from 2.47×10^5 diesel pump will be 300 million kg in Odisha.

7. Annual CO_2 emissions from 1.38×10^5 electric pump sets to be 250 million kg in Odisha.

8. Total annual CO_2 emissions can be mitigated by 550 million kg with the replacement of existing diesel and electric pump sets in Odisha by the adoption of solar photovoltaic powered system in irrigation sector.

9. Total annual electrical energy consumption from 1.38×10^5 electric pump sets can be saved in the tune of 15×10^7 kWh (saving around 150 million units of electricity costing about Rs. 750 million/annum).

10. Total annual diesel consumption from 2.47×10^5 diesel pump sets can be saved in the tune of 10×10^7 liters of diesel (saving of Rs. 6,000 million/annum).

8.5.3 ESTIMATION FOR MITIGATION OF CH_4 EMISSION BY SHIFTING FROM ANAEROBIC TO AEROBIC RICE CULTIVATION

Rice fields have been identified as a major source of atmospheric methane. The global methane emission rate from rice fields was recently estimated as 40 Tg/year (1 Tg = 10^{12} g), which accounts for about 8% of the total methane emission. Predictions based on population growth rate in countries, where rice is the main food crop, indicate that rice production must be increased to 65% by 2020 to meet the rice demand for the growing population and thus making rice cultivation a potential major cause of increasing atmospheric methane. Rice fields are most significant contributors of atmospheric methane accounting for 13–15% of the world's total anthropogenic methane emission. Anaerobic decomposition of organic material in the flooded rice fields produces methane, which escapes to the atmosphere primarily by diffuse transport through the rice plants during the growing season. Up land rice fields, which are not flooded, do not therefore produce significant quantities of methane. As aerobic rice production system is the "Improved upland rice" in terms of yield potential and "improved low land rice" in terms of drought tolerance, it favors remission of methane emission due to the growth of plants in non-flooded and unsaturated soil condition. Aerobic rice technology may therefore be a suitable remedy for future climate change under water-short environment with lesser methane emission.

The reduction in the amount of methane emission from the conventional method of rice cultivation is estimated as follows:

1. On an average, 50 kg methane is emitted from 1 hectare of transplanted rice crops in one crop growing season.
2. Annual Area under rice cultivation in Odisha is 4.0 million hectares (both kharif and rabi).
3. Annual Area under rice cultivation in India is 43.0 million hectares (both kharif and rabi).
4. Annual CH_4 emission from 4 million hectares rice fields is 0.2 million tons in Odisha.
5. Annual CH_4 emission from 43 million hectares rice fields is 2.15 million tons in India.
6. Mitigation of CH_4 emissions through aerobic rice cultivation is 0.2 million tons and 2.15 million tons per annum in Odisha and India, respectively.

   a. Reliable source of energy through solar photovoltaic system in irrigation sector for water pumping.
   b. Popularization of solar photovoltaic powered water pumping system among the farming community.
   c. Water saving technology in rice cultivation.
   d. Coverage of more area under rice cultivation with the available water resources and non-irrigated areas particularly during rabi season.
   e. Less dependence on fossil fuel powered pump sets in irrigation sector and thus mitigating CO_2 emissions to the atmosphere.
   f. Mitigating CH_4 emission to the atmosphere by switching gradually from flooded to non-flooded rice cultivation.
   g. Preventing global warming through reduced anthropogenic greenhouse gas emissions and thus controlling climate change.
   h. Affordable and sustainable rice production technology for small and marginal farmers for favoring food security.

8.6 CONCLUSIONS

A portable solar photovoltaic powered drip-irrigation system for aerobic rice cultivation in warm and humid climate of Odisha can be a viable proposition looking into the present day's concerns of water scarcity and energy crisis in agricultural sector. The findings of the study would definitely help the farming community to go for cultivating rice in non-irrigated areas at the individual's level in water deficient areas and off-grid remote locations without depending upon the grid electricity and community based irrigation set-up. The set-up developed also seems to be economically viable for resource poor farmers of the state because of its estimated short pay-back period of about 4 years. The farmers may also achieve livelihood security through this technology by cultivating rice in aerobic method during rabi season in off-grid rural and non-irrigated areas and also in kharif season due to irregular rainfall. It is therefore a way to achieve energy security, improved water use efficiency, livelihood security, environment protecting ability and thereby attaining food security for the fast growing population of our state and country. The wide spread propagation of this technology would certainly help and encourage the small and marginal farmers of our state to become energy independent for irrigating the land and self dependent on in-time and precise as well as judicious application of water to the crops causing increased productivity per drop of water use and ultimately protecting environment against flooded method of rice cultivation. The following conclusions were drawn from the study.

1. Wide popularization of solar photovoltaic powered water pumping system for aerobic rice cultivation through sustainable energy source.

2. Monthly income of *Rs. 4000/-* throughout the year may be possible by adopting aerobic rice cultivation in 1 acre of land both during rabi and summer seasons.

3. The small and marginal farmers of the state may be attracted to adopt solar photovoltaic powered water pumping system as the hourly operating cost is *Rs. 71/hour* and *Rs. 44/hour* for electric pump set and *Rs. 100/hour* for diesel pump set.

4. The existing area under rice cultivation in the state may be enhanced by adopting aerobic rice cultivation in the water-deficient and non-irrigated areas mostly during rabi season by ensuring required water supply through solar photovoltaic system.

5. The proposed set up may also be utilized for irrigating land in rainy season in case of irregular rainfall.
6. The proposed set up may also be utilized for home lighting in case of power cuts in grid connected locations or off-grid area as well.
7. Pay-back period of the proposed set up is *4 years*, due to which it may be easily accepted by the small and marginal farmers of the state in-spite of its high initial cost.
8. Total annual CO_2 emissions can be mitigated by *0.55 million tons* with the replacement of existing diesel and electric pump sets in our state by the adoption of a reliable solar photovoltaic powered system in irrigation sector.
9. Total annual CH_4 emissions can be mitigated by *0.2 million tons* from 4.0 million hectares rice fields in Odisha.
10. Total annual electrical energy consumption from 1.38×10^5 electric pump sets can be saved in the tune of 15×10^7 kWh (saving around 150 million units of electricity costing about *Rs. 750 million/annum*).
11. Total annual diesel consumption from 2.47×10^5 diesel pump sets can be saved in the tune of 10×10^7 liters of diesel (saving around *Rs. 6000 million/annum*).

8.7 SUMMARY

A water saving technology and less water consuming rice production system without any compromise with the decline in yield are the urgent necessity of the present scenario of increasing water scarcity and achieving food security for the fast growing population of our country. One such crop growing practice, introduced recently is through aerobic method of rice cultivation, which has been developed as a promising and viable technology for the situation where uncertainty in assured irrigation and irregularity of rainfall prevail. Submergence of rice field continuously with water for longer period in a crop-growing season in traditional method of rice cultivation is now a major concern for global warming due to the emission of most potent greenhouse gas i.e., methane to the atmosphere. Shifting from flooded to non-flooded method of rice cultivation may therefore be the need of the hour to address the above issues. Efficient water management through micro- irrigation, particularly by drip irrigation in non-flooded, unsaturated or un-puddled rice cultivation system may be a viable option looking into the present day's major constraints of water stress and increasing concentration of atmospheric methane. Water use in drip irrigation system is mostly from the underground source through mechanically or electrically operated pump sets. The erratic grid supply of electricity and increasing cost of diesel/petrol for use in pump sets are becoming a great problem for the resource poor farmers of the country like India in achieving assured irrigation. Hence, a sustainable and environment-friendly power source through solar photovoltaic system may be a feasible proposition for energy security and environmental protecting approach of the water pumping devices in irrigation sector. Nature's free gift of sunlight and portability mode of the proposed solar photovoltaic water pumping system may also be the added advantages looking into less dependence of fossil fuel, safety, less possibility of threats for theft and the durability of the device in the long run. An attempt is therefore made to develop a portable solar photovoltaic powered drip irrigation system for aerobic rice cultivation in order to face the today's challenges of energy crisis, water scarcity, global warming and ultimately climate change for achieving sustainable production and productivity of rice mostly in the water-deficient, non-irrigated and off-grid areas. The attempts for growing more crop per drop of water use, reliable source of power availability and adoption of location-specific cropping pattern in the study may be the profitable, remunerative and crop-diversified approach to avoid yield decline and sustainability in aerobic rice cultivation. There may be the saving of 40–45 % of water for irrigation purpose compared to the conventional method, mitigation of 0.55 million tons of CO_2 with the replacement of existing diesel and electric pump sets and 0.2 million tons of CH_4 from 4.0 million hectares of rice fields in the state of Odisha, India, through the system, developed by adopting aerobic rice cultivation. The pay- back period of the setup is estimated to be 4 years and total annual saving of Rs. 6750 million due to reduction in the use of electrical energy and petroleum fuels through the existing pump sets in the state. Monthly income of Rs. 4000/- throughout the year may also be possible by adopting aerobic rice cultivation in 1 acre (0.4 ha) of land.

KEYWORDS

- **Aerobic rice cultivation**
- **Climate change**
- **Cost analysis**
- **Global warming**
- **Greenhouse gas mitigation**
- **India**
- **Irrigation**
- **Kharif season**
- **Methane**
- **Micro irrigation**
- **Monsoon**
- **Odisha**
- **Pay back period**
- **Rabi season**
- **Rice**
- **Solar photovoltaic system**
- **Water use efficiency**

REFERENCES

1. Alberto, Ma Carmelita R., Reiner Wassmann, Takashi Hirano, Akira Miyata, Ryusuke Hatano, Arvind Kumar, Agnes Padre, & Modesto Amante (2011). Comparisons of energy balance and Evapotranspiration between flooded and aerobic rice fields in the Philippines. *Agricultural Water Management*, 98, 1417–1430.
2. Anand, Shalini, Dahiya, R. P., Vikash Talyan, & Prem Vrat (2005). Investigations of methane emissions from rice cultivation in Indian context. *Environment International*, 31, 469–482.
3. Datta, A., Santra, S. C., & Adhya, T. K. (2013). Effect of inorganic fertilizers (N, P, K) on methane emission from tropical rice field of India. *Atmospheric Environment*, 66, 123–130.
4. Epule, E. T., Peng, C., & Mafany, N. M. (2011). Methane emission from paddy rice fields: strategies towards achieving a win-win sustainability scenario between rice production and methane emission reduction. *Journal of Sustainable Development*, 4(6), 188–196.

5. Garg, Amit, Bhushan Kankal, & Shukla, P. R. (2011). Methane emissions in India: Sub-regional and sectoral trends. *Atmospheric Environment*, 45, 4922–4929.

6. Gopal, C., Mohanraj, M., Chandra Mohan, P., & Chandrasekhar, P. (2013). Renewable energy source water pumping systems-A literature reviews. *Renewable and Sustainable Energy Reviews*, 25, 351–370.

7. Hamidat, A., Benyoucef, B., & Hartani, T. (2003). *Renewable Energy*, 28, 1081–1096.

8. Jain, N., Pathak, H., Mitra, S., & Bhatia, A. (2004). Emission of methane from rice fields: Review. *Journal of Scientific and Industrial Research*, 63, 101–115.

9. Kato, Yoichiro, Midori Okami, & Keisuke Katsura (2009). Yield potential and water use efficiency of aerobic rice (*Oryza sativa* L.) in Japan. *Field Crops Research*, 113, 328–334.

10. Katsura, Keisuke, & Yukie Nakaide (2011). Factors that determine grain weight in rice under high-yielding aerobic culture: The importance of husk size. *Field Crops Research,* 123, 266–272.

11. Kelley, Leah C., Eric Gilbertson, Anwar Sheikh, & Steven D. Eppinger and Steven Dubowsky (2010). On the feasibility of solar-powered irrigation. *Renewable and Sustainable Energy Reviews*, 14, 2669–2682.

12. Meah, Kala, Steven Fletcher, & Sadrul Ula (2008). Solar photovoltaic water pumping for remote locations. *Renewable and Sustainable Energy Reviews*, 12, 472–487.

13. Narale, P. D., Rathore, N. S., & Kothari, S. (2013). Study of solar PV water pumping system for irrigation of horticulture crops. *International Journal of Engineering Science Invention*, 2(12), 54–60.

14. Parthasarathi, T., Vanitha, K., Lakshamanakumar, P., & Kalaiyarasi, D. (2012). Aerobic rice-mitigating water stress for the future climate change. *International Journal of Agronomy and Plant Production*, 3(7), 241–254.

15. Patel, D. P., Anup Das, G. C., Munda, P. K., Ghosh, Juri Sandhya Bordoloi, & Manoj Kumar (2010). Evaluation of yield and physiological attributes of high-yielding rice varieties under aerobic and flood-irrigated management practices in mid-hills ecosystem. *Agricultural Water Management*, 97, 1269–1276.

16. Pathak, H., & Aggarwal, P. K. (2012). *Low carbon technologies for agriculture: A study on rice and wheat systems in Indo-Gangetic plains. Division of Environmental Science*, IARI, New Delhi.

17. Singh, Jay Shankar (2010). Capping methane emission. *Science Reporter*, September, pages 29–30.

18. Singh, Ravinder, Kundu, D. K., & Bandyopadhya, K. K. (2010). Enhancing agricultural productivity through enhanced water use efficiency. *Journal of Agricultural Physics*, 10, 1–15.

CHAPTER 9

MANAGEMENT OF IRRIGATION SYSTEM: QUALITY PERFORMANCE OF EGYPTIAN WHEAT

HANI A. MANSOUR,[1] ABDELGAWAD SAAD,[2]
AYMAN A. A. IBRAHIM,[3] and MOHAMED E. EL-HAGAREY[4]

[1]*Researcher, National Research Centre (NRC), Agricultural Division, Water Relations and Field Irrigation Department, El-Buhouth St., El-Dokki, Giza, Cairo, Egypt; E-mail: mansourhani2011@gmail.com; Corresponding author*

[2]*Researcher, Agricultural Research Center (ARC), Agricultural Engineering Research Institute (AEnRI), 12311 Dokki, Giza, Cairo, Egypt. E-mail: en_gawad2000@yahoo.com*

[3]*E-mail: eng_ayman288@yahoo.com*

[4]*Researcher, Desert Research Center (DRC), Soil Conservation and Water Resources Dept., Irrigation and Drainage Unite, Cairo, Egypt. E-mail: elhagarey@gmail.com*

Edited version of, *"Hani A. Mansour, Abdelgawad Saad, Ayman A. A. Ibrahim, and Mohamed E. El-Hagarey, 2015. Management of Sprinkler Irrigation System and Different Egyptian Wheat Varieties for II: Technological Quality Properties. www.euroessays.org, Open Access Article in European Journal of Academic Essays, 2(1), 7–13."*

CONTENTS

9.1   INTRODUCTION

Wheat quality is defined in terms of specific properties that determine suitability for milling and bread production [6]. Wheat crop (*Triticum spp.*) is an important cereal crop in the world in terms of area and production. It is a stable food for more than one third of the world population. In Egypt, wheat is the main winter cereal crop that is used as a stable food grain for urban and rural societies. The wheat area over the last 10 years (2004–2014) has expanded from (0.18–0.25 million ha) and the average productivity per ha has increased from 6.4 to 8.8 million tons during that period.

Protein is a primary quality component that influences the most of wheat grain baking quality characteristics. In hard wheat, the majority of the variation in loaf volume of bread can be attributed directly to differences in protein concentration [8]. Flour protein percentage is a good predictor of loaf volume, which itself is a function of the environmental conditions under which the crop is grown [24]. Reese et al. [23] found that determination of grain protein is only one test of flour quality and additional information is needed. The physical properties of wheat-flour dough such as extensibility and resistance to extension influence its mixing behavior very strongly. These rheological properties are highly heritable [24]. Rheological tests (farinorgam and extensio-gram) are carried out on unfermented dough and can be subdivided into tests, which give information about water absorption, mixing requirements and dough behavior. Water absorption is an important quality factor to the baker, as it is related to the amount of bread what can be produced from a given weight of flour [3]. It also has a profound influence on crumb softness and bread keeping characteristics [27]. The baking test is therefore the most useful test available for determining the practical value of a particular flour sample.

Traditionally, loaf volume has been considered as the most important property for the bread - making quality. Bread-making quality usually reacts like other quantitative characteristics to favorable or unfavorable environmental conditions and varies its performance. It is unrealistic to expect the same level of performance in all environments [10]. For the milling and baking industry, it is desirable that quality traits should be maintained as stable as possible through all environments. There exits different concepts of stability definition. According to static concept (called also as biological), stable genotypes possess unchanged or con-stant performances regardless of any variation of environmental condi-tions. A genotype is considered to be stable if its among-environment variance is small [14]. The coefficient of variability (CV) describes this type of stability [8]. This CV depends on the diversity of the environ-ments in the experiments.

In the terms of relative stability, we compare genotype quality trait with other genotypes in certain environment for using Pi (cultivar perfor-mance) value. The Pi of genotype is the mean square distance between genotype one and the genotype with the maximum response [13]. The smaller the estimated value of Pi, the lesser is its distance to the geno-type with maximum value, and thus the better is the genotype [7]. The genotypes response to environment is multivariate, yet the parametric approach tries to transform it to univariate problem via stability charac-ters. There is possible to cluster genotypes according to their response structure. This represents shifts from ranking stability by a quantitative measure to assigning genotypes into qualitatively homogeneous stability subset [14].

The Farinograph test is applied to winter wheat. The resistance of dough is evaluated by the Farinograph test, which means the evaluation of behavior of dough against mixing at a specified constant speed with specified water addition (ISO 5530–1:2013). Parameters determined by 5530 standard are consistency, Farinograph unit (FU), water absorption capacity of flour, dough development time, stability, mixing tolerance index and Farinograph quality number (FQN) [4]. The different baking products require wheat flours with different quality.

Winter wheat (*Triticum aestivum L.*) and maize (*Zea mays L.*) are considered main crops in the region of Nubaria. Available irrigation water for crop yields has been pumped from the Nile River or from Ground water, and average crop yield has substantially increased. However, recently there has been a rapid decline in available water resources from the ground water. There is an urgent need for more efficient water use in order to sustain agriculture in the area or Nubaria region [18, 20, 21]. Scheduling of irrigation is then more complex because irrigation operations must be based on the relationships between climate conditions, crop-growing stages, water requirements (ETc) and crop gross water applied. Alternative irrigation intervals must be assessed to determine which irrigation scheduling will give highest crop yield and water productivity (WP) for a given amounts of irrigation water [16, 17, 19].

Use of sprinkler irrigation, where smaller amounts of water can be uniformly applied to field, further helps to achieve high water use efficiencies [25]. Crops sprinkled with low quality water are exposed in two ways: Salts can affect plant growth and yield; direct salt adsorption through the leaves as well as increased soil salinity [15]. The use of the line source sprinkler method has been advocated for obtaining salinity production functions under such conditions [21]. With this method, it is possible to determine the separate and interactive effects of the quantity and salinity of applied water on crop yields. Therefore, the double-line sprinkler source was used for two growing seasons to determine the water–yield relations of wheat (*Triticumaestivum L.*) for two different types of marginal quality waters i.e., saline and alkali waters.

This chapter discusses effects of different water amounts based on crop water consumptive use (ETc) and four wheat varieties (Giza 168; Sids 12, Misr 1, and Misr 2 on some quality properties.

9.2   MATERIALS AND METHODS

The present investigation was conducted at National Research Center, El-Noubaria Research Station El-Behaira Governorate, during the two successive seasons of 2012/2013 and 2013/2014 to study the effects of different water amounts (100, 75, 50% of ETc) and four wheat varieties (Giza 168; Sids 12, Misr 1, and Misr 2) on some quality properties.

Some soil physical, chemical and water properties of the soil were determined [11, 22]. Soils of both investigated sites were sandy loam in texture. Analysis of farmyard manure was as follows: 4.85 dSm^{-1} (EC, 1:20), 7.77 (pH, 1:20), 11.2% (OM), 5.4, 0.85 and 1.12% total (N, P and K) and 1:16.5 (C:N ratio).

9.2.1   TECHNOLOGICAL QUALITY PROPERTIES

Moisture, fat, protein and ash contents were determined according to the standard methods [2]. The total sugar was calculated by difference method [5]:

$$\text{Total sugar (\%)} = [100 - (\text{moisture (\%)} + \text{crude protein (\%)} + \text{total lipids (\%)} + \text{ash (\%)}] \qquad (1)$$

Farinograph (Brabender GmbH and Co. Duisburg, Germany) was used to estimate dough quality parameters [1]. A 300 g sample at 14 % moisture content was weighed and placed into the corresponding Farinograph mixing bowl. Water was added to the flour and mixed to form dough, then the dough is mixed. The Farinograph records the curve on a graph paper. The amount of added water (absorption) affects the position of the curve on the graph paper. Less water increases dough consistency and moves the curve upward. The curve is centered on the 500-Brabender unit (BU) line 20 BU by adding the appropriate amount of water and is run until the curve leaves the 500-BU line. Dough Period (min), dough stability (min), mixing tolerance (B.U.) and water absorption (%) were analyzed.

Main factors and treatments mean were compared using analysis of variance (ANOVA) and the least significant difference (LSD) between systems at P + 1% [26].

9.3   RESULTS AND DISCUSSION

9.3.1   EFFECTS OF WATER APPLIED AMOUNT AND WHEAT VARIETIES ON WHEAT QUALITY

The data in Table 9.1 and Figures 9.1 and 9.2 illustrate the effects of three water levels (50, 75 and 100% of ETc) on quality parameters: % net flour, % of grain protein, % of fat, and % total sugar in flour. Regarding ETc, means values of % net flour, % of grain protein, % of fat, % total sugar and % wet gluten in flour and % dry gluten in flour, can be arranged in the following ascending order: 50 < 75 < 100. Table 9.1 and Figures 9.1 and 9.2 also illustrate the effects of wheat varieties (Giza 168, Sids 12, Misr 1 and Misr 2) on quality parameters: % net flour, % of grain protein, % of fat, and % total sugar in flour. In respect to wheat varieties, % net flour % of fat and % total sugar can be arranged in the following descending order: Giza 168 > Sids 12 > Misr 1 and > Misr 2. While % grain protein, % wet gluten in flour and % dry gluten in flour can be arranged in the following ascending order: Giza 168 < Sids 12 < Misr 1 and < Misr 2. The maximum values of % net flour, % of grain protein, % of fat, and % total sugar in flour were recorded under interactions of 120 X Giza 168, 120 X Misr 2,

TABLE 9.1   Effects of Evapotranspiration (ETc) Levels and Wheat Varieties on Grain Quality Properties

Water use ETc (%)	Wheat cultivars	% of net flour	% of grain protein	% of fat	% of total sugar
100	Giza 168	69.82	8.59	2.20	1.10
	Sids 12	69.04	8.86	1.92	1.02
	Misr 1	70.48	11.64	1.75	0.81
	Misr 2	74.02	11.87	1.64	0.69
Mean		**73.00**	**10.24**	**1.88**	**0.91**
75	Giza 168	71.80	9.81	2.55	1.34
	Sids 12	70.99	10.05	2.35	1.21
	Misr 1	72.45	13.02	1.97	0.91
	Misr 2	76.14	13.27	1.79	0.79
Mean		**74.60**	**11.54**	**2.16**	**1.06**
50	Giza 168	73.65	10.66	3.51	1.56

TABLE 9.1   Continued

Water use ETc (%)	Wheat cultivars	% of net flour	% of grain protein	% of fat	% of total sugar
	Sids 12	72.98	11.07	3.16	1.45
	Misr 1	74.35	13.75	2.79	1.33
	Misr 2	74.03	14.04	2.37	1.04
Mean		**72.91**	**12.38**	**2.96**	**1.34**
Mean	Giza 168	71.76	9.69	2.75	1.33
	Sids 12	71.01	9.99	2.48	1.23
	Misr 1	71.25	12.80	2.17	1.02
	Misr 2	71.01	13.06	1.93	0.89
LSD$_{0.05}$ for ETc means		**1.24**	**0.70**	**0.27**	**0.13**
LSD$_{0.05}$ for cultivars means		**0.64**	**0.21**	**0.10**	**0.06**
LSD$_{0.05}$ for interaction		**1.24**	**1.12**	**0.12**	**0.06**

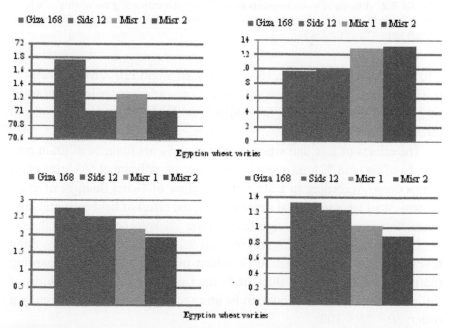

FIGURE 9.1   Effects of wheat varieties on grain quality properties. Y-axis: Top left – % of net flour; Top right – % of protein; Bottom left – % of fat in grain; Bottom right – % of total sugar.

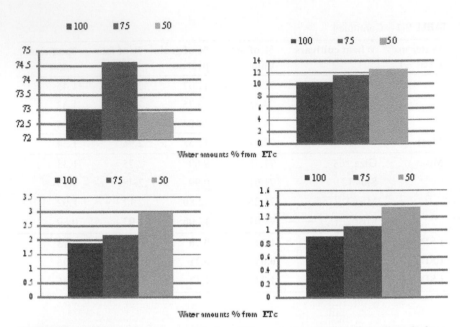

FIGURE 9.2 Effects of water amounts from ETc on properties of grain quality of wheat. Y-axis: Top left – % of net flour; Top right – % of protein; Bottom left – % of fat in grain; Bottom right – % of total sugar.

120 X Giza 168, 120 X Giza 168, 120 X Misr 2 and 120 X Misr 2, respectively. Whereas the minimum values were recorded under interactions of 60 X Misr 2, 60 X Giza 168, 60 X Misr 2, 60 X Misr 2, 60 X Giza 168 and 60 X Giza 168, respectively.

The effects of ETc and wheat varieties on % net flour, % of grain protein, % of fat, and % total sugar in flour were significant at 5% level of all interactions. According to LSD = 0.05, values of % net flour, % of grain protein, % of fat, and % total sugar in flour, the effect of field capacities and wheat varieties on all were significant at the 5% level between all values.

Data in Table 9.1 and Figures 9.1 and 9.2 show that the effect of three levels 50, 75, 100% from ETc on wheat quality parameters. Regarding main factor ETc, means values of % net flour, % of grain protein, % of fat, and % total sugar in flour can be arranged in the following ascending order: 50 < 75 < 100.

The interaction between field capacities and wheat varieties had a significant effect on % net flour. The maximum values of % net flour

(75.69 and 76.58 %) were obtained by Giza 168 wheat variety irrigated by ETc. The % of grain protein increases gradually by increasing ETc, where application of 50% ETc achieved the maximum % of grain protein. The highest % of grain protein (variety Giza 168) significantly exceeded the results of varieties (Sids 12, Misr 1, and Misr 2).

The interaction between ETc and wheat varieties had a significant effect on % of grain protein. The maximum values of % of grain protein were obtained by Giza 168 wheat variety irrigated by 50% ETc. The increasing of protein with increasing nitrogen fertilizer levels may be due to nitrogen element in the formation of amino acid structure.

In Table 9.1 for LSD = 0.05 of % net flour, % of grain protein, % of fat, and % total sugar, the effect of ETc and wheat varieties on all, there are significant differences at 5% level between all values. Concerning the effect of field capacities and wheat varieties on % net flour, % of grain protein, % of fat, and % total sugar in flour, there were significant differences at the 5% level at all interactions. The increasing of quality parameters with increasing nitrogen fertilizer levels may be due to nitrogen element in formation of amino acid structure. The maximum values in (% net flour, % of grain protein, % of fat, and % total sugar in flour) were recorded under interactions of 50% ETC X Giza 168, 50% ETc X Misr 2, 50% ETc X Giza 168, 50% ETc X Giza 168, 50% ETc X Misr 2 and 50% ETc X Misr 2, respectively. Whereas the minimum values were recorded (69.93, 9.15, 1.75 and 0.86 %) under interactions of 100% ETc X Misr 2, 100% ETc X Giza 168, 100% ETc X Misr 2, 100% ETc X Misr 2, 100% ETc XGiza 168 and 100% ETc X Giza 168, respectively.

In Table 9.1, it is noticed that % net flour, % of grain protein, % of fat, and % total sugar), the increasing of quality parameters with decreasing ETc levels may be due to concentrations of applied fertilizers of NPK and other elements in the sandy soils with lowest ETc and more leaching.

All measured values of dough quality evaluation (Farinograph) of wheat flour samples are shown in Table 9.2 and Figures 9.3 and 9.4, We can be clearly observe differences between values under ETc levels and wheat varieties. The values for water absorption increased from 58.97% at 50% ETc to 64.32% at 100% ETc, and maximum value with Misr1 variety. The same phenomenon was also observed for mixing tolerance (B.U.) values, which were 133.5, 137 and 139.5 B.U at ETc levels at 50,

TABLE 9.2   Effects of Evapotranspiration (ETc) Levels and Wheat Varieties on Dough Quality

ETc (%)	Varieties	Dough period (min)	Dough stability (min)	B.U. mixing tolerance	% of water absorption
100	Giza 168	2.6	2.2	147	59.92
	Sids 12	2.3	2.2	183.5	64.46
	Misr 1	1.7	2.1	118	69.92
	Misr 2	1.3	2.0	108.5	62.97
Mean		**1.78**	**2.13**	**139.5**	**64.32**
75	Giza 168	2.6	2.2	144	58.42
	Sids 12	2.3	2.1	179.5	61.01
	Misr 1	1.7	2.1	116	68.38
	Misr 2	1.3	2.0	108.5	61.17
Mean		**1.74**	**2.14**	**137**	**62.25**
50	Giza 168	2.4	2.2	139.5	54.15
	Sids 12	2.3	2.1	177	59.41
	Misr 1	1.7	2.1	112	65.42
	Misr 2	1.2	2.0	105	56.87
Mean		**1.66**	**2.08**	**133.5**	**58.97**
Mean	Giza 168	2.5	2.2	143.5	57.51
	Sids 12	2.3	2.1	180	61.65
	Misr 1	1.7	2.1	115	67.91
	Misr 2	1.3	2.1	108	60.33
$LSD_{0.05}$ for ETc Means		**0.03**	**0.008**	**2**	**0.73**
$LSD_{0.05}$ for cultivars Means		**0.25**	**0.012**	**3.5**	**2.04**
$LSD_{0.05}$ for Interaction		**0.025**	**0.01**	**1.5**	**1.06**

75 and 100%, respectively. On the other hand dough period (min) and dough stability (min) values of the wheat flour sample were (1.66 and 2.08), (1.74 and 2.14) and (1.78 and 2.13) min at 50,75 and 100% of ETc levels, respectively. The minimum values of dough period (min) and dough stability (min) were recorded with Misr 2 variety and maximum values with Giza 168 variety.

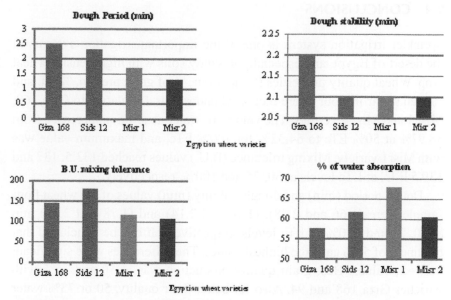

FIGURE 9.3   Effects of wheat varieties on Farinograph properties. Y-axis: Top left – Dough period in min; Top right – Dough stability in min; Bottom left – B.U. mixing tolerance; Bottom right – % of water absorption.

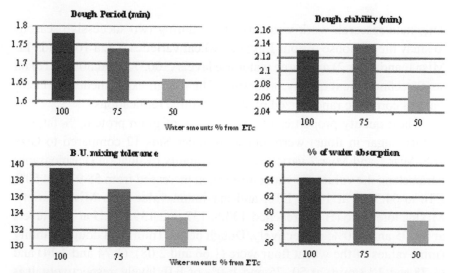

FIGURE 9.4   Effects of water amounts (% from ETc) on Farinograph properties of wheat. Y-axis: Top left – Dough period in min; Top right – Dough stability in min; Bottom left – B.U. mixing tolerance; Bottom right – % of water absorption.

9.4 CONCLUSIONS

Sprinkler irrigation system is one of the important irrigation systems in the desert of Egypt, and especially in sandy soils with high-density wheat crop. Wheat quality properties (% net flour, % of grain protein, % fat, and % total sugar in flour) were increased under Sids 12 relative to Giza 168, Misr 1 and Misr 2 verities. The values for water absorption increased from 58.97% at 50% ETc to 64.32% for 100% ETc, and maximum value was with Misr1 variety. Mixing tolerance (B.U.) values reached 133.5, 137 and 139.5 B.U. at ETc levels at 50, 75 and 100%, respectively.

Dough period (min) and dough stability (min) values of the wheat flour sample were (1.66 and 2.08), (1.74 and 2.14) and (1.78 and 2.13) min at 50, 75 and 100% of ETc levels, respectively. It can be concluded that treatment of 50% gave the highest values. The differences were significant between values. Best grain quality production was at 50 and 75% with varieties Giza 168 and 94. Also for better flour quality, 50 or 75% water levels and Sids 12 and Misr 1 varieties are recommended.

9.5 SUMMARY

This field research work was carried out during two successive seasons to study the response of four Egyptian wheat varieties (Giza 168, Sids 12, Misr 1, and Misr 2) and three water use levels (100, 75, and 50%, ETc) on some technological quality properties. The design experiment was factorial in s complete randomized blocks with three replications.

Wheat quality properties (% net flour, % of grain protein, % fat, and % total sugar in flour) were increased under Sids 12 compared to Giza 168, Misr 1 and Misr 2 varieties. Dough quality evaluation (Farinograph) parameters: the values for water absorption increased from 58.97% at 50% ETc to 64.32% at 100% ETc, and maximum value with Misr1 variety. Mixing tolerance (B.U.) reached 133.5, 137 and 139.5 B.U at ETc levels at 50, 75 and 100%, respectively. Dough period (min) and dough stability (min) values of the wheat flour were (1.66 and 2.08), (1.74 and 2.14) and (1.78 and 2.13 min) at 50, 75 and 100% of ETc levels, respectively. It is concluded that treatment of 50% gave the highest values. The best grain quality production was at 50 and 75% for Giza 168 and Sids 12 varieties. For better flour, it is recommended to use 50 or 75% applied water and Sids 12 and Misr 1 varieties.

KEYWORDS

- % fat
- % total sugar
- Cereal
- Dough period
- Dough stability
- Egyptian varieties
- Evapotranspiration
- Farinograph
- Grain protein
- Net flour
- Production
- Sprinkler irrigation
- Technological quality properties
- Wheat

REFERENCES

1. AACC International (2011). Method 54–21.01: Farinograph method for flour. In *Approved methods of the American Association of Cereal Chemists*. 11th ed. Am. Assoc. Cereal Chem., St. Paul, MN.
2. AOAC (2000). *Official Methods of Analysis*. Association of Official Analytical Chemists Washington, DC.
3. *Agronomy Research Monograph 8* (Special Issue III), pp. 637–644, 2010.
4. D'Appolonia, B. L., & Kunerth, W. H. (1984). *The Farinograph Handbook*. American Association of Cereal Chemistry, St. Paul, MN. VII-64.
5. Egan, H. I., Kirk, R. S., & Sawyer, R. (1981). *Person's Chemical Analysis of Food*. 8th ed., Churchill Livingstogne, Edintugh.

6. Finney, K. F., Yamazaki, W. T., Youngs, V. L., & Rubenthaler, G. L. (1987). *Quality of Hard, Soft, and Durum Wheats. Wheat and Wheat Improvement.* Agronomy Monograph 13, 677– 748.

7. Flores, F., Moreno, M. T., & Cubero, J. I. (1998). A comparison of univariate and multivariate methods to analyze GxE interaction. *Field Crops Research,* 56, 271–286.

8. Fowler, B. (2002). *Winter Cereal Production.* www.usask.ca/agriculture/cropsci/winter_cereals/ (08.03.2007).

9. Francis, T. R., & Kannenberg, L. W. (1978). Yield stability studies in short-season maize, I: A descriptive method for grouping genotypes. *Can. J. Plant Sci.,* 58, 1029–2034.

10. Grausgruber, H., Oberforster, M., Werteker, M., Ruckenbauer, J., & Vollmann, J. (2000). Stability of quality traits in Austrian-grown winter wheats. *Field Crop Research,* 66, 257–267.

11. Klute, A. (1986). *Water Retention: Laboratory Methods.* In: *A. Klute (ed.), Methods of Soil Analysis, Part 1, Physical and Mineralogical Methods.* 9 ASA and SSSA, Madison, WI. Pages 635–662.

12. Levy, Y., Columbus, D., & Sadan, D. (1999). Trickle linear gradient for assessment of the salt tolerance of citrus rootstocks in the orchard. *Irrig. Sci.,* 18, 181–184.

13. Lin, C. S., & Binns, M. R. (1998). A method of analyzing cultivar x location x year experiments: A new stability parameter. *Theor. Appl. Genet.,* 76, 425–430.

14. Lin, C. S., Binns, M. R., & Lefkovitch, L. P. (1986). Stability analysis: Where do we stand. *Crop Sci.,* 26, 894–900.

15. Maas, E. V., & Grattan, S. R. (1999). Crop yields as affected by salinity. In: *Skaggs, R.W., van Shilfgaarde, J. (Eds.), Agricultural Drainage.* Agron. Monog. 38, ASA CSSA SSSA, Madison, Wisconsin, pp. 55–108.

16. Mansour, H. A. (2006). *The Response of Grape Fruits to Application of Water and Fertilizers Under Different Localized Irrigation Systems.* M.Sc. Thesis, Faculty of Agriculture, Agric., Ain Shams University, Egypt.

17. Mansour, H. A. (2012). *Design Considerations for Closed Circuits of Drip Irrigation System.* PhD Thesis, Faculty of Agriculture, Agric., Ain Shams University, Egypt.

18. Mansour, H. A., Abdallah, E. F., Gaballah, M. S., & Gyuricza, Cs. (2015a). Impact of bubbler discharge and irrigation water quantity on 1- Hydraulic performance evaluation and maize biomass yield. *Int. J. of GEOMATE,* 9, 1538–1544.

19. Mansour, H. A., & Abd El-Hady, M. (2014). Performance of Irrigation Systems under Water Salinity in Wheat Production. *IOSR-JAVS Journal of Agriculture and Veterinary Science,* 7(7), 19–24.

20. Mansour, H. A., Abdel-Hady, M., Ebtisam I. El-Dardiry, & Bralts, V. F. (2015b). Performance of automatic control different localized irrigation systems and lateral lengths, 1-emitters clogging and maize (*Zea mays* L.) growth and yield. *Int. J. of GEOMATE,* 9, 1545–1552.

21. Mansour, H. A., Gaballah, M. S., Abd El-Hady, M., & Ebtisam I. Eldardiry (2014). Influence of different localized irrigation systems and treated agricultural wastewater on distribution uniformities, potato growth tuber yield and water use Efficiency. *International Journal of Advanced Research,* 2, 143–150.

22. Rebecca, B. (2004). *Soil Survey Laboratory Methods Manual*. Soil Survey Investigations Report 42 by Natural Resources Conservation Services.

23. Reese, C. L., Clay, D., Beck, D., & England, R. (2007). Is protein enough for assessing wheat flour quality? *Western Nutrient Management Conference*, Salt Lake City, UT, 7, 85–90.

24. Simmonds, D. H. (1989). Fundamental aspects of wheat quality: the protein fraction. In: *CSIRO Editorial Services (Eds.). Wheat and Wheat Quality in Australia*, 9, 183–208.

25. Singh, H. P., Kaushish, S. P., & Kumar, A. (2000). Micro-Irrigation. In: *Proceedings of the International Conference on Micro and Sprinkler Irrigation Systems*, 2000, Central Board of Irrigation and Power, New Delhi, February 8–10, p. 794.

26. Snedecor, G. W., & Cochran, W. G. (1980). *Statistical Methods*. 7th Edition. Iowa State Univ. Press, Iowa, U.S.A.

27. Tipples, K. H. (1986). Testing candidate wheat cultivars for quality. *Wheat Production in Canada, Proceedings of the Canadian Wheat Production Symposium*.

CHAPTER 10

MAXIMIZING PROFITS BY USING DIFFERENT PLANTING GEOMETRY UNDER MICRO IRRIGATION

AJAI SINGH

Associate Professor and Head, Centre for Water Engineering and Management, Central University of Jharkhand, Brambe, Ranchi–834205, India. E-mail: ajai_jpo@yahoo.com; ajai.singh@cuj.ac.in

CONTENTS

10.1　INTRODUCTION

Water is now one of the natural resources, which needs to be used very carefully for any purpose. A judicious water management approach should be inculcated in the habit of human being. Water is important for every sphere of our life and nowadays approach is shifting from global water management to local level. The world's six billion people are already appropriating 54% of all accessible freshwater reserves. It is predicted that by 2025 humankind's share will be 70%. The UNDP's World Water Development Report 2003 accounts for only 8% of global water consumption.

The agricultural sector is the largest user of water globally and accounts for about 70% of the total freshwater abstraction. However, it is estimated that industrial water requirement will increase and grab the share of household and agriculture sector. In fact, in high-income countries, industrial water use already accounts for as much as around 60% of the total fresh water consumption that is almost twice the amount used in agriculture. It is likely, then, that this will become a global trend even as more and more nations begin to choose industry over agriculture, as a key to economic growth. This estimate reflects how we need to allocate more water for human being. Agricultural use of water accounts for nearly 70% of the water used throughout the world, and the majority of this water is used for irrigation.

Traditionally, efforts to address water supply problems have focused on major and medium irrigation projects. However, use of water in India is characterized by an increasing dependence on groundwater for irrigation. The annual extraction of groundwater in India (210 billion m^3) is by far the highest in the world. Groundwater today provides for more than 60% of the net irrigated area. It accounted for over 85% of the addition to the irrigated area in the last 30 years. The area irrigated by canals and tanks has actually undergone a decline even in absolute terms since the 1990s. Mainly groundwater is used to irrigation crop under drip irrigation.

India has the second largest net irrigated area in the world, after China. The irrigation efficiency under canal irrigation is not more than 40% and for ground water schemes, it is 69%. The net irrigated area in the country is 53.5 M-ha, which is about 38% of the total area sown. Although considerable area has been brought under irrigation since independence, yet there is much scope for its expansion in the future. Irrigation water for agriculture finds competition from domestic use, industrial and hydroelectric projects. At present, the efficiency of the irrigation systems adopted is less than 30%. As such as 50% of the water release at the project head is lost in transmission of the canal outlet. Additional losses occur in water courses which is directly proportional to their length and duration of water flow. Considerable scope exists for enhancing the water use efficiency (WUE) to bring additional area under irrigation. Scientific management of irrigation water is necessary to improve crop productivity and alleviate irrigation related problems such as shortage of irrigation water, water logging, salinity etc.

This chapter deals with innovation. If we can manipulate the planting geometry i.e., changing number of plants per unit area without affecting the yield or number of plants per hectare, considerable amount of water can be saved. If number of plants is increased by changing the planting geometry, of course more water will be needed, but yield from unit area will be increased. It depends how we plan the line of action. If water availability is not a limiting factor, it is possible that we can go for higher number of plants per unit area and thus could fetch more yields.

10.2 MICRO IRRIGATION TECHNOLOGY

Micro irrigation systems are gradually getting encouragement from the government in the form of subsidy. This system has achieved good coverage in water scarce areas. The advantages and disadvantages of micro irrigation systems are now well known and enough documents are available for the readers.

Drip irrigation is an effective way to supply water and nutrients to the plants. The major drawback of the drip irrigation system is its high initial investment. However, the cost can be recovered earlier if proper nutrients, water management and design principles of drip irrigation system are followed. Physical factors such as field dimensions and shape, and topography of lands will influence the layout of pipe networks. Among the various components of drip irrigation system, the cost of lateral is the major factor, which influence the total system cost.

Any effort made to reduce the length of lateral required per unit area of the field will result in reduction of the system cost. The benefit-cost ratios for various fruits and vegetable crops have been worked out. It is seen that for vegetable crops, the benefit-cost ratio was in the range of 2–3 excluding the water savings and 3 to 3.5 including water saving. The economic feasibility of drip irrigation in combination with different types of mulches for an okra crop was evaluated. The study indicated that

100% irrigation requirement was met through drip irrigation along with black plastic mulch. This irrigation management gave the highest yield with 72% increase in yield as compared to furrow irrigation [2].

Several investigations have shown the selection of most economical size of the pipe in pumped irrigation system. In India, government has initiated several programs to boast the adoption of micro irrigation system. To achieve this, several *Plasticulture Development Project Centers* (PDPP) were established throughout the country. The main focus of the center was to conduct research on advancement of micro irrigation systems and organizing training programs for farmers as well as for officers of the State Governments.

Tiwari and Reddy [1] evaluated the effect of the planting geometry on yield, capital cost, operating cost and net return for banana crop irrigated with trickle irrigation system. It was found that there was variation in the yield with different planting geometries and the maximum yield was recorded for 2 m x 2 m planting geometry. In several studies, trickle irrigation system has been recommended to design based on emission uniformity and economic criteria. It has been recommended an economical design with varying pipe sizes for main lines rather than design with uniform pipe size. It was also reported that the saving in total annual cost was 29% over the criteria, where the emission uniformity was the sole parameter in designing the system. Good amount of work has been carried out to design of drip irrigation system based on crop planting geometry for vegetable crop.

10.3 METHODS AND MATERIALS

In a research work where the author has participated as the co-author, an attempt was made to access the effect of okra crop geometry on yield, initial investment, total annual cost of drip system and net return for crop grown on red lateritic soil, sub-humid tropical climate of Kharagpur, West Bengal. The lateral lines were laid in the middle of four rows of crop and were provided with online emitters of 4 lph discharge. Lateral lines can be laid in between two rows also but the cost becomes high. The plant spacing was 30, 40, 50 and 60 cm. The normal practice is to grow okra with 60 x 60 cm^2 spacing.

The lateral spacing for each crop geometry was calculated while keeping the plant population same for all treatments. It means unplanted area will be more in case of 30 x 30 cm^2 spacing. If the whole area is planted with this spacing, of course the number of plants per hectare will be more. Same plant population was kept for fair comparison among the different plant spacing. Standard agronomical practices were followed in case of seed sowing, fertilizers and pest application.

10.4 RESULTS AND DISCUSSION

Water requirement of okra crop was estimated by Modified Penman's Method [3] as given Table 10.1.

For estimating the Benefit–Cost ratio (BCR), economic analysis was conducted. The seasonal cost of drip irrigation system included: depreciation, prevailing bank rate of interest and repair and maintenance @ 12% per annum and 2% of the fixed cost, respectively. The useful life of drip system was considered as 5 years or 10 seasons.

Author focused mainly on the economic aspect of the study though they also conducted biometrics analysis and effect of market price on BCR also. The salient findings of the study were:

- For irrigating one hectare of okra crop, 665 mm of water was required.
- Maximum yield was obtained in case of plant spacing of 30 x 30 cm², which reduces the lateral length by half as compared to 60 x 60 cm² plant spacing. Good saving of money was realized.
- Highest benefit-cost ratio of 2.18 was obtained for plant spacing of 30 x 30 cm².

In this study, authors fixed the number of plants per hectare. If number of plant are not fixed then it would be much more in case of plant spacing

TABLE 10.1 Estimated Crop Water Requirement for Okra

Months	ET_o, mm/d	Crop coefficient, K_c	Wetted area, 1.2 m x 1.2 m	Volume of water, l/d/8 plants
	(1)	(2)	(3)	(1 x 2 x 3)
February	7.38	0.45	1.44	4.80
March	9.06	0.75	1.44	9.78
April	9.16	1.15	1.44	15.16
May	9.16	0.87	1.44	11.50

of 30 x 30 cm^2 as compared to 60 x 60 cm^2. Optimal spacing of plants is required to be determined by conducting experiments to maximize profits and to utilize our natural resources more judiciously.

10.5 SUMMARY

This chapter deals with innovative method of planting okra. Author claims that: If we can manipulate the planting geometry (changing number of plants per unit area without affecting the yield or number of plants per hectare), considerable amount of water can be saved. If number of plants is increased by changing the planting geometry, of course more water will be needed, but yield from unit area will be increased. It depends how we plan the line of action. If water availability is not a limiting factor, it is possible that we can go for higher number of plants per unit area and thus could fetch more yields.

In this study, okra water requirement was 665 mm/ha. The benefit cost ratio was 2.18 for a plant spacing of 30 x 30 cm^2. Maximum yield was obtained under plant spacing of 30 x 30 cm^2 compared to 60 x 60 cm^2 plant spacing. Good saving in the high investment was realized.

KEYWORDS

- Benefit–cost ratio
- Drip irrigation
- High initial investment
- India
- Micro irrigation
- Okra
- Online emitters
- Plant spacing
- Plastic mulch
- Water requirement
- Water saving

REFERENCES

1. Tiwari, K. N., & Reddy, K. Y. (1997). Economic analysis of trickle irrigation system considering planting geometry. *Agricultural Water Management*, 34, 95.
2. Tiwari, K. N., Mal, P. K., Singh, R. M., & Chattopadhyay, A. (1998). Response of okra (*Abelmoschus esculentus M.*) to drip irrigation under mulch and non-mulch conditions. *Agricultural Water Management*, 38, 91–102.
3. Tiwari, K. N., Singh, A., Mal, P. K., & Pandey, A. (2001). Effect of crop geometry on yield and economics of okra (*Abelmoschus exculentus L. Moench*) under drip irrigation. *Jr. of the Institution of Engineers, Division of Agricultural Engineering, India*, 82, 9–12.

CHAPTER 11

DESIGN OF LATERAL LINES

VICTOR A. GILLESPIE,[1] ALLAN L. PHILLIPS,[2] and I. PAI WU[3]

[1]*Retired Professor and Extension Specialist, Department of Molecular Biosciences and Bioengineering, University of Hawaii at Mañoa, Honolulu, Hawaii; Mailing address: 3050 Maileway, Honolulu – Hawaii – 96822, USA*

[2]*Former Professor and Director, Department of Agricultural and Biosystems Engineering, University of Puerto Rico - Mayaguez Campus, Mayaguez, Puerto Rico 00681, USA. Current Address: 7602 Kilgore Rd, Avoca, MI – 48006. E-mail: allan@greatlakes.net*

[3]*Retired Professor, Department of Molecular Biosciences and Bioengineering, University of Hawaii at Mañoa, Honolulu, Hawaii; Mailing address: 3050 Maileway, Honolulu – Hawaii – 96822, USA. Tel. 808-956-8809. E-mail: ipaiwu@hawaii.edu*

CONTENTS

Reprinted with permission from, "Victor A. Gillespie, Allan L. Phillips, and I. Pai Wu, (2013). *Design of lateral lines*. Chapter 13, pp. 247–259. In: *Management of Drip/Trickle or Micro Irrigation*, edited by Megh R. Goyal. Apple Academic Press, Inc."

11.1   INTRODUCTION

In a drip irrigation system, a major design criteria is a minimization of the discharge (or emitter flow) variation along a drip irrigation line, either a lateral or a submain [1]. A typical drip irrigation system is shown in Figure 11.1. The discharge variation can be kept within acceptable limits in laterals or submains of a fixed diameter by designing a proper length for a given operating pressure. The discharge (or emitter flow) variation is controlled by the pressure variation along the line which results from the combined effect of friction drop and line slopes. When the kinetic energy is considered to be small and neglected in a drip irrigation line, the pressure variation will be simply a linear combination of the friction drop and energy gain or loss due to slopes as suggested by Wu and Gitlin [6] and Howell and Hiler [2].

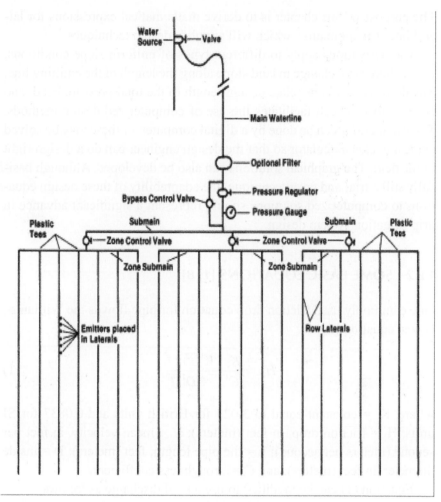

FIGURE 11.1   Layout of lateral lines and components of a drip system for the design.

A lateral length (or submain) can be designed by using a step by step calculation using a computer. The computer program can be used to simulate different situations to develop design charts as shown by Wu and Fangmeier [5]. Simplified design procedures were developed [3, 7] by using a general shape of the energy gradient line and line slopes. Design charts for lateral line design were introduced by Wu and Gitlin [7], however trial and error techniques are required in the design procedure.

The purpose of this chapter is to derive mathematical expressions for lateral lines (or submains), which will simplify design techniques.

The derivations apply to different types of uniform slope conditions, where there is no change in land slope along the length of the emitting line. The derived equations relate design length to the total pressure head, and are in a form which facilitates the use of computerized design methods. The calculations can be done by a digital computer, or these may be solved using a pocket calculator so that the design engineer can do a design right in the field. The graphical solutions can also be developed. Although basically still a trial and error technique, the adaptability of these design equations to computerized solutions should represent a significant advance in drip irrigation system design.

11.2   SOME BASIC EQUATIONS [1, 8]

One commonly used friction drop equation for pipe flow is the Williams–Hazen equation [4].

$$H_f = \frac{K_1 * V^{1.852} * L}{C^{1.852} * D^{1.167}} \tag{1}$$

where, K_1 = constant equal to 3.023 for British units and 0.0837 for SI units; H_f = friction drop, in feet (meters); V = mean velocity, in feet per second (meters per second); L = the pipe length, feet (meters); D = inside diameter, in feet (meters); and C = a roughness coefficient.

Equation (1) can be modified to use a total discharge as follows:

$$H_f = \frac{K_2 * Q^{1.852} * L}{C^{1.852} * D^{4.871}} \tag{2}$$

where, K_2 = constant equal to 10.45 for British units and 2.264×10^7 for SI units; Q = expressed as gallons per minute (liters per second); and D = inside diameter, inches (centimeters).

Equation (2) calculates the friction drop using total discharge. For lateral line or submain, the discharge in the line decrease with respect to the length of the line. The total friction drop at the end of the line can be calculated by applying a correction factor of, [1/2.852], indicated by Wu and Gitlin [8]. The total energy drop due to friction at the end of a lateral line or submain can be expressed as:

$$\Delta H = \frac{K_3 * Q^{1.852} * L}{C^{1.852} * D^{4.871}}$$ (3)

where, K_3 = constant equal to 3.6642 for British units and 7.94×10^6 for SI units; and ΔH = the total friction drop at the end of a lateral line or submain, in feet (meters).

Assuming the emitter flow q is uniform or is designed with a certain variation, one can rearrange Eq. (3) into:

$$\Delta H = \frac{K_3 * q^{1.852} * L^{2.852}}{C^{1.852} * S_p^{1.852} * D^{4.871}}$$ (4)

where, q = average emitter flow, in gallons per minute (liter per second); and Sp = emitter spacing, in feet (meter).

In a drip irrigation design, the terms q, C, S_p, and D are usually known, therefore:

$$\Delta H = K * L^{2.852}, \text{ where } K = \frac{K_3 * q^{1.852}}{C^{1.852} * S_p^{1.852} * D^{4.871}} = \text{Constant}$$ (5)

The total friction drop divided by total length is a dimensionless term S [$=\Delta H/L$]. Equation (5) can be expressed as:

$$S = K * L^{1.852}$$ (6)

The total friction drop shown in Eq. (5) is the total friction drop over the full length of the line. The friction drop along the line can be determined from a dimensionless energy gradient line as derived by Wu and Gitlin [8]. It can be expressed as follows:

$$\Delta Hp = [1 - \{1 - P/L\}^{2.852}] * \Delta H$$ (7)

where, ΔH_p = the total friction drop at a distance P, from the inlet.

When a lateral line or submain is laid on uniform slopes, the total energy gain (down slope situation) or loss (up slope situation) due to change in elevation can be expressed as:

$$\Delta H' = S_o * L \tag{8}$$

where, $\Delta H'$ = the total energy gain or loss due to uniform slope at the end of the line, in feet (meters); and S_o = the line slope.

The energy gain or loss at a point along the line due to uniform slopes can be shown as:

$$\Delta H'_p = S_o * p \tag{9}$$

$$\Delta H'_p = P/L * \Delta H' \tag{10}$$

where, $\Delta H'_p$ = the energy gain or loss due to slopes at a length P measured from the inlet; S_o = the land slope; and $\Delta H'$ = the energy gain or loss due to slope over the total length of the line.

11.3   PRESSURE PROFILES

The pressure head profiles along the lateral or submain can be determined from the inlet pressure, friction drop, and energy change due to slopes. It can be expressed as:

$$H_p = H - \Delta H_p \pm \Delta H'_p \tag{11}$$

where, H = the inlet pressure or operating pressure expressed as pressure head, in feet (meters); the plus (+) sign means down slope; and the minus (−) sign means up slope.

Substituting equations (7) and (10) into equation (11), we have:

$$H_p = H - [1 - \{1 - P/L\}^{2.852}] \Delta H \pm P/L * \Delta H' \tag{12}$$

The Eq. (12) describes pressure profiles along a lateral line or submain. The shape of profiles will depend on the inlet pressure (initial pressure), total friction drip, and total energy change by slopes. There are five typical pressure profiles as shown in Figure 11.2.

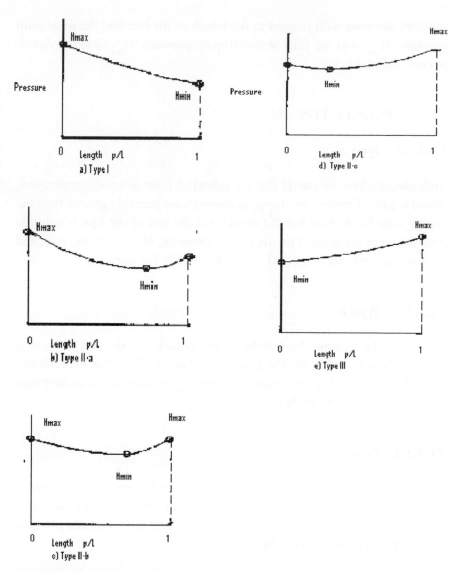

FIGURE 11.2   Profiles of pressure head.

11.3.1   PROFILE TYPE I

This occurs when the lateral line (or submain) is on zero or uphill slope. Energy is lost by both elevation change due to up slope and friction. The pressure decrease with respect to the length of the line and the maximum pressure, H_{max} is at the inlet and minimum pressure, H_{min} is at the downstream end of the line.

11.3.2 PROFILE TYPE II

11.3.2.1 Type a

This occurs when the lateral line (or submain) is on down slope situation, where a gain of energy by slopes at downstream points is greater than the energy drop by friction but the pressure at the end of the line is still less than the inlet pressure. The maximum pressure, H_{max} is at the inlet and a minimum pressure is located somewhere along the line.

11.3.2.2 Type b

This is similar to Type II-a but the profile is such that the end pressure is equal to the inlet pressure. The maximum pressure, H_{max} is at the inlet and the end of the line. The minimum pressure, H_{min} is located somewhere near the middle section of the line.

11.3.2.3 Type c

This occurs when the line slope is even steeper so the pressure at the end of line is higher than the inlet pressure. In this condition, the maximum pressure, H_{max} is at the downstream end of the line and the minimum pressure is located somewhere along the line.

11.3.3 PROFILE TYPE III

This occurs when the lateral line (or submain) is on steep down slope conditions where the energy gain by slopes is larger than the friction drop for all sections along the line. In this condition, the maximum pressure is at the downstream end of the line and minimum pressure is at the inlet.

The location of the minimum pressure along the pressure profile II-a-b-c, can be determined by differentiating Eq. (12) with respect to the length P and setting the derivative equal to zero.

$$\{2.852*(1 - P/L)^{1.852} * \Delta H/L\} - \Delta H'/L = 0 \qquad (13)$$

If the term $\Delta H/L$, the ratio of total friction drop to length, is set as energy slope S, Eq. (13) becomes:

$$\{2.852*(1 - P/L)^{1.852} * S\} - S_o = 0 \qquad (14)$$

Simplifying:

$$P/L = 1 - [0.3506 \, S_o/S]^{0.54} \qquad (15)$$

Equation (15) shows the location of the point of minimum pressure when both S_o and S are known.

11.4 DESIGN EQUATIONS

Since the five pressure profiles are smooth curves as shown in Figure 11.2, pressure variation can be used as design criteria. The pressure variation is defined as:

$$H_{var} = \frac{H_{max} - H_{min}}{H_{max}} \qquad (16)$$

where, H_{var} = pressure variation. The maximum and minimum pressure along the line are H_{max} and H_{min}, respectively.

The emitter flow variation can also be expressed as follows:

$$q_{var} = \frac{q_{max} - q_{min}}{q_{max}} \qquad (17)$$

where, q_{max} = maximum emitter flow along the emitting line; and q_{min} = minimum emitter flows along the emitting line produced by H_{max} and H_{min}.

For the orifice type of emitter flow, the relationship between q_{var} and H_{var} is given by Eq. (18):

$$H_{var} = 1 - [1 - q_{var}]^2 \tag{18}$$

Using pressure variation as design criteria, the design equations for each pressure profile type are derived.

11.4.1 PROFILE TYPE I

The inlet pressure is the maximum pressure and the minimum pressure is at the end of the line:

$$H_{min} = H - (\Delta H + \Delta H') \tag{19}$$

The pressure variation can be expressed as:

$$H_{var} = \frac{H - [H - (\Delta H + \Delta H')]}{H} \tag{20}$$

$$H_{var} = \frac{\Delta H + \Delta H'}{H} \tag{21}$$

Both sides of Eq. (21) can be multiplied by H/L, to obtain:

$$\frac{H_{var}}{L} * H = \frac{\Delta H}{L} + \frac{\Delta H'}{L} \tag{22}$$

$$H_{var} * H = (S + S_o)L \tag{23}$$

$$L = \frac{H_{var} * H}{S + S_o} \tag{24}$$

The values of H_{var} and H are selected by the designer; S_o can usually be obtained from field measurements; S and L are unknown. If S is as a function of L [Eq. (6)], it is possible to substitute Eq. (6) for S and derive a computational form of Eq. (24) that contains only one unknown variable, L. It is the same equation as given by Howell and Hiler [2].

$$L = \frac{H_{var} * H}{K * L^{1.852} + S_o} \tag{25}$$

11.4.2 PROFILE TYPE II

11.4.2.1 Type a

The inlet pressure is the maximum pressure and the minimum pressure is somewhere along the line. The line slope is downhill and there is energy gain due to slope. The pressure variation can be expressed as:

$$H_{var} = \frac{H - [H + (\Delta H'_p + \Delta H_p)]}{H} \tag{26}$$

$$H_{var} = \frac{\Delta H_p - \Delta H'_p}{H} \tag{27}$$

$$\frac{H_{var} * H}{L} = \frac{\Delta H_p}{L} - \frac{\Delta H'_p}{L} \tag{28}$$

Substituting Eqs. (7) and (10) into Eq. (28) and simplifying, we obtain:

$$L = \frac{H_{var} * H}{[1 - \{1 - P/L\}^{2.852}] * S - \dfrac{P * S_0}{L}} \tag{29}$$

Substituting Eq. (15) into Eq. (29) and simplifying, we obtain:

$$L = \frac{H_{var} * H}{S + S_o[0.3687 * (S_o / S)^{0.54} - 1]} \tag{30}$$

The computerized form of Eq. (30) is obtained by substituting the Eq. (6) into Eq. (30):

$$L = \frac{H_{var} * H}{K * L^{1.852} + S_0[0.3687 * (S_o / K * L^{1.852})^{0.54} - 1]} \qquad (31)$$

11.4.2.2 Type b

This is similar to Type II-a, the only difference is that S and S_o are equal as defined by Eqs. (6) and (8). It is therefore possible to substitute S_o for S in both Eqs. (30) and (31). Equation (30) can be shown as:

$$L = \frac{H_{var} * H}{S + S_o[0.3687 * (S_o / S)^{0.54} - 1]} \qquad (32)$$

Simplifying:

$$L = \frac{H_{var} * H}{0.3687 * S_o} \qquad (33)$$

11.4.2.3 Type c

The maximum pressure is located at the downstream end of line and the minimum pressure is somewhere along the line. The pressure variation can be expressed as:

$$H_{var} = \frac{[H + (\Delta H' - \Delta H) - [H + (\Delta H_p - \Delta H'_p)]}{H + (\Delta H' - \Delta H)} \qquad (34)$$

$$H_{var} = \frac{\Delta H' - \Delta H + \Delta H_p - \Delta H'_p)]}{H + (\Delta H' - \Delta H)} \qquad (35)$$

$$H_{var}H = [\Delta H'/L - \Delta H/L] + \Delta H_p/L - \Delta H'_p/L] -$$
$$H_{var}/L * [\Delta H'/L - \Delta H/L] \qquad (36)$$

Substituting Eqs. (7), (10) and (15), and simplifying, we obtain:

$$L = \frac{H_{var} * H}{S_o[0.3687(S_o / S)^{0.54}] - [(S_o - S) * H_{var}]} \qquad (37)$$

$$L = \frac{H_{var} * H}{S_o[0.3687 * (S_o / KL^{1.852})^{0.54}] - H_{var} * [S_o - KL^{1.852}]} \quad (38)$$

11.4.3 PROFILE TYPE III

The derivation of the Type III profile is simpler than the other down slope situations because there is no minimum point along the pressure profile. The value H_{min} is H at the head of the emitting line, and H_{max} is at the end of the emitting line. The pressure variation can be expressed as:

$$H_{var} = \frac{H + (\Delta H' - \Delta H) - H}{H + (\Delta H' - \Delta H)} \quad (39)$$

$$H_{var}H + H_{var} * (\Delta H' - \Delta H) = (\Delta H' - \Delta H) \quad (40)$$

$$\frac{H_{var} * H}{L} = (S_o - S) - H_{var} * (S_o - S) \quad (41)$$

$$L = \frac{H_{var}H}{(S_o - S) * (1 - H_{var})} \quad (42)$$

The design length can be expressed as:

$$L = \frac{H_{var}H}{(S_o - KL^{1.852}) * (1 - H_{var})} \quad (43)$$

11.5 CRITERIA FOR THE SELECTION OF THE APPROPRIATE DESIGN EQUATION

The criteria for selecting which of the five design equations to use for a given land slope and flow situations is dependent on the relationship between S and S_o. The criteria for the Type I profile is simplest, Eq. (25) is used when there is zero slope or for uphill slopes. The criteria for choosing which of the four down slope design equations to use are based on the magnitude of S and S_o and on Eq. (15). The Type II-a profile is characterized by S being greater than S_o:

$$S > S_o, S/S_o > 1 : \frac{KL^{1.852}}{S_o} > 1 \qquad (44)$$

The profile Type II-b is characterized because S is equal to S_o:

$$S = S_o, S/S_o = 1 : \frac{KL^{1.852}}{S_o} = 1 \qquad (45)$$

The profiles Types II-c and III are characterized, because S is smaller than S_o:

$$S < S_o, S/S_o < 1 : \frac{KL^{1.852}}{S_o} < 1 \qquad (46)$$

If the land slope and flow conditions satisfy this inequality, it is possible to use equation (15) to determine which design equation to use for the Type II-c pressure profile. The minimum point occurs at P/L greater than zero and less than one. This occurs if the following inequality holds true:

$$0 < \left(0.3506 \frac{S_o}{S} \right)^{0.54} < 1 \qquad (47)$$

$$0 < \left(0.3506 \frac{S_o}{S} \right) < 1 \qquad (48)$$

$$0 < \frac{S_o}{S} < 2.852 \qquad (49)$$

$$0 < \frac{S_o}{KL^{1.852}} < 2.852 \qquad (50)$$

Conversely, if the land slope and flow situations is such that $S < S_o$, but S_o/S 2.852, then it is appropriate to use the Type III profile design equation. When the design procedure begins, S is usually unknown, and it is not possible to solve the down slope criteria equations until L is either known or calculated. If S is not known, it is necessary to intuitively select one of the four down slope design equations and to solve it for a value of S. This value can then be tested using the appropriate criteria equation. If the criteria are not satisfied, it is necessary to solve for another value of S using another of the downhill design equations. In most down slope situations, it is appropriate to first use the Type II-a profile design equation. Generally the designer will try to design a lateral line to have a gentle down slope condition, which usually will result in a Type II-a, b, or c profile.

11.6 DESIGN EXAMPLES

In the developed design equations, the design length cannot be solved directly. One can use a calculator and use a trial and error method to determine the length, or use Newton's method of approximation iteratively to determine the length using a computer program. Two design examples are shown in the following subsections.

11.6.1 EXAMPLE 1

A lateral line is on a 1% uphill slope. The following data are given, and it is necessary to determine the maximum L for the land slope and flow conditions using Type I profile design equation: Emitter spacing, $S_p = 2.0$ ft. (0.61 m); Emitter diameter, $d = 0.010$ in (0.254 mm); Design emitter flow, $q = 0.0047$ gal/min (2.1 x 10^{-5} L/s); Lateral line diameter, $D = 0.56$ in (14.2 mm); Inlet pressure, $H = 10.4$ ft. (3.17 m); Pressure variation magnitude, $H_{var} = 0.19$; Land slope uphill, $S_o = 0.010$; and Roughness coefficient, $C = 150$. The Eq. (5) is written as:

$$K = \frac{7.94 x 10^6 q^{1.852}}{C^{1.852} * S_p^{1.852} * D^{4.871}}$$
(51a)

$$K = \frac{7.94 x 10^6 (2.1 x 10^{-5})^{1.852}}{(150)^{1.852} * (0.61)^{1.852} * (1.42)^{4.871}}$$
(51b)

$$K = 7.82 \times 10^{-8}$$
(51c)

From the Eq. (25):

$$L = \frac{Hvar * H}{K * (L^{1.852}) + So} = \frac{(0.19)(3.17)}{(7.82x10^{-8})(L^{1.852}) + (0.010)} \tag{52a}$$

$$L = 178 \text{ feet (54.27 meters)} \tag{52b}$$

We can also obtain a graphical solution solving H for various L and tracing a graph to determine "L." The particular type of line can extend approximately to 180 feet (55 m) for an up hill slope of 1%, before H_{var} will exceed 19% that corresponds to a q_{var} of 10%.

11.6.2   EXAMPLE 2

A lateral line is on a 1.5% downhill slope. The first design equation used is that for the Type II-a profile. Emitter spacing, $S_p = 8$ ft. (2.4 m); Emitter diameter, $d = 0.019$ in (0.48 mm); Design emitter flow, $q = 0.026$ gal/min (1.14×10^{-4} L/s); Emitting line diameter, $D = 0.612$ in (15.5 mm); inlet pressure, $H = 28.37$ ft. (8.65 m); Pressure variation magnitude, $H_{var} = 0.19$; Land slope downhill, $S_o = 0.015$; and Roughness coefficient, $C = 137$. Using these values, we obtain:

$$K = 7.98 \times 10^6 (1.14 \text{ x } 10^{-14})^{1.852} \tag{53a}$$

$$K = 9.93 \times 10^{-7} \tag{53b}$$

From Eq. (31):

$$L = 9.93 \times 10^{-7} L^{1.852} + 0.015[0.3687(0.015/9.93 \times 10^{-7} L^{1.852})^{0.54} - 1] \tag{54a}$$

$$L = 659 \text{ ft. (201 meters)} \tag{54b}$$

The solution can also be obtained, graphing L and H as shown in Example 1. The answer need to be verified to determine, if the right equation was used.

$$\frac{S}{S_o} = \frac{9.93x10^{-7}(201)^{1.852}}{0.015} = 1.22 \qquad (55)$$

11.7 SUMMARY

Five pressure profiles are presented and considered in this chapter. These represent design conditions which result from a lateral line (or submain) laid on uniform slopes. Procedures are developed to identify pressure profiles by land slope and total friction drop at the end of the line. Equations for designing lateral length (or submain) based on a given criteria, pressure variation, are derived. These equations cannot be solved directly, but solutions can be obtained using trial and error technique on a pocket calculator or by using Newton's method of approximation in a computer program. The developed mathematical equations can be useful in the future development of computerized drip irrigation system design.

KEYWORDS

- Conduit
- Discharge variation
- Down hill slope
- Drainage
- Emitter
- Friction drop equation
- Head loss
- Main line
- Pressure head profiles
- Pressure variation
- Submain line
- Term
- Up hill slope

REFERENCES

A. Literature Cited

1. Gillespie, V. A., Phillips, A. L. & Pai Wu, I. (1979). Drip irrigation design equations. *Journal Irrigation and Drainage, ASCE*, 105 (IR3), 247–257 Paper # 14819.
2. Howell, T. A., & Hiler, E. A. (1974). Trickle irrigation lateral design. *Transactions American Society of Agricultural Engineers*, 17(5), 902–908.
3. Howell, T. A., & Hiler, E. A. (1974). Designing trickle irrigation laterals for uniformity. *Journal of the Irrigation and Drainage Division*, ASCE, 100 (IR4): 443–454, Paper 10983.
4. Williams, G. S., & Hazen, A. (1960). *Hydraulic Tables*, 3rd Ed. New York: John Wiley & Sons.
5. Wu, I. P., & Fangmeir, D. C. (1974). *Hydraulic design of twin-chamber trickle irrigation laterals*. Technical Bulletin No. 216, The Agricultural experiment station, Tucson, Ariz.
6. Wu, I. P., & Gitlin, H. M. (1973). Hydraulics and uniformity for drip irrigation. *Journal of the Irrigation and Drainage Division*, ASCE, 99 (IR3), 157–168. Paper 9786.
7. Wu, I. P., & Gitlin, H. M. (1974). *Design of drip irrigation lines*. Technical Bulletin No. 96, Hawaii Agricultural Experiment Station, University of Hawaii, Honolulu, Hawaii.
8. Wu, I. P., & Gitlin, H. M. (1975) December. Energy gradient line for drip irrigation laterals. *Journal of the Irrigation and Drainage Division*, ASCE, 101 (IR4), 323–326. Paper 11750.

B. Books/Bulletins/Journal and Proceedings/Reports

1. Abreu, V. M., & Luís S. P., (2002). Sprinkler irrigation systems design using. Paper number 02–2254, St. Joseph, MI: ASAE Annual Meeting.
2. Assouline, S., (2002). The effects of micro drip and conventional drip Irrigation on water distribution and uptake. *Soil Science Society of America Journal*, 66: 1630–1636.
3. Brandt R. A., & William, W. B. (1988). Reliability in design. ASAE Distinguished Lecture No. 13, pages 1–27. Paper #913C0888. Winter Meeting of the American Society of Agricultural Engineers, December, Chicago-Illinois .
4. Camp, C. R., Sadler, E. J., & Busscher, W. J. (1997). A Comparison of uniformity measures for drip irrigation systems. *Transactions of the ASAE*, 40(4), 1013–1020.
5. Camp, C. R., Bauer, P. J. & Hunt, P. G. (1997). Subsurface drip irrigation lateral spacing and management for cotton in the southeastern coastal plain. Transactions of the ASAE, 40(4), 993–999.
6. Chieng, S., & Ghaemi, A. (2003). Uniformity in a micro irrigation with partially clogged emitters. Paper number 032097, 2003 ASAE Annual Meeting.
7. Manoliadis, O. G., (2002). Analysis of irrigation systems using sustainability-related criteria. *Journal of Environmental Quality*, 30, 1150–1153.
8. Moore, S., Han, Y. J., Khalilian, A., Owino, T. O., & Niyazi, B. (2005). Instrumentation for variable-rate lateral irrigation system. Paper number 052184 at ASAE Annual Meeting.

9. Zhu, H., Butt, C. L., Lamb, M. C., & Blankenship, P. D. (2004). An implement to install and retrieve surface drip irrigation laterals. *Applied Engineering in Agriculture*, 20(1), 17–23.

C. Web Page Links

1. Evans, R., & Sneed, R. E. (1996). Selection and management of efficient low volume irrigations systems. North Carolina Cooperative Extension Service. EBAE-91–153. http://www.bae.ncsu.edu/programs/extension/evans/ebae-91–153.html
2. Fisher, G. W., & Turf, S. (2001). Innovations in Irrigation. http://grounds-mag.com/irrigation/grounds_maintenance_innovations_irrigation/
3. Gallion, G. B., 2005. Irrigating difficult spaces. http://grounds-mag.com/irrigation/grounds_maintenance_irrigating_difficult_spaces/
4. Haman, D. Z., & Smajstrla, A. G. Design tips for drip irrigation of vegetables. University of Florida. http://edis.ifas.ufl.edu/AE093
5. Hill, R. W., (2000). Management of sprinkler irrigation systems. http://ucanr.org/alf_symp/2000/00–119.pdf
6. Kizer, M. A., (1990). Drip (trickle) irrigation systems. Oklahoma Cooperative Extension Fact Sheet. <http://pods.dasnr.okstate.edu/docushare/dsweb/Get/Document-1443/f-1511%20web.pdf>
7. McGuirk, S., (2001). January. Irrigating steep-sloped landscapes. Grounds Maintenance. http://grounds-mag.com/irrigation/grounds_maintenance_irrigating_steepsloped_landscapes/
8. Solomon, K. H., (1988). Irrigation system selection. Center for Irrigation Technology Irrigation Notes. California State University, Fresno, California 93740–0018. http://www.wateright.org/site2/publications/880105.asp
9. Stryker, J. Drip irrigation design guidelines. http://www.irrigationtutorials.com/dripguide.htm
10. Subsurface Drip Irrigation Systems, (2000). Geo-flow surface drip systems. http://www.geoflow.com/landscape%20(inclu%20golf)/design_general.htm
11. Szolosi (2005). *Drip/Trickle*. Pratt Water Pvt. Ltd. http://www.irrig8right.com.au/Irrigation_Methods/Micro_Irrigation/Drip/Drip_Trickle/Details_DR.htm

PART IV

FIELD EVALUATION OF
MICRO IRRIGATION SYSTEMS

CHAPTER 12

POTENTIAL PROBLEMS IN DRIP IRRIGATION SYSTEMS: THEIR MANAGEMENT AND TROUBLE SHOOTING

ASHOK MHASKE, R. N. KATKAR, and R. B. BINIWALE

Associate Professor, Soil Water Conservation Engg. Agril College, Dr. PDKV, Nagpur. Mobile: +91 9422767788; E-mail: mhaskear@gmail.com

CONTENTS

12.1   INTRODUCTION

Micro irrigation systems can deliver water and nutrients in metered quantity and at measured frequencies directly to the root zone of plants. An extensive network of pipes is used to distribute water to emitters which discharge it in droplets in micro irrigation systems. In the past decade, the use of micro irrigation to provide water to horticultural crops has increased dramatically. Micro irrigation, if properly managed, offers several potential advantages over other irrigation methods, as follows:

a.   Enhanced water use efficiency.

b.   More uniformity in water application.

c.   Minimum deep percolation and runoff.

d.   Enhanced weed control.

e.   Land with different or too steep slope can irrigated being the pressurized irrigation system.

f.   Judicious delivery of fertilizer and other chemicals through the irrigation system.

g.   Less humid environment reduced bacteria, fungi, disease, and other pests.

h.   Some of the problems which interfere with efficiency of micro irrigation are discussed in this chapter.

12.2   EMITTER CLOGGING

Emitter plugging can result from physical (grit), biological (bacteria and algae), or chemical (scale) causes. Frequently, plugging is caused by a combination of more than one of these factors. Emitter plugging is the biggest problem faced by operator of a micro irrigations system, especially when these systems utilize treated wastewater effluent stored in surface reservoirs. The reason behind that is there are very small water passages in most emitters, therefore the emitters gets easily plugged by suspended particles of the mineral, algae, zooplankton and other organisms in the reservoirs. This can reduce the emission rates, leading to non-uniform water distribution and thereby causes water stresses and damage to crops.

In most of the cases, some contaminants enter the system during the installation of micro-irrigation system, which are not sufficiently flushed out of the system. The contaminants included in this category are insect, shavings of PVC pipes and fittings, Teflon tape, and soil particles.

Some of the contaminants are present in the irrigation water delivered to the users are not sufficiently filtered out through appropriate filtration unit. Therefore, important factors to consider in selecting a filtration method are emitter design and quality of the water source. Filters should be sized according to the emitter manufacturer's recommendations or, in the absence of manufacturer's recommendations, to remove any particles larger than one-tenth the diameter of the smallest opening in the emitter flow path. The contaminant present in the water may include soil particles, leaving or dead organic material and scales from rusty pipes.

Finally, as the water stagnates in the polytube lines or evaporates from the emitters or orifice between irrigations, the contaminants may grow, aggregate or precipitate in water. Under certain circumstances iron oxide, manganese dioxide, calcium carbonate, bacterial and algae slimes can form in micro irrigation system.

The appropriate solution to a particular plugging problem must be based upon the nature of the problem. Acid injection can remove scale deposits, reduce or eliminate mineral precipitation, and create an environment unsuitable for microbial growth. Calcium precipitate and chemical treatment is often required to prevent emitter plugging due to microbial growth and/or mineral precipitation. The attachment of inorganic particles to microbial slime is a significant source of emitter plugging. Chlorination is an effective measure against microbial activity [11–14, 22]. Drying of the system thoroughly will serve to shrink organic materials enough to partially open clogged emitters or orifice, which can be treated more effectively with the chemicals.

12.3 SALINITY MANAGEMENT

The salts deposited by the irrigation water can be frequently leached out in the area where significant rainfall occurs during the year. However, in the area where there is scanty rainfall, considerable deposits of salts may get accumulate in the soil due to absence of the natural leaching. The deposited salts are directly proportional to the salinity of the irrigation water. When saline irrigation water is used in a arid region salt be frequently deposited on the soil surface. Salt will also concentrate below the soil surface at the perimeter of the soil volume wetted by the each emitter. Properly designed and managed irrigation system can tolerate the buildup of the salt in the soil. Erratic and uncertain heavy rains may wash accumulated salt back in to the root zone, which may cause the salinity of the micro irrigation system if the system is not operating. Therefore, when there is rain after salt accumulation, irrigation water should be applied until about 5 cm of the rainfall, to prevent the salts moving into root zone and to help leaching of the salt below the root zone.

When the saline water is used for irrigation through drip irrigation/ micro irrigation, salt will move radially with the soil water and will concentrate at the center edges of the wetted perimeter. So in this case one consideration is addressed of keeping the emitting device in such a way as to result water movement away from the plant.

12.4 BACTERIAL PRECIPITATION OF SULFUR

When air-water contact is avoided until water is discharged from the system then sulfur bacteria problem can be minimized. Defective valves or pipefittings on the suction side of the irrigation pump are common causes of sulfur bacteria problems. Water containing more than 0.1 ppm of the total sulfides then bacteria can produce organic sulfur slime (Table 12.1). These bacteria produce the white cottony masses of the slime due to which emitter may completely blocked. In the system if the iron and sulfur is interacted then the problem can be created in the micro-irrigation system. Filters screens manufactured with the stainless steel can cause the precipitation of iron sulfides if used to filter the water with high concentration of sulfides. When the iron and sulfides are present in dissolved form may produce the chemical reaction in which the insoluble iron sulfides are

TABLE 12.1 Criteria for Plugging Potential of Drip Irrigation System Water Sources [2]

Factor	Plugging hazards		
	Slight	Moderate	Severe
	Parts per million (ppm) except pH		
Physical			
Suspended solids (Filterable)	<50	50–100	>100
Chemical			
Hardness	<150	150–300	>300
Hydrogen sulfide	<0.5	0.5–2.0	>2.0
Iron	<0.1	0.1–1.5	>1.5
Manganese	<0.1	0.1–1.5	>1.5
pH	<7.0	7.0–7.5	>7.5

formed. Under certain conditions, the injection of the nutrients containing iron into a sulfides bearing water will also cause precipitation. The use of an air bladder or diaphragm may be made to separate the air from the water which will minimize this problem.

The following are several possible chlorine injection recommendations:

- Inject continuously at a low level to obtain 1 to 2 ppm of free chlorine at the ends of the laterals (Figure 12.3).
- Inject at intervals (once at the end of each irrigation cycle) at concentrations of 20 ppm and for duration long enough to reach the last emitter in the system.
- Inject a slug treatment in high concentrations (50 ppm) weekly at the end of an irrigation cycle and for duration sufficient to distribute the chlorine through the entire piping system.
- To control the sulfur slime bacteria, intermittent treatment of chlorination at 1 ppm free available chlorine which is measured in the field for 30 to 50 minutes daily.

12.5 BACTERIAL PRECIPITATION OF IRON

Soluble ferrous iron is a primary energy source for certain iron-precipitating bacteria [15]. These bacteria can attach to surfaces and oxidize ferrous iron to its insoluble ferric iron form. In this process, the bacteria create a slime that can form aggregates called ochre, which may combine with other materials in the micro irrigation tubing and cause emitter plugging [14].

There are some bacteria which can oxidize the soluble ferrous oxide and can produce the precipitation of insoluble ferric oxide. The problem with the iron bacteria may occurred with iron concentration as low as 0.1 ppm. Attachment of iron precipitate which is of red filamentous sludge's to PVC and polyethylene tubing may completely block the emitters.

The most appropriate treatment for bacterial precipitation of iron is chlorination of water to kill the bacterial activity. Chlorination may be performed continuously at the rate of 1 ppm free available chlorine, or intermittently at the rate of 10 to 20 ppm for 30 to 60 minutes daily.

If the presence of bacteria is due to contamination of the well then the super chlorination of the well greatly reduces the problem. Super chlorination of the well is done by injecting 200 to 500 ppm into well. The volume of the water to be treated can be estimated by knowing the depth and diameter of the well.

12.6 CHEMICAL PRECIPITATION OF IRON

In terms of the iron solubility, the temperature, pressure, pH and the relative concentrations of other substances in solution are the most important variables. Many changes occur in the physical and chemical environment of the water when the water is pumped out of an aquifer and into an irrigation system. The dissolved iron in the aquifer water will generally be in the form of ferrous oxide. When the water is in the aquifer beneath the ground, this condition may favors the solubility of the iron, while the precipitation will probably be favored once the water entered in the irrigation system, particularly between irrigations. This leads to the precipitation of the iron in the system. The injection of the phosphates or calcium salt will accelerate this process and should be avoided in iron bearing waters. Clogging of the emitters may occur even If the very small quantity of iron is present (0.2 ppm or more). Experiments have shown that a ferrous iron content as low as 0.2 ppm can contribute to iron deposition [15]. The process of the iron bearing waters where the chemical precipitation is a problem may be achieved either by removing the iron from the water or by retaining the iron in the solution. Following are the some recommended techniques.

12.6.1   AERATION AND SETTLING

Removal of the iron from the irrigation water with aeration and settling is the most reliable and practicable method. The water is aerated thoroughly at the inlet of the settling basin, either by run over a series of the baffles or allowing it to fall through or by spraying the water through the air. Both the method will add large amount of the oxygen into the water which oxidizes the ferrous oxides to ferric oxide. The precipitation of the ferric oxide out of the water will settle at the bottom of the basin, provided the sufficient settling time is allowed.

Aeration and settling is the most appropriate method of iron removal which is less expensive, requires low maintenance and little training. Existing ponds can sometimes be used as settling basins. They need not be elaborate structures; however, settling basins should be accessible for cleaning, and large enough that the velocity of the flowing water is sufficiently slow for particles to settle out. Most of the iron bearing water is well waters and they are pumped out of the well into a pressurized system.

12.6.2   CHLORINE PRECIPITATION

Normally chlorine is used to precipitate iron in the micro-irrigation system when aeration and settling is not possible then. Ferrous oxide will be instantly oxidized to ferric oxide with free chlorine which will precipitate out of the solution. The iron concentration must be decided and accordingly chlorine must be injected. Chlorine should be injected continuously at a low level to obtain 1 to 2 ppm of free chlorine at the ends of the laterals or Inject at intervals (once at the end of each irrigation cycle) at concentrations of 20 ppm and for duration long enough to reach the last emitter in the system. The method used will depend on the growth potential of microbial organisms, the injection method and equipment, and the scheduling of injection of other chemicals.

Some additional chlorine may be required to kill any iron bacteria and to control the growth of the bacterial slime. When chlorine is injected, a test kit should be used to check to see that the injection rate is sufficient. Color test kit (D.P.D.), that measures 'free residual' chlorine, (the primary bactericidal agent) should be used. The orthotolidine-type test kit, which is often used to measure total chlorine content in swimming pools, is not satisfactory for this purpose. The turbulence of the water should be created within the system for achieving the complete mixing of the chlorine otherwise iron will pass into submain and lateral lines where it will precipitate. After complete mixing iron, it is removed by the filtration.

A sand filter is the most appropriate filter. It should frequently backwashed with automatic backwash system (Figures 12.1 and 12.2). Precipitation of the iron with use of chlorine is effective only when the proper procedure is followed. But in this process serious attention for the operation and maintenance of the filtration unit and chlorine injection equipments is required. Similarly cost factor of the chlorine and the filtration unit must be significant in large systems with high levels of the iron in the water source. Serious care should be taken in those cases where the manganese is present in irrigation water. Oxidation of the manganese by chlorine proceeds at a much slower speed than that of the oxidation of the iron. When manganese is present in the irrigation water there will be clogging problem in the irrigation system due to use of chlorine, which result in manganese precipitation after the filter.

12.6.3 pH CONTROL

The pH of the water is very important because it can determine the solubility of iron. When the pH of the water is lowering, i.e., acidic then iron is most soluble. Soluble iron is also called as clear water iron. The pH

FIGURE 12.1   Filtration process: automatic backwash gravel filters.

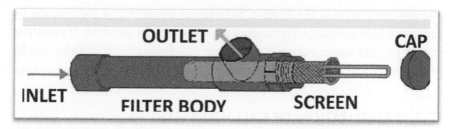

FIGURE 12.2   Screen filter.

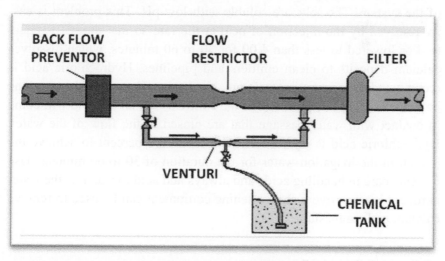

FIGURE 12.3   Injection of chemicals through venturi tube.

of water may rise when the water is pumped out of aquifer and the iron will precipitate out. Iron in the solution may be maintained by injection of the acid. Similarly acid may be used to dissolve the iron sediment, which have deposited over the period of time. Before using one of the above preventive measures, partially clogged drip irrigation system with iron can be cleared by using acid injection.

12.6.4 IRON SULFIDE PRECIPITATION

In most sedimentary rocks sulfur-bearing minerals are present and a soluble form of sulfate is carried by water. Sulfates are difficult to precipitate and generally remain in solution. Sulfate can be used as a food source by bacteria which produces hydrogen sulfide gas as a by-product. If sufficient iron is present under moderate reducing conditions, iron sulfides can be precipitated, and a sand media filter is suggested to remove the precipitate.

12.7   PRECIPITATION OF CALCIUM SALTS

The precipitation of the calcium salts in micro irrigation is common with some waters, and appears as a white film or plating on the inner surfaces of the system. The salts are soluble with low pH. This problem is easily resolved by the injection of acid at the rate so that the pH is equal to 4 or lowered to less than 4.00 for 30 to 60 minutes, which dissolves calcium deposits to clean emitters and pipelines. Hydrochloric acid is recommended for treating calcium blockages. However, this is done before occurring full blockage of the emitters, so that the acid will come in contact with water passage that are closed to the flow of the water. Hydrochloric acid is injected at the rate of five percent to achieve the pH 4.0 in the irrigation water for the duration of 30 to 60 minutes. Use extreme care in handling acids and always add acid to water. If the water hardness is excessive water softening equipment can be used to remove calcium and magnesium.

12.8   ANT DAMAGE

Insect damage to thin-walled polyethylene drip tubing or "tape" is a major problem. Damage to the thin walled polyethylene tube by ant has been observed in various part of the world and is the major damage in some areas. Ant damage is nothing but chewing of holes through the sides of the strip tubing and due to these orifices are enlarged in the strip tubing. This damage destroys the integrity of the tape or drip tubing, resulting in small to massive leaks that may result in poor moisture distribution and soil erosion, which ultimately results in economic losses to the growers. Ant damage to drip tape is most severe in drip tubing and tapes having wall thicknesses of less than 0.015 inches.

Sufficient wall thickness with ant resistant tubing or tapes can protect the tubing or tapes from making holes through the tubing wall to prevent the orifice enlargement. Similarly ant damage can be controlled with the chlorinated hydrocarbons.

12.9 BORON TOXICITY

Boron in small amount is an essential element for plant growth. Boron concentration slightly above the critical level is toxic to the plant. Boron problems originating from the water are probably more frequent than those originating in the soil. Sensitivity to boron varies greatly between crops. Perennial crops such as trees and vines are usually more prone to boron toxicity than vegetable crops. Boron toxicity symptoms include the yellowing of leaves and marginal burns. Tree crops may not show leaf symptoms but may have twig dieback or gum on the limbs and trunk. The usual range of the boron in the natural water is from 0.01 ppm to 10 ppm. Boron content which can be tolerated by the crop has no countable effect on physical properties of the soil. It is difficult to remove the boron from the soil as chloride and nitrate but it can be removed by successive leaching.

Boron in a small amount is essential for plant growth. Some plants are more sensitive to an excess than others. Plants grown on sandy soils which have been irrigated for several years by water exceedingly low in boron (less than 0.02 ppm) may develop boron deficiency. Plants grown in soils high in lime may tolerate more boron than grown in the non-calcareous soils. The ranges in Table 12.2 provide satisfactory guide to the boron hazard in the irrigation water.

TABLE 12.2   Different Levels of Boron Hazard in the Irrigation Water [3, 4]

Level	Effect for the crop
Below 0.5 ppm	Satisfactory for all crops
0.5 to 1.00 ppm	Satisfactory for most crops, sensitive crops may show injury (may show leaf injury but yield may not be affected)
1.00 to 2.00 ppm	Satisfactory for the semi- tolerant crops. Yield and vigor may be reduced in the sensitive crops
2.00 ppm and over	Only tolerant crop can produce satisfactory yield

12.10   CHLORINE AND COPPER SULFATE FOR ALGAE CONTROL

Chlorination of irrigation water used for drip irrigation is required to prevent the agglomeration of fine organic and inorganic particles with bacterial byproducts. A chlorination treatment program is regulated by the severity of the plugging problem. Weekly or even bi-weekly treatments may have to be employed depending on the severity. In order to obtain optimum results the chlorine must be injected at the right concentration for the required interval of time. The amount of copper sulfate or chlorine necessary to control various types of algae are shown in Tables 12. 3 and 12. 4. Some engineers use a standard dose for all types say 1.20 mg per liter.

12.11   DAMAGE TO INFIELD DISTRIBUTION SYSTEMS BY ANIMALS

There is no serious animal damage to the micro-irrigation system because of frequent tillage and farming practices destroying the animal's habitats. Borrowing animals like squirrels, gophers, rats, mice cause damage to the tubing laid on surface or underground. These animals are able to damage the micro-irrigation system by chewing holes in the lateral lines. To solve this problem to some extent following solutions can be used:

   a.   biological control;
   b.   trapping to control the animal population;
   c.   to stop animal food supply;
   d.   extermination.

One common biological method for controlling rodent populations is to attract rodent predators, most frequently by using owl boxes. Because they are relatively inexpensive and can be populated for a long time, owl boxes are being employed in greater numbers as part of a rodent management program. The principle is simple, the higher the owl population the fewer the rodents. Before installation of drip irrigation system immediate surrounding areas should be cleared of rodents prior to installing drip line. To inhibit rodents from invading a clean field placing approved bates around the perimeter of the field prior to irrigation system installation is a good practice to follow.

TABLE 12.3 Copper Sulfate and Chlorine for Algae Control [3, 4]

Organism	Odor	CuSO$_4$ (ppm)	CuSO$_4$ (lb)	Chlorine (ppm)
Diatomaceae				
Asterionella	Aromatic	0.10	0.8	0.5–1.0
Melosira		0.30	2.5	2.0
Synedra	Earthy	1.00	8.3	1.0
Navicula		0.07	0.6	
Chlorophyceae				
Conferva		1.00	8.3	2.5
Scenedesmus		0.30	1.7	
Spirogyra		0.20	1.7	0.7–1.5
Ulothrix		0.20	1.7	
Volvox	Fishy	0.25	2.1	0.3–1.0
Xygnema		0.70	5.8	
Coelastrum		0.30	2.5	
Cyanophyceae				
Anabaena	Moldy	0.10	0.8	0.5–1.0
Clathrocystis	Grassy	0.10	0.8	0.5–1.0
Oscillaria		0.20	1.7	1.1
Aphanizomenon	Moldy	0.15	1.2	0.5–1.0
Protozoa				
Euglena		0.50	4.2	
Uroglena	Fishy	0.05	0.4	0.3–1.0
Peridinium	Fishy	2.00	16.6	—
Chlamydomonas	—	0.50	4.2	
Dinobryon	Aromatic	0.30	2.5	0.3–1.0
Synura	Cucumber	0.10	0.8	0.3–1.0
Schizomycetes				
Beggiatoa	Putrefactive	5.00	41.5	—
Crenothrix	Putrefactive	0.30	2.5	0.5

TABLE 12.4 Common Chlorine Compounds Used in Micro Irrigation [2]

Compound	Form	Percent available
Calcium hypochlorite	Dry	65–70
Sodium hypochlorite	Liquid	5.26–15
Chlorine gas	Gas	100

Population reduction of large rodents such as pocket gophers on small to medium sized fields is effectively done by trapping. It is also effective to clean up remaining animals after a poison control program. In smaller rodent such as mice where there is rapid production rates, trapping method is not a cost effective.

Body-gripping traps work exceptionally well for capturing pocket gophers. Traps can be set in the main tunnel or in a lateral, preferably near the freshest mound. Gophers usually visit traps within a few hours of setting so newly placed traps should be checked twice daily. If a trap has not been visited within 48 hours, move it to a new location. Trapping is usually most effective in the spring and fall when the gophers are actively building mounds. The information section at the end of this document lists several sources for purchasing and placement of these traps.

Burrowing animals using the food generally will either be weed or crop being grown. Weed control will reduce the problem if the food supply material is like weeds and if food supply is the crop then animal population control will minimize the problem.

In extermination several rodenticides including toxicants and anti-coagulants are in current use for managing rodent populations in and around agricultural fields. To keep the animal away from the lateral lines some chemicals which tastes or smell bad to the animal either may be injected through the system or laid down with lateral during installation. In general, chemical injection through the drip irrigation system is the better technique.

12.12   FIELD FAILURE OF POLYETHYLENE HOSE AND TUBING

There are two types of failures and very different in their causes and symptoms, and must be carefully distinguished in order to make an accurate diagnosis. The laterals are subject to failure due to either exposure to ultra-violet lights, and environmental stress cracking.

12.12.1   ULTRA-VIOLET DEGRADATION

When ultra-violet light penetrates the plastic then degradation of polyethylene hose and tubing occurs and causes the molecular changes, therefore it results in progressive degradation of the plastic. Cracking of the hose tubing which is exposed to sunlight are symptoms of U.V. degradation. Because of outdoor applications UV stabilizers have been developed to protect these plastics from degradation. The ideal UV stabilizer should absorb the UV radiation and dissipate it in a harmless manner. Carbon black has been found to be one such stabilizer. This carbon black colors the clear the polyethylene black and absorbs the ultra-violate light. There are many bodes which suggest concentrating of 2–3 CB for best results The second variable introduced is cross linking of the polymer which is achieved by irradiating the finished product.

Shorter wavelength UV light is very susceptible to absorption by air. If sun is low in the sky it passes through more air than if sun is directly overhead. S plastics exposed near equator degrade faster. Altitude has also some effect. Sunlight found at high altitude passes through less atmosphere than at lower altitude, thus exposed to more intensive radiation.

12.12.2   ENVIRONMENTAL STRESS CRACKING

Drip irrigation tubing which has been exposed to the environment, start splitting and cracking is called the environmental stress cracking. Frequent failure will occurs where the material has been stressed. Inline emitter or fitting points on pipes and tubing are the typical example of environmental stress cracking. The environmental stress cracking is due to the use of inappropriate types of polyethylene, inferior quality grades polyethylene, adding of polyethylene crack obtained from the toy makers. Extruded polyethylene hose and tubing is tested for its environmental stress crack resistance by folding a 12-inch sample in two. The sample is checked daily for signs of failure, which generally appear as cracks or splits at the stress points. The times to failure will typically range from the few minutes to several hours.

12.13   SUMMARY

This chapter discusses potential problems in the drip irrigation systems under topics such as: emitter clogging, salinity management bacterial precipitation of iron, bacterial precipitation of sulfur, chemical precipitation of iron, aeration and settling, chlorine precipitation, pH control, precipitation of the calcium salts, ant damage, boron toxicity, copper sulfate and chlorine for algae control, animal damage to infield distribution systems, field failure of polyethylene hose and tubing, ultraviolet degradation, and environmental stress cracking.

KEYWORDS

- **Acid treatment**
- **Algae**
- **Application uniformity**
- **ASAE**
- **Bacterial slime**
- **Boron**
- **Calcium**
- **Calcium carbonate**
- **Chlorination**
- **Chlorine**
- **Clogging**
- **Cracking**
- **Design**
- **Distribution uniformity**
- **Drip tube**
- **Emitter**
- **Environmental stress**
- **Ferric iron**
- **Ferrous oxide**
- **Filters**
- **Fresh water**
- **Inline emitter**
- **Iron**
- **Iron oxide**
- **Irrigation system**

- Irrigation water
- Line source
- Manganese
- Manganese dioxide
- Micro irrigation
- Organic material
- Polyethylene
- Polyethylene hose
- Precipitate
- Precipitation
- Preventative maintenance
- Reclaimed waste water
- Root zone
- Saline water
- Salinity
- Sewage effluent
- Stress crack
- Stress point
- Teflon tape
- Trickle irrigation
- Under treated waste water
- Wastewater
- Water distribution
- Water quality

REFERENCES

1. ASAE Standard (1996). Amer. Soc. of Agric. Eng. St Joseph, MI, 792–797.
2. Bill, Lamont (2012). Maintaining drip irrigation systems, Penn State Horticulture and Extension. *The Vegetable & Small Fruit Gazette, Penn State University Extension, Vol. 16.*
3. Boswell, M. J. (1985). Design characteristic of line source drip tubes. Proc. of the 3rd Int. Drip/Trickle Irr. Cong., Vol.1, CA, USA, 306–312.
4. Boswell, M. J. (1985). *Micro irrigation design manual.* Second edition, USA.

5. Braud, H. J., & Soon, A. M. (1980). Trickle irrigation design for improved application uniformity. ASAE and CSAE Mtg. on Trickle (drip) Irri., Paper 79–2571, Winnipeg, Canada.

6. Bucks, I. V., Nakayama, F. S., & Gilbert, I. G.(1979). Trickle irrigation water quality and preventative maintenance. *Agric. Water Manage.,* 2(2), 149–162.

7. Capra, A., & Scicolone, B. (1998). Water quality and distribution uniformity in drip/ trickle irrigation systems. *J. Agric. Eng. Res.,* 70, 355–365.

8. Dorota, Z. Haman (2014). Causes and prevention of emitter plugging. In: *Micro-irrigation Systems.* The EDIS website at http://edis.ifas.ufl.edu.

9. EI-Tantawy, M. T., Matter, M. A., & Arafa Y. E. (2009). Filters and emitters performance under treated waste water. *Misr. J. Ag. Eng.,* 26(2), 886–904.

10. Feigin, A., Ravina, I., & Shalhevet, J. (1991). Irrigation with treated sewage effluent. In: *Management for Environmental Protection.* Springer, Berlin, Heidelberg, New York.

11. Ford, H. W. (1977). Controlling Certain Types of Slime Clogging in Drip/Trickle Irrigation Systems. Proceedings of the 7th International Agricultural Plastics Congress, San Diego, California.

12. Ford, H. W. (1979). *A Key for determining the use of sodium hypochlorite (liquid chlorine) to inhibit iron and slime clogging of low pressure irrigation systems in Florida.* Lake Alfred, CREC Research Report CS 79–3. IFAS, University of Florida.

13. Ford, H. W. (1987). *Iron Ochre and Related Sludge Deposits in Subsurface Drain Lines.* Extension Cir. 671. IFAS, University of Florida.

14. Ford, H. W., & Tucker, D. P. H. (1975). Blockage of drip irrigation filters and emitters by iron-sulfur-bacterial products. *Hort Science,* 10(1), 62–64.

15. Gilbert, R. G., & Ford, H. W. (1986). Operational Principles. Chapter 3. *Trickle Irrigation for Crop Production. (Eds. Nakayama and Bucks).* Elsevier Science Publishers. Amsterdam, Netherlands.

16. Keller, J., & Bliesner, R. D. (1991). *Sprinkle and Trickle Irrigation.* AVI Book, New York, USA, p. 652.

17. Liu, H., & Huang, G. (2009). Laboratory experiment on drip emitter clogging with fresh water and treated sewage effluent. *Agric. Water Manage.,* 96, 245–756.

18. Nakayama, F. S., & Bucks, D. A. (1991). Water Quality in Drip/Trickle Irrigation. A VI Book. New York. USA. pp. 187–102.

19. Oswald, W. J. (1989). Use of wastewater effluent in agriculture. *Desalination,* 72, 67–80.

20. Ravina, I., Paz, E., Sofer, Z., Marcu, A., Shisha, A., & Sagi, G. (1992). Control of emitter clogging in drip irrigation with reclaimed wastewater. *Irrig. Sci.,* 13, 129–139.

21. Ravina, I., Paz, E., Sofer, Z., Marcu, A., Shisha, A., & Sagi, G. (1997). Control of emitter clogging in drip irrigation with stored treated municipal sewage. *Agric. Water Manage.* 33(2), 127–137.

22. Tyson, A. W., & Harrison, K. A. (1985). *Chlorination of Drip Irrigation Systems to Prevent Emitter Clogging.* Misc. Publ. 183. Cooperative Extension Service, University of Georgia.

CHAPTER 13

EMITTERS: PRESSURE AND DISCHARGE RELATIONSHIP

RAJESH GAJBHIYE,[1] DNYANESHWAR TATHOD,[1,2]
and YOGESH MAHATALE[3]

[1]Former Student, Department of Irrigation and Drainage
Engineering, Dr. Panjabrao Deshmukh Krishi Vidyapeeth,
Akola 444104 (MS), India

[2]Assistant Professor, College of Agricultural Engineering and Tech.,
Warwat Road, Jalgaon Jamod, Dist. Buldhana, Maharashtra, India.
Mobile: +91 9604818220; E-mail: dnyanutathod@gmail.com

[3]Former Student at Soil Water Conservation Engg. Agril College,
Dr. PDKV, Akola; Postal address: A T Mahendra Colony,
Near Sharda Agency, P.O. VMV Amravati – 444604
(MS), India. Mobile: +91 9326279798 or 9420124699;
E-mail: yrmahalle1@gmail.com

CONTENTS

13.1   INTRODUCTION

Water is scarce, precious natural resources and most crucial element, which must be planned, developed, conserved and managed in a sustainable manner. Optimum management of available water resources at farm level is needed because of increasing demands, limited resources, water table variation in space and time, and soil combination [15]. It is important to increase the crop yield under limited water sources for optimum crop production to meet future need of food production.

Limited water supply should be used efficiently to irrigate more areas with same quantity of water. Pressurized irrigation techniques are very reliable solution for obtaining higher uniformity and application efficiency.

Drip irrigation can potentially provide high application efficiency and application uniformity. Both are important in producing uniformly high crop yields and preserving water quality when both water and chemicals are applied through the irrigation system. In many irrigation methods, drip irrigation is popular for a plant crop because water can be applied directly to the root zone of the crop [30].

Trickle irrigation is gaining importance in the world, especially in areas with limited and expensive water supplies, since it allows efficient use of limited resources. Ideally, all emitters in the system should discharge same amount of water, but flow differences between two supposedly identical emitters exist due to: manufacturing variations, pressure differences, emitter plugging, aging, friction head losses throughout the pipe network, emitter sensitivity to pressure and irrigation water temperature changes. Accurate emitter manufacturing is necessary in order to achieve a high degree of system uniformity. However, the complexity of emitter and their individual components make it difficult to maintain precision during production. Along with above factors, variation in pressure causes change in discharge rate of two identical drippers and non-uniformity of water application, ultimately leads to reduction in crop yield. Variation in discharge causes uneven wetting pattern of the soil due to which plant root development restricted.

In most micro irrigation systems, the plant root development is restricted to the wetted soil volume near each emitter or along each lateral because water is supplied to a concentrated portion of the total soil volume. Excessive restriction of plant root development has the potential to decrease plant growth and yield. Therefore, it is more rational if irrigation water is applied according to the change of required soil wetted patterns during plant growing period.

Modern irrigation technologies have high water savings under well management. Due to high irrigation efficiency, size of the irrigated land with current water supply is higher in comparison with surface irrigated areas and it is possible to obtain high crop yield as well as more income with better management. A poorly managed pressurized irrigation system results in non -uniform water distribution. In such systems, the most valuable outcome of evaluation process is irrigation uniformity. Uniformity of drip irrigation system is usually a combination of measuring the variability of emission from individual emitter and pressure variation within the entire system.

Considering the importance of effect of change in pressure on discharge rate of drip irrigation, the study was undertaken entitled "*Evaluation of pressure and discharge relationship in drip irrigation*" at the research farm of Irrigation and Drainage Engineering at C.A.E.T. Jalgaon (Jamod) with following objectives.

- To evaluate emitter discharge uniformity for a drip irrigation system.
- To study effect of variations in pressure on discharge rate of emitters.

13.2   LITERATURE REVIEW

13.2.1   IMPORTANT OF DRIP IRRIGATION

Bucks and Davis [6] stated that the drip irrigation is characterized by high efficiency/low volume, localize over a long period application of water, low-pressure requirement and application of water near root zone.

Selvaraj et al. [21] reported that the fresh rhizome yield of turmeric under drip irrigation scheduled at 80% of surface irrigation was superior over surface irrigation schedule at 0.90 IW/CPE ratio. Singandhupe et al. [23] revealed that majority of tropical and subtropical fruit crops require very frequent irrigation during fruit bearing and fruit development stages and it is only possible through drip irrigation since limited wetted area is irrigated unlike surface flood method.

Tiwari [28] evaluated the economic feasibility of drip irrigation and observed that the application of drip irrigation was found to increased yield compared to that of furrow irrigation. Troy [29] studied the performance of drip for home garden and stated that efficiency of drip irrigation is 90–95%. Only 5–10% water loss comes from water evaporation from the small portion of the soil surface.

Bhardwaj [3] reported 100% yield increase in banana, 40–50% in sugarcane, pomegranate and 25% in grapes and cotton under drip irrigation. Tiwari et al. [27] reported that the productivity of horticultural crop can be enhanced to the tune of about 60–70% by adopting drip irrigation. Ibragimov et al. [11] compared drip and furrow irrigation under cotton and obtain 18–42% saving of irrigation water with drip system in comparison with furrow.

13.2.2 UNIFORMITY OF DRIP IRRIGATION

Camp [7] stated that uniformity will depend on emitter manufacturing variation, land slope-induced hydraulic variability of the irrigation unit, head losses in pipes, emitter sensitivity to pressure and temperature variation and emitter clogging.

Tsay and Ju [30] analyzed uniformity of drip irrigation by numerical simulation and concluded that the uniformity analysis could be modified by restricting the irrigation domain to the most effected boundary in order to get more significant comparing results. Pereira [17] indicated that improvement of irrigation system requires considerations of the factors influencing the hydraulic processes, the water infiltration and the uniformity of water application to the entire field. Priyanjith et al. [19] determined the field evaluation of drip irrigation system for small-scale banana orchards and showed that emission uniformity and statistical uniformity were 53% and 88% respectively.

Soccol et al. [26] studied the performance analysis of a trickle irrigation subunit installed in an apple orchard and revealed that the performance of system was average and emission uniformity, statistical uniformity and coefficient of global variations were 74.51, 77.69, and 22.31%, respectively; and the efficiency parameters were below expectations. Safi et al. [20] studied water application uniformity of sub-surface drip irrigation and observed that for used and unused tapes, emitter performance at 150 kPa was better than at 50, 90 and 200 kPa. The maximum uniformity coefficient values of the unused and used tapes of 34 m long were 96.9% and 91.8%, respectively. Changade et al. [8] determined the emission uniformity of emitter in gravity-fed drip irrigation system and revealed that the emission uniformity of system was 90.58% and was good based on ASAE standard. The manufactures coefficient of emitter was 0.0428, which was good.

Acar et al. [1] studied the impact of pressurized irrigation technologies on efficient water resources and found that the average UC and DU values of drip irrigated areas were 80.9% and 68.9%, respectively. Umara et al. [31] reported that average Christian uniformity coefficient values of 88 and 96.4% was obtained from the 1.5 and 2.2 mm orifice emitters, respectively and were considerably satisfactory for convectional micro system.

13.2.3 PRESSURE AND DISCHARGE RELATIONSHIPS

Zur and Tal [32] studied the emitter sensitivity to pressure/temperature and found that the discharge sensitivity to pressure was dependent upon flow regime and cross-sectional area of flow of emitter. Little et al. [16] evaluated 258 irrigation systems by mobile field laboratories and observed that higher uniformity distribution was observed only for the farms with >100 acres. They also accounted that variation in pressure is responsible for poor distribution and average uniformity was relatively low.

Anyoji and Wu [2] calculated the normal distribution of water for drip under turbulent flow and stated that it is less effected by temperature. If the water is filtered by filters before irrigation, the emitter variation will be only effected by pressure and manufacturing variation. Shani et al. [22] showed that the flow rate decreases for non-compensating emitter and this reduction is greater in loamy than in sandy soil. The over pressure at the discharge point is higher in the first where dripper has lower infiltration.

Sinobas et al. [25] measured flow rate under controlled conditions on sample of six commercial emitters by applying pressure of 100, 150 and 200 kPa and observed that the discharge rate of pressure-compensating emitters was decreased over the operating time. The variation over 6% was observed. Kang [13] reported that water application uniformity increases or slightly decreases as operating pressure head increases in a range when emission exponent $x \leq 0.5$ in most cases. The water application uniformity decreases as operating pressure head increases in a range when emission exponent $x \geq 0.5$. Priyanjith et al. [19] observed that the mean discharge rate of all emitters was 2.8 lph. The most emitter operated close to the mean discharge rate, although in last lateral the variation of discharge rate was > 20% and other three laterals showed almost uniform discharge rate.

Gil et al. [10] studied emitter discharge variability of subsurface drip irrigation and reported that the flow rate portability of non-compensating emitters in homogeneous soils with high infiltration is more or less the same as for surface drip irrigation. Kadale et al. [12] observed that discharge sensitivity to pressure varied from 0.50–0.62 depending on the flow regime and cross-sectional area of flow of each emitter type. Singh and Kumar [24] evaluated the hydraulic performance of drip irrigation and revealed that at a particular spacing, the uniformity coefficient and emission uniformity increased while coefficient of variation decreased as the operating pressure head increased for all emission devices.

Dogan [9] studied the effects of pressure fluctuations on drip lateral emitter flow rate and diameter change and reported that an increase in pressure resulted in increased emitter flow rates limited to 10%. Popale et al. [18] examined the hydraulic performance of drip irrigation system and showed that the uniformity coefficient and emission uniformity increased while coefficient of variation decreased as operating pressure increased for all emission devices.

13.3 MATERIALS AND METHODS

13.3.1 EXPERIMENTAL SITE

The study was conducted at the research farm of Irrigation and Drainage Engineering in Jalgaon (Jamod), MS, India. The 15 m laterals with two diameters (12 mm and 16 mm) were selected with two discharge rates of 4 lph and 8 lph.

13.3.2 MATERIALS

The 99 aluminum moisture boxes of 7.5×2.5 cm^2 were used for the collection of soil samples. In drip irrigation, providing an appropriate pressure is prime requirement and required pressure can be maintained by using of pressure gauge. The pressure gauge was fitted in lateral to maintain the pressure of 0.5, 0.75, 1.0, and 1.5 kg/cm^2.

The lateral was made up of low-density polyethylene chloride tube of 12 mm and 16 mm in diameter; and it was connected to the underground submain pipe to supply water. The lateral was used with 4 lph and 8 lph drippers at 15 cm and 20 cm dripper spacing. The turbo key plus online NPC drippers of 4 lph and 8 lph were used. The connector was used for connecting the lateral to field the underground submain. Two types of connectors were 16×16 mm for 16 mm diameter lateral and 16×12 mm for 12 mm lateral. End cap was at the end of the lateral. Two types of end caps were 16 mm and 12 mm in diameter.

The measuring cylinder of 100 mL capacity was used to collect the flow. Appendix I shows the photos during the study.

13.3.3 PRESSURE AND DISCHARGE RELATIONSHIP

The emitter is the important part of drip irrigation. An emitter with a high degree of pressure compensation (x = 0) is technically possible, although the ideal emitter has not yet been invented. Emitter flow rates may fluctuate as pressure along the lateral line varies due to friction, elevation, and/or accidental restrictions, resulting in a non-uniform water application [5]. Emitter discharge rate is a function of operating pressure.

The pressure sensitivity of an emitter discharge depends mainly on the values of x, which determines how sensitive is the discharge to pressure. The value of x typically falls between 0.1 and 1.0, depending on the make and design of the emitter (hydraulic characteristics). For a fully laminar flow regime, emitters must be very sensitive to pressure head changes and the value of x must be 1.0. This means that a pressure variation of 20% may result in ± 20% emitter flow rate variation. Most non-compensating emitters are always fully turbulent with an x level of about 0.5, indicating that a pressure variation of 20% will result in a flow variation of approximately 10%. On the other hand, for compensating emitters, pressure variation causes little discharge variation. Compensating emitters have an x level ranging from 0.1 to 0.4. An ideal pressure compensating emitter would have an x equal to 0.

The study was conducted considering the different pressures (0.5, 0.75, 1 and 1.5 kg/cm^2) by using two laterals of diameter 12 mm and 16 mm. Effect of pressure on flow rate of emitter was determined by using the equation developed by Keller and Karmeli [14].

$$q = Ke \ H^x \qquad (1)$$

where, q = emitter discharge (lph), K_e = emitter discharge coefficient that characterize the emitter dimensions, H = operating pressure at the emitter in kPa, and x = emitter discharge exponent.

Small differences between among identical emitters may result in significant discharge variations. The manufacturer's coefficient of emitter variation is a measure of the variability of discharge of a random sample of a given make, model and size of emitter, as operation or aging has taken place. The manufacturer's variation is caused by pressure and heat instability during emitter production. In addition, a high CV_m could occur due to a heterogeneous mixture of the materials used in the production of emitters. Typical values for CV_m range from 2 to 15%, although higher values are possible [4]. Classification of CV_m values based on ASAE standards is shown in Table 13.1.

The effects of pressure on discharge rate of emitter were evaluated. Two different drippers (NPC 4 lph and 8 lph) to determine CV x, and k values at 4 different pressures (0.5, 0.75, 1.0, 1.5 kg/cm²) were used. Keller and Krameli [14] introduced the coefficient of variation as a statistical measure for emitter manufacturing variation. This coefficient of manufacturer's variation was included in design equations for emission uniformity.

$$CV = 100 \ x \ [S_q/q_{avg}] \tag{2}$$

where, CV = discharge coefficient of variation (%), Sq = standard deviation of discharge rates of the emitters in the sample (lph), and q_{avg} = mean of emitter discharge rate (lph).

To determine the effect of different pressures on the variation of discharge rate, the discharge variation along with pressure was determined considering diameter of lateral and spacing between two drippers. Two diameters of 16 mm and 12 mm were selected along with dripper spacing of 15 cm and 12 cm. The system was operated for one hour considering

TABLE 13.1 ASAE Recommended Classification of Manufacturer's Coefficient of Variation (CV_m)

CV_m (%)	Classification
<5	Excellent
5–7	Average
7–11	Marginal
11–15	Poor
>15	Unacceptable

the irrigation operation time for major crop in Jalgaon. The moisture boxes were kept below each emitter and flow were collected by varying the pressure from 0.5 to 1.5 kg/cm^2.

13.3.4 DETERMINATION OF EMISSION UNIFORMITY

The uniformity of water application is of major concern in design procedures. It is typically used as a primary measure of the potential performance of the system. To determine if water and chemicals are applied uniformly, it is necessary to evaluate emitter discharge uniformity and system performance. Application uniformity of drip irrigation system can be expressed by several uniformity parameters. However, emitter discharge must be for a representative sample of emitters in a system. For determination of emission uniformity of drip irrigation, the system was operated for one hour. And discharge was collected. Emission uniformity of drip irrigation system was determined by equation:

$$EU, \% = 100 \times [q_n/q_a] \tag{3}$$

where, EU = emission uniformity of emitters (%), q_n = average discharge from emitters in the lowest 25% of the discharge range (lph), q_a = average discharge of all emitters in lph.

13.4 RESULTS AND DISCUSSION

13.4.1 PRESSURE AND DISCHARGE RELATIONSHIPS

The experiment was conducted at the field of Department of Irrigation and Drainage Engineering. The observations for the discharge rate at various pressures were recorded. For studying the effect of pressure on discharge rate, four pressures of 0.5, 0.75, 1, and 1.5 kg/cm^2 were applied to two lateral sizes. The Discharge variation was observed for 12 mm and 16 mm laterals with dripper spacing of 15 cm and 20 cm. The system was operated for one hour at 0.5, 0.75, 1, and 1.5 kg/cm^2 pressure.

Table 13.2 shows the discharge of 4 lph and 8 lph dripper for 12 mm lateral with dripper spacings of 15 cm and 20 cm. At 15 cm dripper spacing:

TABLE 13.2 Variation in Discharge with Pressure for 12 mm Lateral

Pressure	Size of Dripper	
(kg/cm²)	4 *lph* Q$_{ave}$	8 *lph* (Q$_{ave}$)
15 cm dripper spacing		
0.5	2.67	4.85
0.75	2.89	6.05
1.0	3.29	7.41
1.5	4.22	7.91
20 cm dripper spacing		
0.5	2.25	5.80
0.75	3.34	7.00
1.0	3.36	7.75
1.5	4.29	8.45

it was observed that at pressure 0.5 kg/cm² the discharge of 4 lph dripper was 2.67 lph and 4.85 lph for 8 lph dripper, which was too low than the recommended value. The discharge was increased as the pressure increases from 0.5 to 1.5 kg/cm². At 1 kg/cm² pressure, the discharge rate was 3.29 near the recommended value for 4 lph, but at 1.5 kg/cm² pressure, the discharge rate was 4.22 more than recommended value for 4 lph dripper. While for 8 lph dripper, the discharge rate was 7.91 near the recommended value at 1.5 kg/cm² pressure.

At dripper spacing of 20 cm in Table 13.2: it was observed that at pressure 0.5 kg/cm² the discharge of 4 lph dripper was 2.25 lph and 5.80 lph for 8 lph dripper. The discharge was increased as the pressure increases from 0.5 to 1.5 kg/cm². For 1 kg/cm² pressure, the discharge rate was 3.66 and 7.75 near the recommended discharge for both 4 lph and 8 lph, but at 1.5 kg/cm² pressure, the discharge rate was 4.29 and 8.45 more than recommended value for 4 lph and 8 lph.

Table 13.3 shows the discharge of 4 lph and 8 lph drippers for 16 mm lateral with dripper spacing of 15 cm and 20 cm. At 15 cm dripper spacing, it can be observed that at pressure 0.5 kg/cm² the discharge of 4 lph dripper was 2.88 lph and 5.27 lph for 8 lph dripper. The data shows increasing trend of discharge with increase in pressure from 0.5 to 1.5 kg/cm². For 1 kg/cm²

TABLE 13.3 Variation in Discharge With Pressure for 16 mm Lateral

Pressure	Size of Dripper	
(kg/cm²)	4 *lph* Q_{ave}	8 *lph* (Q_{ave})
15 cm dripper spacing		
0.5	2.88	5.27
0.75	3.62	5.85
1.0	3.86	6.74
1.5	4.42	7.66
20 cm dripper spacing		
0.5	3.02	5.44
0.75	3.56	5.86
1.0	3.92	6.74
1.5	4.75	7.87

pressure, the discharge rate was 3.86 near the recommended discharge for 4 lph, but at 1.5 kg/cm² the discharge rate was 4.42 more than recommended value for 4 lph dripper. While for 8 lph dripper the discharge rate was 7.66 near the recommended value at 1.5 kg/cm² pressure. Therefore, it can be concluded that the results were more accurate for 8 lph dripper as it gives more uniform discharge than 4 lph.

At 20 cm dripper spacing: Table 13.3 shows that at 0.5 kg/cm² the discharge of 4 lph dripper was 3.02 lph and 5.44 lph for 8 lph dripper. Table 13.3 shows increasing trend in discharge with pressure increase from 0.5 to 1.5 kg/cm². For 1 kg/cm², the discharge rate was 3.92 near the recommended discharge for 4 lph, but at 1.5 kg/cm² the discharge rate was 4.75 more than the recommended value for 4 lph. While for 8 lph dripper, the discharge rate was 7.87 near the recommended value at 1.5 kg/cm². Therefore, it can be concluded that the results were more accurate for 8 lph dripper. In Table 13.3, it was observed that better results can be obtained for 8 lph dripper having the spacing 20 cm.

From Tables 13.2 and 13.3, it can be observed that there was considerable effect of pressure on discharge variation as the drippers are of non-pressure compensating types. When the pressure increases above 1.0 kg/cm² for 4 lph dripper, the discharge was beyond the recommended value. There was

TABLE 13.4 Values of K_e and x for 4 lph and 8 lph Dripper with 15 cm and 20 cm Dripper Spacings

Dripper size	Dripper Spacing			
	15 cm		20 cm	
	K_e	X	K_e	x
12 mm lateral				
4 lph	2.209	0.150	1.985	0.194
8 lph	4.241	0.167	5.279	0.123
16 mm lateral				
4 lph	2.606	0.134	2.614	0.145
8 lph	4.605	0.126	4.694	0.124

also impact of dripper spacing. It can be concluded that 16 mm lateral with 20 cm spacing between drippers gave most accurate results.

Table 13.4 shows the values of K_e and x. The values of K_e and x were determined by linear regression analysis of a logarithmic plot of the measured discharge against applied pressure (Figures 13.1–13.4).

Table 13.4 shows that the values of x for 4 lph and 8 lph dripper were 0.150 and 0.167, respectively for 15 cm spacing. For 20 cm spacing the value of x was 0.194 for 4 lph dripper and 0.123 for 8 lph dripper. The values of x for 4 lph and 8 lph dripper were 0.134 and 0.126, respectively for 15 cm spacing. For 20 cm spacing the value of x was 0.145 for 4 lph dripper and 0.124 for 8 lph dripper. It can be concluded that the value of x ranged from 0.1 to 0.2.

The Figures 13.1–13.4 show the relationships between discharge and pressure for determining the value of k and x. Discharge was plotted on y axis and pressure was plotted on x axis and exponential relationships were obtained for 4 lph and 8 lph drippers. These graphs were used to determine the corresponding values of k and x in each case (Table 13.4).

Graphs 1 to 4 show increasing trend in discharge with pressure. Graphs shows that for 1.5 kg/cm^2 pressure, the discharges were above the recommended value as the drippers are NPC. Hence it can be concluded that the drip system should be operated up to 1.0 kg/cm^2 pressure in clayey soil for drippers spacings of 15 cm and 20 cm.

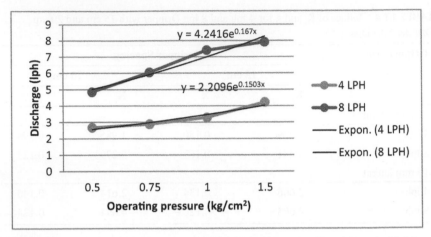

FIGURE 13.1   Regression values for 4 lph and 8 lph drippers on 12 mm lateral with 15 cm dripper spacing.

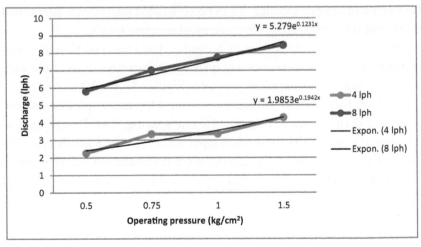

FIGURE 13.2   Regression values for 4 lph and 8 lph dripper on 12 mm lateral with 20 cm dripper spacing.

13.4.2   DETERMINATION OF EMISSION UNIFORMITY AND COEFFICIENT OF VARIATION

The Table 13.5 shows values of emission uniformity and coefficient of variation for 4 lph and 8 lph at different operating pressure. The Figures 13.5–13.8 indicate effects of variation in pressure on coefficient of variation for 4 lph and 8 lph drippers.

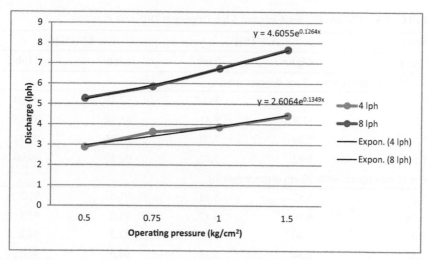

FIGURE 13.3 Regression values for 4 lph and 8 lph dripper on 16 mm lateral with 15 cm dripper spacing.

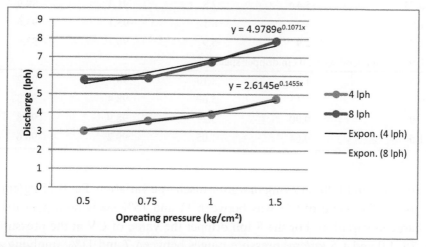

FIGURE 13.4 Regression values for 4 lph and 8 lph dripper on 16 mm lateral with 20 cm dripper spacing.

13.4.2.1 For 12 mm Lateral with 15 cm Dripper Spacing

From Table 13.5, it can be concluded that value of CV for 4 lph dripper at 1.5 kg/cm^2 pressure was between 5 and 7%, which shows an average variation in discharge; and for 1.0 and 0.5 kg/cm^2 pressure the CV ranges

TABLE 13.5   CV and EU for 4 lph and 8 lph Dripper at Different Pressure

Pressure	CV (%)		EU (%)	
(kg/cm^2)	4 lph	8 lph	4 lph	8 lph
For 12 mm lateral with 15 cm dripper spacing				
0.5	9.82	11.67	87.38	86.69
0.75	11.81	8.42	85.78	87.45
1.0	7.39	7.70	94.85	87.72
1.5	5.92	7.27	92.55	89.90
For 12 mm lateral with 20 cm dripper spacing				
0.5	5.52	5.75	91.4	91.2
0.75	6.10	6.17	91.2	91.1
1.0	4.40	3.45	93.3	94.8
1.5	3.34	2.78	94.5	96.1
For 16 mm lateral with 15 cm dripper spacing				
0.5	2.7	8.39	96.6	85.04
0.75	5.4	7.8	91.3	89.9
1.0	3.7	6.1	95.1	93.6
1.5	2.9	5.9	96.1	93.5
For 16 mm lateral with 20 cm dripper spacing				
0.5	1.3	2.10	98.1	97.2
0.75	1.13	5.60	98.4	92.3
1.0	0.90	3.21	98.9	95.8
1.5	0.64	1.05	99.2	98.6

between 7 and 11% showing marginal discharge variation. For 0.75 kg/cm^2 pressure the value of CV was between 11 and 15% showing a poor discharge in variation. For the 8 lph dripper the value of CV at the pressure 0.75, 1.0 and 1.5 kg/cm^2 pressure ranges between 7 and 11%, implying a marginal variation in discharge and for 0.5 kg/cm^2 pressure the value of CV was between 11 and 15%, showing poor variation in discharge.

The maximum emission uniformity for 4 lph dripper was 94.85% at 1.0 kg/cm^2 pressure and for 8 lph dripper was 89.90% at 1.5 kg/cm^2 pressure. So it can be concluded that 1.5 kg/cm^2 was a better for 8 lph dripper of 12 mm lateral with 15 cm spacing and for 4 lph dripper 1.0 kg/cm^2 pressure was better for 12 mm lateral with 15 cm spacing.

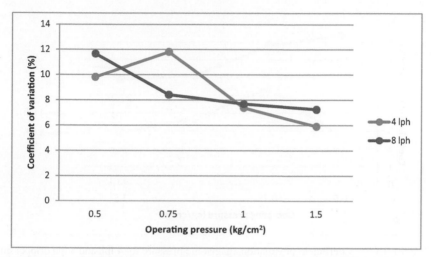

FIGURE 13.5 Coefficient of variation against the pressure for 4 lph and 8 lph dripper on 12 mm lateral with 15 cm dripper spacing.

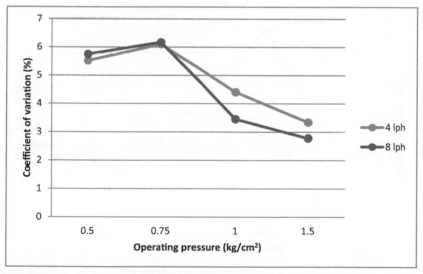

FIGURE 13.6 Coefficient of variation against the pressure for 4 lph and 8 lph dripper on 12 mm lateral with 20 cm dripper spacing.

13.4.2.2 For 12 mm Lateral with 20 cm Dripper Spacing

Table 13.5 shows that coefficient of variation for 4 lph dripper at 1.0 kg/cm^2 and 1.5 kg/cm^2 pressure was less than 5%. So it was an excellent variation

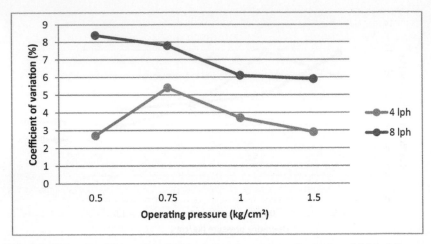

FIGURE 13.7 Coefficient of variation against the pressure for 4 lph and 8 lph dripper on 16 mm lateral with 15 cm dripper spacing.

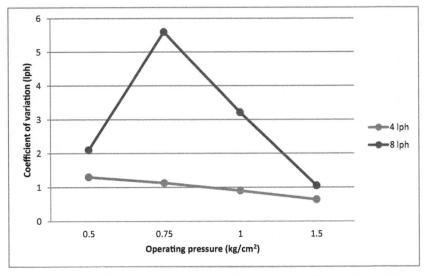

FIGURE 13.8 Coefficient of variation against the pressure for 4 lph and 8 lph dripper on 16 mm lateral with 20 cm dripper spacing.

in discharge. For 0.5 and 0.75 kg/cm² pressure the discharge variation ranges between 5 to 7%, which shows an average variation in discharge. For 8 lph dripper the coefficient of variation at 1.0 and 1.5 kg/cm² pressure was less than 5%, which indicates the excellent variation in discharge and for 0.5, 0.75 kg/cm² pressure the value of CV ranges between 5 to 7%, which shows an average variation in discharge.

The maximum emission uniformity of 4 lph dripper and 8 lph dripper was 94.5% and 96.1%, respectively at 1.5 kg/cm² pressure. So it can be concluded that the 1.5 kg/cm² pressure was a better for 12 mm lateral with 20 cm spacing for 4 lph and 8 lph drippers.

13.4.2.3 For 16 mm Lateral with 15 cm Dripper Spacing

In Table 13.5, it can be observed that value of CV for 4 lph dripper at 0.5, 1 and 1.5 kg/cm² pressure was less than 5%. According to Table 13.1, it represents excellent variation in discharge. For 0.75 kg/cm² the value of CV is ranges between 5 and 7%, which shows average variation in discharge. For 8 lph dripper the value of CV ranges between 5 and 7% at 0.75, 1.0 and 1.5 kg/cm² pressure, which indicates an average variation in discharge and for 0.5 kg/cm² pressure the value of CV ranges between 7 and 11%, indicating marginal variation in discharge.

The emission uniformity for 4 lph dripper was 96.1% and 96.6% at 1.5 and 0.5 kg/cm² pressure and for 8 lph dripper was 93.6% and 93.5% at 1 kg/cm² and 1.5 kg/cm² pressure. Therefore, it can be concluded that the 1.0 kg/cm² pressure can be considered better for 16 mm lateral with 15 cm dripper spacing for 8 lph dripper and 1.5 kg/cm² was better for 4 lph dripper.

13.4.2.4 For 16 mm Lateral with 20 cm Dripper Spacing

From Table 13.5, it can be concluded that value of CV for 4 lph dripper at 0.5, 0.75, 1.0 and 1.5 kg/cm² pressure was less than 5%, indicating excellent variation in discharge. The value of CV was less than 5% at 0.5, 1, and 1.5 kg/cm² pressure for 8 lph dripper, which shows an excellent variation in discharge and for 0.75 kg/cm² pressure the value of CV range from 5 to 7%, which shows an average variation in discharge.

The maximum emission uniformity for 4 lph and 8 lph dripper was 99.2% and 98.6% at the pressure of 1.5 kg/cm². Therefore, it was concluded that the better operating pressure was 1.5 kg/cm² for 4 lph and 8 lph drippers of 16 mm lateral with 20 cm spacing. From Table 13.5, it can be concluded that EU was better for the 16 mm lateral with 20 cm dripper spacing for 4 lph dripper and 8 lph drippers.

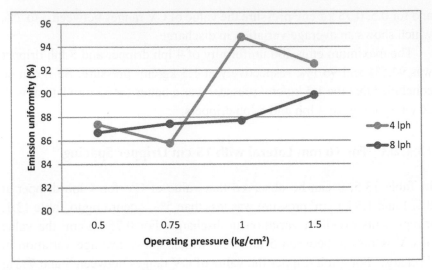

FIGURE 13.9 Emission uniformity against the pressure for 4 lph and 8 lph dripper on 12 mm lateral with 15 cm dripper spacing.

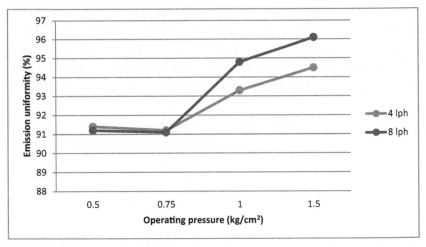

FIGURE 13.10 Emission uniformity against the pressure for 4 lph and 8 lph dripper on 12 mm lateral with 20 cm dripper spacing.

The Figures 13.9–13.12 show the effects of pressure on emission uniformity. The emission uniformity was plotted on y-axis and pressure was plotted on x-axis.

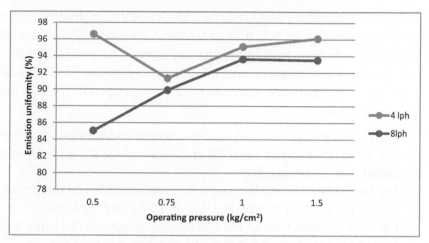

FIGURE 13.11 Emission uniformity against the pressure for 4 lph and 8 lph dripper on 16 mm lateral with 15 cm dripper spacing.

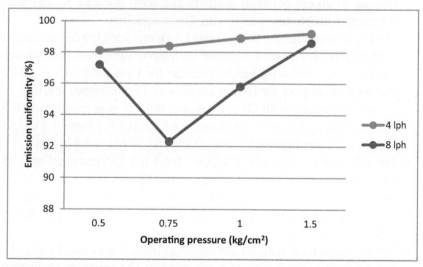

FIGURE 13.12 Emission uniformity against the pressure for 4 lph and 8 lph dripper on 16 mm lateral with 20 cm dripper spacing.

13.5 CONCLUSIONS

The study was conducted at the research farm of Irrigation and Drainage Engineering Department under the title *"Determination of Pressure and*

Discharge Relationships for Emitters." The variation in discharge was observed at pressures at 0.5, 0.75, 1.0, and 1.5 kg/cm². The values of k and x were calculated by using Keller and Karmeli's formula. The efficiency of drip irrigation in terms of emission uniformity and coefficient of variation was also calculated using formula. The highlights of findings are summarized below:

- The discharge rate was 3.86 near the recommended value at 1.0 kg/cm² for 4 lph, compared to 7.66 at 1.5 kg/cm² for 8 lph for 16 mm lateral with 15 cm spacing.
- For 16 mm lateral with 20 cm spacing, recommended discharge was 3,92 at 1.5 kg/cm² for 4 lph and 7,87 at 1.5 kg/cm² for 8 lph.
- For 12 mm lateral with 15 cm spacing, the acceptable discharge was 3.29 at 1.0 kg/cm² for 4 lph and 7.91 at 1.5 kg/cm² for 8 lph.
- The discharge rate gives most accurate results at a pressure of 1.0 kg/cm² for 4 lph and 8 lph: 3.36 and 7.75 lph for 12 mm lateral for the spacing of 20 cm between drippers and most accurate results were obtained from 16 mm lateral with 20 cm spacing.
- The CV was excellent at 0.5, 1.0 and 1.5 kg/cm² for 4 lph dripper and an average at 0.75, 1.0 and 1.5 kg/cm² for 8 lph dripper. The emission uniformity was 96.6% at 0.5 kg/cm² pressure for 4 lph and 93.6% at 1.0 kg/cm² for 8 lph dripper for 16 mm lateral with 15 cm dripper spacing.
- For 16 mm lateral with 20 cm spacing: the CV was excellent at 0.5, 0.75, 1.0 and 1.5 kg/cm² for 4 lph and for 8 lph at 0.5, 1.0 and 1.5 kg/cm² and the emission uniformity was 99.2% and 98.6% at 1.5 kg/cm².
- The CV was average at 1.5 kg/cm² for 4 lph and marginal at 0.75, 1.0 and 1.5 kg/cm² for 8 lph dripper for 12 mm lateral with 15 cm spacing. The emission uniformity for 4 lph was 94.85% at 1.0 kg/cm² pressure and for 8 lph was 89.90% at 1.5 kg/cm² for 12 mm lateral with 15 cm spacing.
- For 12 mm lateral with 20 cm spacing: the CV was excellent at 1.0 and 1.5 kg/cm² for 4 lph and 8 lph dripper. The emission uniformity of 4 lph and 8 lph drippers was 94.5% and 96.1% at 1.5 kg/cm².

13.6 SUMMARY

The experiment was conducted to examine effect of pressure on discharge variation along with the efficiency of drip irrigation in term of emission uniformity. For the study, authors considered two dripper spacings and two diameters of laterals (12 mm and 16 mm). The system was operated at 0.5, 0.75, 1.0 and 1.5 kg/cm². The system was operated for one hour, considering water requirement of major crops in Jalgaon (Jamod). The study revealed that the discharge was directly related to the pressure, showing exponential relationships. Out of two diameters and two spacings, best results were obtained for 16 mm lateral with a drpper spacing of 20 cm. For 4 lph dripper, the recommended discharge was obtained at 1.0 kg/cm² pressure. The values of x ranged between 0.1 and 0.2. The value of CV was <5% in most of the cases and was excellent. The EU was maximum for 16 mm lateral with 20 cm dripper spacing at 1.5 kg/cm².

KEYWORDS

- **Coefficient of variation**
- **Discharge**
- **Drip irrigation**
- **Dripper**
- **Dripper spacing**
- **Emission**
- **Emitter**
- **Lateral**
- **Pressure**
- **Pressure gauge**
- **Regression analysis**
- **Uniformity**

REFERENCES

1. Acar, B., Ramazan, T., & Mithat, D. (2010). Impact of pressurized irrigation technologies on efficient water resources. Int. J. of Sustainable Water and Environmental System, I, 1–4.
2. Anyoji, H., & Wu, L. P. (1994). Normal distribution water applications for drip irrigation schedules. Transaction of the ASAE, 37, 159–164.

3. Bhardwaj, S. K. (2001). Importance of drip irrigation in Indian Agriculture. Kissan World, 28, 32–33.
4. Boswell, M. J. (1985). Design characteristics of line source drip tubes. Proceeding of the third international trickle irrigation congress, volume 1, California, USA, Pages 306–312.
5. Braud, H. J., & Soon, A. M. (1980). Trickle irrigation design for improved application uniformity. ASAE and CSAE national meeting o trickle irrigation. Paper No. 1979–85.
6. Bucks, D. A., & Davis, S. (1986). Historical development. In: Trickle Irrigation for Crop Production. Eds.: F. S. Nakayama and D. A. Bucks. Dev. in Agr. Eng. 9, 1–26.
7. Camp, C. R. (1998). Subsurface drip irrigation: a review. Trans ASAE, 41, 1353–1367.
8. Changade, N. M., Chavan, M. L., Jadhav, S. B., & Bhagyavant, R. G. (2009). Determination of emission uniformity of emitter in gravity fed drip irrigation system. International Journal of Agricultural Eng., II No.1, 88–91.
9. Dogan, E. (2010). Effects of drip irrigation system pressure fluctuations on drip lateral emitter flow rate and diameter change. Journal of Agricultural Science, 16, 235–241.
10. Gil, M., Sinobas, L. R., Juana, L., Sanchez, R., & Losada, A. (2007). Emitter discharge variability of sub surface drip irrigation in uniform soils.
11. Ibragimov, N., Evett, S. R., Esanbekov, Y., Kamilov, B. S., Mirzaev, L., & Lamers, J. P. (2007). Water use efficiency of irrigated cotton in Uzbekistan under drip irrigation. Agric. Water Manage., 90, 112–120.
12. Kadale, A. S., Shah, N. M., & Mallik, M. (2007). Determination of permissible lateral pressure variation for various types of emitter-A Case study. Karnataka J. Agric. Sci., 20(2), 444–445
13. Kang, Y. (2000). Effect of operating pressures on micro irrigation uniformity. http://www.biotechnology.uni-koeln.de/inco2-dev/.../41_kangy.com
14. Keller, J., & Karmeli, D. (1974). Trickle irrigation design parameters. Trans. ASAE, 17, 678–784.
15. Kumar, R., & Singh, J. (2003). Regional water management modeling for decision support in irrigated agriculture. J. Irrig. Drain. Eng., 129, 432–439.
16. Little, G. E., Hills, D. J., & Hanson, B. R. (1993). Uniformity in pressurized irrigation systems depends on design, installation. California Agri., 47(3), 18–21.
17. Pereira, L. (1999). Higher performance through combined improvements in irrigation methods and scheduling: A discussion. Agric. Water Manage., 40, 153–169.
18. Popale, P. G., Bombale, V. T., & Magar, A. P. (2011). Hydraulic performance of drip irrigation system. Engineering and Technology in India, 2, 24–28.
19. Priyanjith, K. J. K. T., Kuruppuarachchi, D. P. S., & Gunathilaka, H. A. (2002). Field evaluation of drip irrigation system for small-scale banana orchard.
20. Safi, B., Mohammad, R. N., & Hossein, A. (2007). Water application uniformity of drip irrigation system at various operating pressure and tape lengths. Indian Farming, 31, 275–285.
21. Selvaraj, P. K., Krishnamurthi, V. V., Manickasundaram, P., Martin, G. J., & Ayyaswamy, M. (1997). Drip irrigation for sugarcane. Indian Farming, 46, 17–20.
22. Shani, U., Xue, S., Katz, R. G., & Warrck, A. W. (1996). Soil limiting from subsurface emitters. I: Pressure measurements. Journal of Irrigation and Drainage Engineering, 122, 291–295.

23. Singandhupe, R. B., Singh, S. R., Chaudhari, K., & Patil, N. G. (1998). Micro irrigation and sprinkler irrigation system proceedings, 28–30 April, New Delhi, II, 54.
24. Singh, P., & Kumar, S. (2007). Evaluation of hydraulic performance of drip irrigation system. Journal of Agricultural Engineering, 44, 106–108.
25. Sinobas, L. R., Juana, L., & Losada, A. (1999). Effect of temperature change on emitter discharge. Journal of Irrigation and Drainage Engineering, pp. 64–73.
26. Soccol, O. J., Ullmann, M. N., & Frizzone, J. A. (2002). Performances analysis of trickle irrigation subunit installed in an apple orchard. Brazilian Archives of Biology And Technology, 45, 525–530.
27. Tiwari, K. N., Kannan, N., & Mal, P. K. (2002). Enhancing productivity of horticultural crops through drip irrigation. Progress in micro irrigation research and development in India. Water Technology Centre for Eastern Region (ICAR). Pages 44–49.
28. Tiwari, K. N. (1998). Response of okra to drip irrigation under mulch and non-mulch conditions. Agril. Water Management, 38, 91–102.
29. Troy, P. R. (1999). Drip irrigation for the yard and gardens. http://www.igreenproduct.com
30. Tsay, T. S., & Ju, S. H. (1998). Analysis of drip irrigation uniformity by numerical simulation. Journal of Chinese Agricultural Engineering, 44(4).
31. Umara, B. G., Audu, I., & Bashir, A. U. (2011). Performance evaluation of bamboo low cost micro irrigation lateral system. ARPN Journal of Engineering and Applied Sciences, 6(5), www.arpnjournals.com
32. Zur, B., & Tal, S. (1981). Emitter discharge sensitivity to pressure and temperature. Journal of the irrigation and drainage division ASCE, 109, 19.

APPENDIX I

Photos of Field Evaluation

Measuring the pressure
by pressure gauge

Collection of water flow
of the emitter

Measurement of discharge at different pressure.

APPENDICES

APPENDIX A   CONVERSION SI AND NON-SI UNITS

To convert the Column 1 in the Column 2,	Column 1	Column 2	To convert the Column 2 in the Column 1,
	Unit	*Unit*	
Multiply by	SI	Non-SI	Multiply by

LINEAR

0.621 _____	kilometer, km (10^3m)	miles, mi _____	1.609
1.094 _____	meter, m	yard, yd _____	0.914
3.28 _____	meter, m	feet, ft _____	0.304
3.94×10^{-2} ___	millimeter, mm (10^{-3})	inch, in _____	25.4

SQUARES

2.47 _____	hectare, he	acre _____	0.405
2.47 _____	square kilometer, km^2	acre _____	4.05×10^{-3}
0.386 _____	square kilometer, km^2	square mile, mi^2 ___	2.590
2.47×10^{-4} ___	square meter, m^2	acre _____	4.05×10^{-3}
10.76 _____	square meter, m^2	square feet, ft^2 _____	9.29×10^{-2}
1.55×10^{-3} ___	mm^2	square inch, in^2 ____	645

CUBICS

9.73×10^{-3} ___	cubic meter, m^3	inch-acre _____	102.8

(Modified and reprinted with permission from: Megh R. Goyal, 2012. Appendices. Pages 317–332. In: *Management of Drip/Trickle or Micro Irrigation* edited by Megh R. Goyal. New Jersey, USA: Apple Academic Press Inc.)

35.3 _____ cubic meter, m³	cubic-feet, ft³ _____ 2.83×10^{-2}
6.10×10^4 ____ cubic meter, m³	cubic inch, in³ _____ 1.64×10^{-5}
2.84×10^{-2} ___ liter, L (10^{-3} m³)	bushel, bu _____ 35.24
1.057 _____ liter, L	liquid quarts, qt ____ 0.946
3.53×10^{-2} ___ liter, L	cubic feet, ft³ _____ 28.3
0.265 _____ liter, L	gallon _____ 3.78
33.78 _____ liter, L	fluid ounce, oz _____ 2.96×10^{-2}
2.11 _____ liter, L	fluid dot, dt _____ 0.473

WEIGHT

2.20×10^{-3} ___ gram, g (10^{-3} kg)	pound, _____ 454
3.52×10^{-2} ___ gram, g (10^{-3} kg)	ounce, oz _____ 28.4
2.205 _____ kilogram, kg	pound, lb _____ 0.454
10^{-2} _____ kilogram, kg	quintal (metric), q __ 100
1.10×10^{-3} ___ kilogram, kg	ton (2000 lbs), ton __ 907
1.10^2 _____ mega gram, mg	ton (US), ton _____ 0.907
1.10^2 _____ metric ton, t	ton (US), ton _____ 0.907

YIELD AND RATE

0.893 _____ kilogram per hectare	pound per acre _____ 1.12
7.77×10^{-2} ___ kilogram per cubic meter	pound per fanega ___ 12.87
1.49×10^{-2} ___ kilogram per hectare	pound per acre, _____ 67.19 60 lb
1.59×10^{-2} ___ kilogram per hectare	pound per acre, _____ 62.71 56 lb
1.86×10^{-2} ___ kilogram per hectare	pound per acre, _____ 53.75 48 lb
0.107 _____ liter per hectare	galloon per acre ____ 9.35
893 _____ ton per hectare	pound per acre _____ 1.12×10^{-3}

893 _____ mega gram per   pound per acre _____ 1.12×10^{-3}
                    hectare
0.446 _____ ton per hectare   ton (2000 lb) per ___ 2.24
                                       acre
2.24 _____ meter per second   mile per hour _____ 0.447

SPECIFIC SURFACE

10 _____ square meter per   square centimeter ___ 0.1
                    kilogram   per gram
10^3 _____ square meter per   square millimeter ___ 10^{-3}
                    kilogram   per gram

PRESSURE

9.90 _____ megapascal, MPa   atmosphere _____ 0.101
10 _____ megapascal   bar _____ 0.1
1.0 _____ megagram per   gram per cubic _____ 1.00
                    cubic meter   centimeter
2.09×10^{-2} ___ pascal, Pa   pound per square ___ 47.9
                                        feet
1.45×10^{-4} ___ pascal, Pa   pound per square ___ 6.90×10^3
                                        inch

TEMPERATURE

1.00   _____ Kelvin, K   centigrade, °C _____ 1.00
(K-273)   (C+273)
(1.8 C + 32) __ centigrade, °C   Fahrenheit,°F _____ (F-32)/
                                        1.8

ENERGY

9.52×10^{-4} ___ Joule J   BTU _____ 1.05×10^3

0.239 _____	Joule, J	calories, cal _____	4.19
0.735 _____	Joule, J	feet-pound _____	1.36
2.387×10^5 ___	Joule per square meter	calories per _____ square centimeter	4.19×10^4
10^5 _____	Newton, N	dynes _____	10^{-5}

WATER REQUIREMENTS

9.73×10^{-3} ___	cubic meter	inch acre _____	102.8
9.81×10^{-3} ___	cubic meter per hour	cubic feet per _____ second	101.9
4.40 _____	cubic meter per hour	galloon (US) _____ per minute	0.227
8.11 _____	hectare-meter	acre-feet _____	0.123
97.28 _____	hectare-meter	acre-inch _____	1.03×10^{-2}
8.1×10^{-2} ___	hectare centimeter	acre-feet _____	12.33

CONCENTRATION

1 _____	centimol per kilogram	milliequivalents ___ per 100 grams	1
0.1 _____	gram per kilogram	percents _____	10
1 _____	milligram per kilogram	parts per million ___	1

NUTRIENTS FOR PLANTS

2.29 _____	P	P_2O_5 _____	0.437
1.20 _____	K	K_2O _____	0.830
1.39 _____	Ca	CaO _____	0.715
1.66 _____	Mg	MgO _____	0.602

NUTRIENT EQUIVALENTS

Column A	Column B	Conversion	Equivalent
		A to B	B to A
N	NH_3	1.216	0.822
	NO_3	4.429	0.226
	KNO_3	7.221	0.1385
	$Ca(NO_3)_2$	5.861	0.171
	$(NH_4)_2SO_4$	4.721	0.212
	NH_4NO_3	5.718	0.175
	$(NH_4)_2HPO_4$	4.718	0.212
P	P_2O_5	2.292	0.436
	PO_4	3.066	0.326
	KH_2PO_4	4.394	0.228
	$(NH_4)_2HPO_4$	4.255	0.235
	H_3PO_4	3.164	0.316
K	K_2O	1.205	0.83
	KNO_3	2.586	0.387
	KH_2PO_4	3.481	0.287
	Kcl	1.907	0.524
	K_2SO_4	2.229	0.449
Ca	CaO	1.399	0.715
	$Ca(NO_3)_2$	4.094	0.244
	$CaCl_2 \times 6H_2O$	5.467	0.183
	$CaSO_4 \times 2H_2O$	4.296	0.233
Mg	MgO	1.658	0.603
	$MgSO_4 \times 7H_2O$	1.014	0.0986
S	H_2SO_4	3.059	0.327
	$(NH_4)_2SO_4$	4.124	0.2425
	K_2SO_4	5.437	0.184
	$MgSO_4 \times 7H_2O$	7.689	0.13
	$CaSO_4 \times 2H_2O$	5.371	0.186

APPENDIX B   PIPE AND CONDUIT FLOW

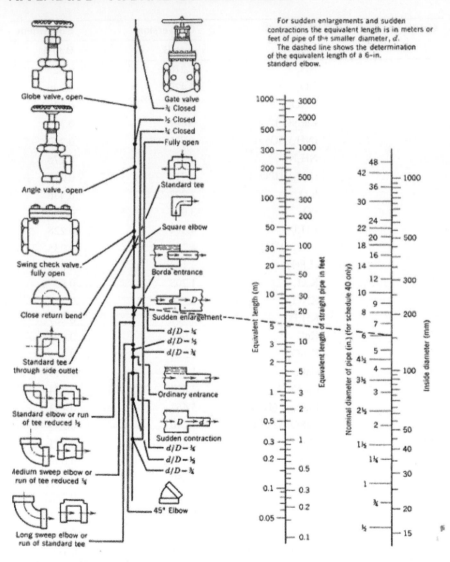

APPENDIX C PERCENTAGE OF DAILY SUNSHINE HOURS: FOR NORTH AND SOUTH HEMISPHERES

Latitude	Jan	Feb	Mar	Apr	May	Jun	Jul	Aug	Sep	Oct	Nov	Dec
NORTH												
0	8.50	7.66	8.49	8.21	8.50	8.22	8.50	8.49	8.21	8.50	8.22	8.50
5	8.32	7.57	8.47	3.29	8.65	8.41	8.67	8.60	8.23	8.42	8.07	8.30
10	8.13	7.47	8.45	8.37	8.81	8.60	8.86	8.71	8.25	8.34	7.91	8.10
15	7.94	7.36	8.43	8.44	8.98	8.80	9.05	8.83	8.28	8.20	7.75	7.88
20	7.74	7.25	8.41	8.52	9.15	9.00	9.25	8.96	8.30	8.18	7.58	7.66
25	7.53	7.14	8.39	8.61	9.33	9.23	9.45	9.09	8.32	8.09	7.40	7.52
30	7.30	7.03	8.38	8.71	9.53	9.49	9.67	9.22	8.33	7.99	7.19	7.15
32	7.20	6.97	8.37	8.76	9.62	9.59	9.77	9.27	8.34	7.95	7.11	7.05
34	7.10	6.91	8.36	8.80	9.72	9.70	9.88	9.33	8.36	7.90	7.02	6.92
36	6.99	6.85	8.35	8.85	9.82	9.82	9.99	9.40	8.37	7.85	6.92	6.79
38	6.87	6.79	8.34	8.90	9.92	9.95	10.1	9.47	3.38	7.80	6.82	6.66
40	6.76	6.72	8.33	8.95	10.0	10.1	10.2	9.54	8.39	7.75	6.72	7.52
42	6.63	6.65	8.31	9.00	10.1	10.2	10.4	9.62	8.40	7.69	6.62	6.37
44	6.49	6.58	8.30	9.06	10.3	10.4	10.5	9.70	8.41	7.63	6.49	6.21
46	6.34	6.50	8.29	9.12	10.4	10.5	10.6	9.79	8.42	7.57	6.36	6.04
48	6.17	6.41	8.27	9.18	10.5	10.7	10.8	9.89	8.44	7.51	6.23	5.86
50	5.98	6.30	8.24	9.24	10.7	10.9	11.0	10.0	8.35	7.45	6.10	5.64
52	5.77	6.19	8.21	9.29	10.9	11.1	11.2	10.1	8.49	7.39	5.93	5.43
54	5.55	6.08	8.18	9.36	11.0	11.4	11.4	10.3	8.51	7.20	5.74	5.18

Appendix C Continued

Latitude	Jan	Feb	Mar	Apr	May	Jun	Jul	Aug	Sep	Oct	Nov	Dec
56	5.30	5.95	8.15	9.45	11.2	11.7	11.6	10.4	8.53	7.21	5.54	4.89
58	5.01	5.81	8.12	9.55	11.5	12.0	12.0	10.6	8.55	7.10	4.31	4.56
60	4.67	5.65	8.08	9.65	11.7	12.4	12.3	10.7	8.57	6.98	5.04	4.22
SOUTH												
0	8.50	7.66	8.49	8.21	8.50	8.22	8.50	8.49	8.21	8.50	8.22	8.50
5	8.68	7.76	8.51	8.15	8.34	8.05	8.33	8.38	8.19	8.56	8.37	8.68
10	8.86	7.87	8.53	8.09	8.18	7.86	8.14	8.27	8.17	8.62	8.53	8.88
15	9.05	7.98	8.55	8.02	8.02	7.65	7.95	8.15	8.15	8.68	8.70	9.10
20	9.24	8.09	8.57	7.94	7.85	7.43	7.76	8.03	8.13	8.76	8.87	9.33
25	9.46	8.21	8.60	7.74	7.66	7.20	7.54	7.90	8.11	8.86	9.04	9.58
30	9.70	8.33	8.62	7.73	7.45	6.96	7.31	7.76	8.07	8.97	9.24	9.85
32	9.81	8.39	8.63	7.69	7.36	6.85	7.21	7.70	8.06	9.01	9.33	9.96
34	9.92	8.45	8.64	7.64	7.27	6.74	7.10	7.63	8.05	9.06	9.42	10.1
36	10.0	8.51	8.65	7.59	7.18	6.62	6.99	7.56	8.04	9.11	9.35	10.2
38	10.2	8.57	8.66	7.54	7.08	6.50	6.87	7.49	8.03	9.16	9.61	10.3
40	10.3	8.63	8.67	7.49	6.97	6.37	6.76	7.41	8.02	9.21	9.71	10.5
42	10.4	8.70	8.68	7.44	6.85	6.23	6.64	7.33	8.01	9.26	9.8	10.6
44	10.5	8.78	8.69	7.38	6.73	6.08	6.51	7.25	7.99	9.31	9.94	10.8
46	10.7	8.86	8.90	7.32	6.61	5.92	6.37	7.16	7.96	9.37	10.1	11.0

APPENDIX D   PSYCHOMETRIC CONSTANT (γ) FOR DIFFERENT ALTITUDES (Z)

$$\gamma = 10^{-3}\ [(C_p.P) \div (\varepsilon.\lambda)] = (0.00163) \times [P \div \lambda]$$

γ, psychrometric constant [kPa C⁻¹]

c_p, specific heat of moist air = 1.013 [kJ kg⁻¹⁰ C⁻¹]

P, atmospheric pressure [kPa].

ε, ratio molecular weight of water vapor/dry air = 0.622

λ, latent heat of vaporization [MJ kg⁻¹] = 2.45 MJ kg⁻¹ at 20°C.

Z (m)	γ kPa/°C	z (m)	γ kPa/°C	z (m)	γ kPa/°C	z (m)	γ kPa/°C
0	0.067	1000	0.060	2000	0.053	3000	0.047
100	0.067	1100	0.059	2100	0.052	3100	0.046
200	0.066	1200	0.058	2200	0.052	3200	0.046
300	0.065	1300	0.058	2300	0.051	3300	0.045
400	0.064	1400	0.057	2400	0.051	3400	0.045
500	0.064	1500	0.056	2500	0.050	3500	0.044
600	0.063	1600	0.056	2600	0.049	3600	0.043
700	0.062	1700	0.055	2700	0.049	3700	0.043
800	0.061	1800	0.054	2800	0.048	3800	0.042
900	0.061	1900	0.054	2900	0.047	3900	0.042
1000	0.060	2000	0.053	3000	0.047	4000	0.041

APPENDIX E   SATURATION VAPOR PRESSURE [e_s] FOR DIFFERENT TEMPERATURES (T)

Vapor pressure function = e_s = [0.6108]*exp{[17.27*T]/[T + 237.3]}							
T °C	e_s kPa	T °C	e_s kPa	T °C	e_s kPa	T °C	e_s kPa
1.0	0.657	13.0	1.498	25.0	3.168	37.0	6.275
1.5	0.681	13.5	1.547	25.5	3.263	37.5	6.448
2.0	0.706	14.0	1.599	26.0	3.361	38.0	6.625
2.5	0.731	14.5	1.651	26.5	3.462	38.5	6.806
3.0	0.758	15.0	1.705	27.0	3.565	39.0	6.991

Appendix E   Continued

| Vapor pressure function = $e_s = [0.6108]*\exp\{[17.27*T]/[T + 237.3]\}$ | | | | | | | |
T °C	e_s kPa	T °C	e_s kPa	T °C	e_s kPa	T °C	e_s kPa
3.5	0.785	15.5	1.761	27.5	3.671	39.5	7.181
4.0	0.813	16.0	1.818	28.0	3.780	40.0	7.376
4.5	0.842	16.5	1.877	28.5	3.891	40.5	7.574
5.0	0.872	17.0	1.938	29.0	4.006	41.0	7.778
5.5	0.903	17.5	2.000	29.5	4.123	41.5	7.986
6.0	0.935	18.0	2.064	30.0	4.243	42.0	8.199
6.5	0.968	18.5	2.130	30.5	4.366	42.5	8.417
7.0	1.002	19.0	2.197	31.0	4.493	43.0	8.640
7.5	1.037	19.5	2.267	31.5	4.622	43.5	8.867
8.0	1.073	20.0	2.338	32.0	4.755	44.0	9.101
8.5	1.110	20.5	2.412	32.5	4.891	44.5	9.339
9.0	1.148	21.0	2.487	33.0	5.030	45.0	9.582
9.5	1.187	21.5	2.564	33.5	5.173	45.5	9.832
10.0	1.228	22.0	2.644	34.0	5.319	46.0	10.086
10.5	1.270	22.5	2.726	34.5	5.469	46.5	10.347
11.0	1.313	23.0	2.809	35.0	5.623	47.0	10.613
11.5	1.357	23.5	2.896	35.5	5.780	47.5	10.885
12.0	1.403	24.0	2.984	36.0	5.941	48.0	11.163
12.5	1.449	24.5	3.075	36.5	6.106	48.5	11.447

APPENDIX F   SLOPE OF VAPOR PRESSURE CURVE (Δ) FOR DIFFERENT TEMPERATURES (T)

$$\Delta = [4098. \, e^0(T)] \div [T + 237.3]^2$$
$$= 2504\{\exp[(17.27T) \div (T + 237.2)]\} \div [T + 237.3]^2$$

T °C	Δ kPa/°C	T °C	Δ kPa/°C	T °C	Δ kPa/°C	T °C	Δ kPa/°C
1.0	0.047	13.0	0.098	25.0	0.189	37.0	0.342
1.5	0.049	13.5	0.101	25.5	0.194	37.5	0.350
2.0	0.050	14.0	0.104	26.0	0.199	38.0	0.358

Appendix F   Continued

T °C	Δ kPa/°C	T °C	Δ kPa/°C	T °C	Δ kPa/°C	T °C	Δ kPa/°C
2.5	0.052	14.5	0.107	26.5	0.204	38.5	0.367
3.0	0.054	15.0	0.110	27.0	0.209	39.0	0.375
3.5	0.055	15.5	0.113	27.5	0.215	39.5	0.384
4.0	0.057	16.0	0.116	28.0	0.220	40.0	0.393
4.5	0.059	16.5	0.119	28.5	0.226	40.5	0.402
5.0	0.061	17.0	0.123	29.0	0.231	41.0	0.412
5.5	0.063	17.5	0.126	29.5	0.237	41.5	0.421
6.0	0.065	18.0	0.130	30.0	0.243	42.0	0.431
6.5	0.067	18.5	0.133	30.5	0.249	42.5	0.441
7.0	0.069	19.0	0.137	31.0	0.256	43.0	0.451
7.5	0.071	19.5	0.141	31.5	0.262	43.5	0.461
8.0	0.073	20.0	0.145	32.0	0.269	44.0	0.471
8.5	0.075	20.5	0.149	32.5	0.275	44.5	0.482
9.0	0.078	21.0	0.153	33.0	0.282	45.0	0.493
9.5	0.080	21.5	0.157	33.5	0.289	45.5	0.504
10.0	0.082	22.0	0.161	34.0	0.296	46.0	0.515
10.5	0.085	22.5	0.165	34.5	0.303	46.5	0.526
11.0	0.087	23.0	0.170	35.0	0.311	47.0	0.538
11.5	0.090	23.5	0.174	35.5	0.318	47.5	0.550
12.0	0.092	24.0	0.179	36.0	0.326	48.0	0.562
12.5	0.095	24.5	0.184	36.5	0.334	48.5	0.574

APPENDIX G   NUMBER OF THE DAY IN THE YEAR (JULIAN DAY)

Day	Jan	Feb	Mar	Apr	May	Jun	Jul	Aug	Sep	Oct	Nov	Dec
1	1	32	60	91	121	152	182	213	244	274	305	335
2	2	33	61	92	122	153	183	214	245	275	306	336
3	3	34	62	93	123	154	184	215	246	276	307	337
4	4	35	63	94	124	155	185	216	247	277	308	338
5	5	36	64	95	125	156	186	217	248	278	309	339
6	6	37	65	96	126	157	187	218	249	279	310	340

Appendix G   Continued

Day	Jan	Feb	Mar	Apr	May	Jun	Jul	Aug	Sep	Oct	Nov	Dec
7	7	38	66	97	127	158	188	219	250	280	311	341
8	8	39	67	98	128	159	189	220	251	281	312	342
9	9	40	68	99	129	160	190	221	252	282	313	343
10	10	41	69	100	130	161	191	222	253	283	314	344
11	11	42	70	101	131	162	192	223	254	284	315	345
12	12	43	71	102	132	163	193	224	255	285	316	346
13	13	44	72	103	133	164	194	225	256	286	317	347
14	14	45	73	104	134	165	195	226	257	287	318	348
15	15	46	74	105	135	166	196	227	258	288	319	349
16	16	47	75	106	136	167	197	228	259	289	320	350
17	17	48	76	107	137	168	198	229	260	290	321	351
18	18	49	77	108	138	169	199	230	261	291	322	352
19	19	50	78	109	139	170	200	231	262	292	323	353
20	20	51	79	110	140	171	201	232	263	293	324	354
21	21	52	80	111	141	172	202	233	264	294	325	355
22	22	53	81	112	142	173	203	234	265	295	326	356
23	23	54	82	113	143	174	204	235	266	296	327	357
24	24	55	83	114	144	175	205	236	267	297	328	358
25	25	56	84	115	145	176	206	237	268	298	329	359
26	26	57	85	116	146	177	207	238	269	299	330	360
27	27	58	86	117	147	178	208	239	270	300	331	361
28	28	59	87	118	148	179	209	240	271	301	332	362
29	29	(60)	88	119	149	180	210	241	272	302	333	363
30	30	—	89	120	150	181	211	242	273	303	334	364
31	31	—	90	—	151	—	212	243	—	304	—	365

APPENDIX H   STEFAN-BOLTZMANN LAW AT DIFFERENT TEMPERATURES (T):

$$[\sigma^*(T_K)^4] = [4.903 \times 10^{-9}], \text{ MJ K}^{-4} \text{ m}^{-2} \text{ day}^{-1}$$
where: $T_K = \{T[°C] + 273.16\}$

T	$\sigma*(T_K)^4$	T	$\sigma*(T_K)^4$	T	$\sigma*(T_K)^4$
			Units		
°C	MJ m^{-2} d^{-1}	°C	MJ m^{-2} d^{-1}	°C	MJ m^{-2} d^{-1}
1.0	27.70	17.0	34.75	33.0	43.08
1.5	27.90	17.5	34.99	33.5	43.36
2.0	28.11	18.0	35.24	34.0	43.64
2.5	28.31	18.5	35.48	34.5	43.93
3.0	28.52	19.0	35.72	35.0	44.21
3.5	28.72	19.5	35.97	35.5	44.50
4.0	28.93	20.0	36.21	36.0	44.79
4.5	29.14	20.5	36.46	36.5	45.08
5.0	29.35	21.0	36.71	37.0	45.37
5.5	29.56	21.5	36.96	37.5	45.67
6.0	29.78	22.0	37.21	38.0	45.96
6.5	29.99	22.5	37.47	38.5	46.26
7.0	30.21	23.0	37.72	39.0	46.56
7.5	30.42	23.5	37.98	39.5	46.85
8.0	30.64	24.0	38.23	40.0	47.15
8.5	30.86	24.5	38.49	40.5	47.46
9.0	31.08	25.0	38.75	41.0	47.76
9.5	31.30	25.5	39.01	41.5	48.06
10.0	31.52	26.0	39.27	42.0	48.37
10.5	31.74	26.5	39.53	42.5	48.68
11.0	31.97	27.0	39.80	43.0	48.99
11.5	32.19	27.5	40.06	43.5	49.30
12.0	32.42	28.0	40.33	44.0	49.61
12.5	32.65	28.5	40.60	44.5	49.92
13.0	32.88	29.0	40.87	45.0	50.24
13.5	33.11	29.5	41.14	45.5	50.56
14.0	33.34	30.0	41.41	46.0	50.87
14.5	33.57	30.5	41.69	46.5	51.19
15.0	33.81	31.0	41.96	47.0	51.51
15.5	34.04	31.5	42.24	47.5	51.84

Appendix H   Continued

T	$\sigma*(T_K)^4$	T	$\sigma*(T_K)^4$	T	$\sigma*(T_K)^4$
			Units		
°C	MJ m^{-2} d^{-1}	°C	MJ m^{-2} d^{-1}	°C	MJ m^{-2} d^{-1}
16.0	34.28	32.0	42.52	48.0	52.16
16.5	34,52	32.5	42.80	48.5	52.49

APPENDIX I   THERMODYNAMIC PROPERTIES OF AIR AND WATER

1. Latent Heat of Vaporization (λ)

$$\lambda = [2.501 - (2.361 \times 10^{-3})\, T]$$

where: λ = latent heat of vaporization [MJ kg^{-1}]; and T = air temperature [°C].

The value of the latent heat varies only slightly over normal temperature ranges. A single value may be taken (for ambient temperature = 20°C): $\lambda = 2.45$ MJ kg^{-1}.

2. Atmospheric Pressure (P)

$$P = P_o\, [\{T_{Ko} - \alpha(Z - Z_o)\} \div \{T_{Ko}\}]^{(g/(\alpha.R))}$$

where: P, atmospheric pressure at elevation z [kPa]

P_o, atmospheric pressure at sea level = 101.3 [kPa]

z, elevation [m]

z_o, elevation at reference level [m]

g, gravitational acceleration = 9.807 [m s^{-2}]

R, specific gas constant = 287 [J kg^{-1} K^{-1}]

α, constant lapse rate for moist air = 0.0065 [K m^{-1}]

T_{Ko}, reference temperature [K] at elevation z_o = 273.16 + T

T, means air temperature for the time period of calculation [°C]

When assuming P_o = 101.3 [kPa] at z_o = 0, and T_{Ko} = 293 [K] for T = 20 [°C], above equation reduces to:

$$P = 101.3[(293-0.0065Z)(293)]^{5.26}$$

3. Atmospheric Density (ρ)

$$\rho = [1000P] \div [T_{Kv} R] = [3.486P] \div [T_{Kv}], \text{ and } T_{Kv} = T_K[1-0.378(e_a)/P]^{-1}$$

where: ρ, atmospheric density [kg m^{-3}]

R, specific gas constant = 287 [J kg^{-1} K^{-1}]

$T_{Kv,}$ virtual temperature [K]

T_K, absolute temperature [K]: $T_K = 273.16 + T$ [°C]

e_a, actual vapor pressure [kPa]

T, mean daily temperature for 24-hour calculation time steps.

For average conditions (e_a in the range 1–5 kPa and P between 80–100 kPa), T_{Kv} can be substituted by: $T_{Kv} \approx 1.01 (T + 273)$

4. Saturation Vapor Pressure function (e_s)

$$e_s = [0.6108]*\exp\{[17.27*T]/[T + 237.3]\}$$

where: e_s, saturation vapor pressure function [kPa]

T, air temperature [°C]

5. Slope Vapor Pressure Curve (Δ)

$$\Delta = [4098.\, e°(T)] \div [T + 237.3]^2$$

$$= 2504\{\exp[(17.27T) \div (T + 237.2)]\} \div [T + 237.3]^2$$

where: Δ, slope vapor pressure curve [kPa C^{-1}]

T, air temperature [°C]

e^0(T), saturation vapor pressure at temperature T [kPa]

In 24-hour calculations, Δ is calculated using mean daily air temperature. In hourly calculations T refers to the hourly mean, T_{hr}.

6. Psychrometric Constant (γ)

$$\gamma = 10^{-3} [(C_p.P) \div (\varepsilon.\lambda)] = (0.00163) \times [P \div \lambda]$$

where: γ, psychrometric constant [kPa C^{-1}]

c_p, specific heat of moist air = 1.013 [kJ kg^{-10} C^{-1}]
P, atmospheric pressure [kPa]: equations 2 or 4
ε, ratio molecular weight of water vapor/dry air = 0.622
λ, latent heat of vaporization [MJ kg^{-1}]

7. Dew Point Temperature (T_{dew})

When data is not available, T_{dew} can be computed from e_a by:

$$T_{dew} = [\{116.91 + 237.3 Log_e(e_a)\} \div \{16.78 - Log_e(e_a)\}]$$

where: T_{dew}, dew point temperature [°C]
       e_a, actual vapor pressure [kPa]

For the case of measurements with the Assmann psychrometer, T_{dew} can be calculated from:

$$T_{dew} = (112 + 0.9T_{wet})[e_a \div (e^0 \, T_{wet})]^{0.125} - [112 - 0.1T_{wet}]$$

8. Short Wave Radiation on a Clear-Sky Day (R_{so})

The calculation of R_{so} is required for computing net long wave radiation and for checking calibration of pyranometers and integrity of R_{so} data. A good approximation for R_{so} for daily and hourly periods is:

$$R_{so} = (0.75 + 2 \times 10^{-5} z)R_a$$

where: z, station elevation [m]
       R_a extraterrestrial radiation [MJ m^{-2} d^{-1}].

Equation is valid for station elevations less than 6000 m having low air turbidity. The equation was developed by linearizing Beer's radiation extinction law as a function of station elevation and assuming that the average angle of the sun above the horizon is about 50°.

For areas of high turbidity caused by pollution or airborne dust or for regions where the sun angle is significantly less than 50° so that the path length of radiation through the atmosphere is increased, an adoption of Beer's law can be employed where P is used to represent atmospheric mass:

$$R_{so} = (R_a) \exp[(-0.0018P) \div (K_t \sin(\Phi))]$$

where: K_t, turbidity coefficient, $0 < K_t < 1.0$ where $K_t = 1.0$ for clean air and

$K_t = 1.0$ for extremely turbid, dusty or polluted air.

P, atmospheric pressure [kPa]

Φ, angle of the sun above the horizon [rad]

R_a, extraterrestrial radiation [MJ m^{-2} d^{-1}]

For hourly or shorter periods, Φ is calculated as:

$$\sin \Phi = \sin \varphi \sin \delta + \cos \varphi \cos \delta \cos \omega$$

where: φ, latitude [rad]

δ, solar declination [rad] (Eq. (24) in Chapter 3)

ω, solar time angle at midpoint of hourly or shorter period [rad]

For 24-hour periods, the mean daily sun angle, weighted according to R_a, can be approximated as:

$$\sin(\Phi_{24}) = \sin[0.85 + 0.3 \, \varphi \sin\{(2\pi J/365) - 1.39\} - 0.42 \, \varphi^2]$$

where: Φ_{24}, average Φ during the daylight period, weighted according to R_a [rad]

φ, latitude [rad]

J, day in the year.

The Φ_{24} variable is used to represent the average sun angle during daylight hours and has been weighted to represent integrated 24-hour transmission effects on 24-hour R_{so} by the atmosphere. Φ_{24} should be limited to > 0. In some situations, the estimation for R_{so} can be improved by modifying to consider the effects of water vapor on short wave absorption, so that: $R_{so} = (K_B + K_D) R_a$ where:

$$K_B = 0.98\exp[\{(-0.00146P) \div (K_t \sin \Phi)\} - 0.091 \{w/\sin \Phi\}^{0.25}]$$

where: K_B, the clearness index for direct beam radiation

K_D, the corresponding index for diffuse beam radiation

$K_D = 0.35 - 0.33 K_B$ for $K_B > 0.15$

$K_D = 0.18 + 0.82 K_B$ for $K_B < 0.15$

R_a, extraterrestrial radiation [MJ m^{-2} d^{-1}]

K_t, turbidity coefficient, $0 < K_t < 1.0$ where $K_t = 1.0$ for clean air and $K_t = 1.0$ for extremely turbid, dusty or polluted air.

P, atmospheric pressure [kPa]

Φ, angle of the sun above the horizon [rad]

W, perceptible water in the atmosphere [mm] = $0.14 \, e_a \, P + 2.1$

e_a, actual vapor pressure [kPa]

P, atmospheric pressure [kPa]

APPENDIX J   PSYCHROMETRIC CHART AT SEA LEVEL

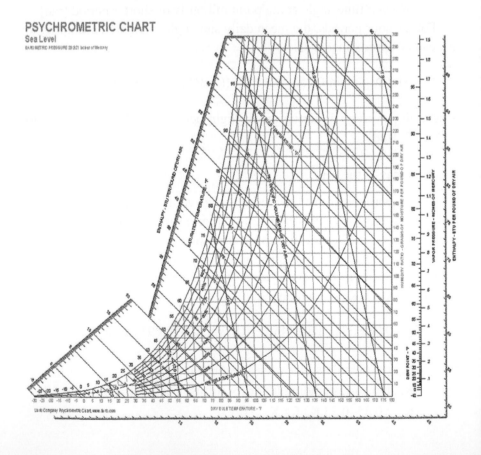

APPENDIX K

(http://www.fao.org/docrep/T0551E/t0551e07.htm#5.5%20field%20
management%20practices%20in%20wastewater%20irrigation)

1. **Relationship between applied water salinity and soil water salinity at different leaching fractions (FAO, 1985)**

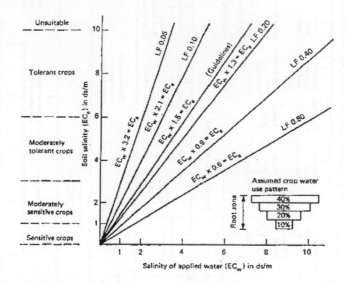

2. **Schematic representations of salt accumulation, planting positions, ridge shapes and watering patterns.**

3.   **Main components of general planning guidelines for wastewater reuse (Cobham and Johnson, 1988)**

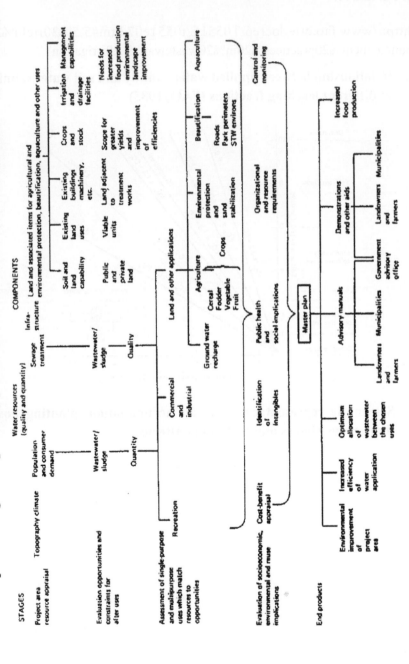

APPENDIX L

From: *Vincent F. Bralts*, 2015. Chapter 3: Evaluation of the uniformity coefficients. In: *Sustainable Micro Irrigation Management for Trees and Vines, Volume 3* by M. R. Goyal (Ed.). Apple Academic Press Inc.

1. Uniformity classification.

Classification	Statistical Uniformity	Emission Uniformity
Excellent	For U = 100–95%	100–94%
Good	For U = 90–85%	87–81%
Fair	For U = 80–75%	75–68%
Poor	For U = 70–65%	62–56%
Not Acceptable	For U < 60%	<50%

2. Acceptable intervals of uniformity in a drip irrigation system.

Type of dripper	Slope	Uniformity interval, %
Point Source: located in planting distance > 3.9 m.	Level*	90–95
	Inclined**	85–90
Point Source: located in planting distance < 3.9 m.	Level*	85–90
	Inclined**	80–90
Drippers inserted in the lines for annual row crops.	Level*	80–90
	Inclined**	75–85

* Level = Slope less that 2%. ** Inclined = Slope greater than 2%.

3. Confidence limits for field uniformity (U).

Field uniformity	18 drippers		36 drippers		72 drippers	
	Confidence limit		Confidence limit		Confidence limit	
	N_{Sum}*	%	N_{Sum}	%	N_{Sum}	%
100%	3	U ± 0.0	6	U ± 0.6%	12	U ± 0.0%
90%	3	U ± 2.9	6	U ± 2.0%	12	U ± 1.4%
80%	3	U ± 5.8	6	U ± 4.0%	12	U ± 2.8%
70%	3	U ± 9.4	6	U ± 6.5%	12	U ± 4.5%
60%	3	U ± 13.3	6	U ± 9.2%	12	U ± 6.5%

*N_{Sum} = 1/6 part of the total measured drippers. This is a number of samples that will be added to calculate T_{max} and T_{min}.

4. Nomograph for statistical uniformity

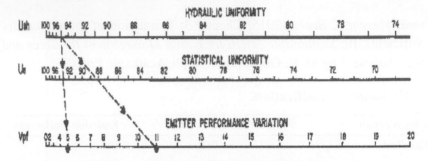

5. The field uniformity of an irrigation system based on the dripper times and the dripper flow rate.

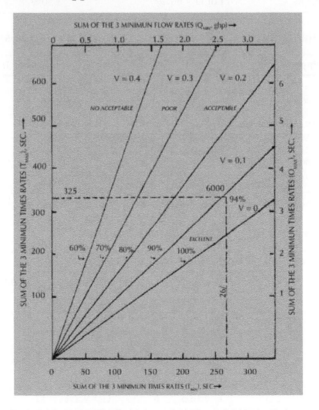

6. **The field uniformity of a drip irrigation system based on the time to collect a known quantity of water or based on pressure for hydraulic uniformity.**

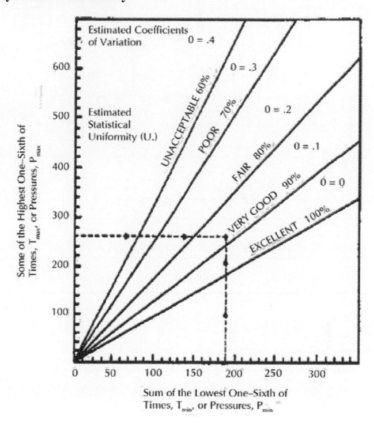

INDEX

A

Acer, 61–66
  platanoides L., 48
  ruburm L., 50, 62–66
Acid treatment, 340
Actual evapotranspiration, 69
Aerobic
  condition, 249, 255, 259
  culture, 255
  method, 253, 262, 274, 275
  rice, 249–255, 258, 259, 265, 269, 272–276
    cultivation, 250–253, 259, 269, 273–277
    field conditions, 252
    production system, 249
    varieties, 249, 253, 255–259
    soil conditions, 249, 253, 254
Aerodynamic factors, 25, 38
Aesthetic ratings, 69
Agricultural
  fields, 338
  productivity, 109, 250
  sector, 155, 156, 159, 160, 246, 248–253, 260, 274, 296
Air
  bladder/diaphragm, 329
  dried soil, 121
  entry value, 141
  pollution, 70
  temperature, 6–10, 20, 25, 155, 167
  water contact, 328
  water interface, 86
Alfisols, 89, 90
Algae, 325–327, 336, 340
Alleviate irrigation, 297
Ambient temperature, 163, 167, 171, 201, 204, 207

Anacardium excelsum, 50
Analysis of variance (ANOVA), 283
Animal damage, 336, 340
Annual time scale, 28–37
Anthropogenic, 246, 257, 272, 273
Anti-evaporants, 80
Aquifer water, 330
Arctostaphylos densiflora, 46
Arid environment, 25, 30, 33–38
Atmospheric
  boundary layer, 86
  concentration, 258
  methane, 248, 257, 258, 272, 276
Automate crop irrigation, 159
Automated irrigation, 208
Autumn seasons, 171

B

Baccharis, 69
Bacterial
  activity, 330
  byproducts, 336
  precipitation, 328, 329, 330, 339
    iron, 329, 330, 339
    sulfur, 328, 339
  slime, 331, 340
Barometric pressures, 157
Belgian Campine region, 50
Belle Fourche Irrigation District (BFID), 157
Benefit-cost ratio (BCR), 88–90, 297, 299
Biomass water pumping systems, 261
Biometrics, 299
Black polyethylene, 106
Blaney-Criddle method, 4
Body-gripping traps, 338
Boom sprayer, 208
Borehole tubing, 144

Printed and bound by CPI Group (UK) Ltd, Croydon, CR0 4YY

23/10/2024

01777701-0002